Biology
Advanced Topics

G.H. Harper, M.SC., PH.D.
Formerly Head of Biology, Watford Boys' Grammar School.

T.J. King, M.A., D.PHIL.
Director of Studies, Abingdon School;
Formerly Head of Biology, Westminster School.

M.B.V. Roberts, M.A., PH.D.
Formerly Head of Biology, Marlborough College and Cheltenham College.

Nelson

Thomas Nelson and Sons Ltd
Nelson House Mayfield Road
Walton-on-Thames Surrey
KT12 5PL UK

51 York Place
Edinburgh
EH1 3JD UK

Thomas Nelson (Hong Kong) Ltd
Toppan Building 10/F
22A Westlands Road
Quarry Bay Hong Kong

Thomas Nelson Australia
480 La Trobe Street
Melbourne Victoria 3000
Australia

Nelson Canada
1120 Birchmount Road
Scarborough Ontario
M1K 5G4 Canada

First published by Thomas Nelson and Sons Ltd 1987

ISBN 0-17-448032-6
NPN 9 8 7 6 5 4

Printed in Hong Kong

Preface

It is now fifteen years since *Biology, A Functional Approach* by M.B.V. Roberts was first published, and considerably longer since it was first conceived; so – having reached puberty – it is not inappropriate for it to reproduce. The result is *Biology, Advanced Topics*.

Biology, Advanced Topics is intended for students who are aiming for high grades in A-level or comparable examinations, or who intend to take special papers for university entrance. In each chapter we have assumed that the relevant groundwork has already been covered and that the student is ready to spread his or her wings in various directions. Though designed as a companion to the fourth edition of *Biology, A Functional Approach*, the book can be used equally well with other A-level materials.

For convenience the chapters follow the same sequence as in *Biology, A Functional Approach*. Each chapter contains between two and four topics which, in our view, merit more discussion than is provided in the parent book. The topics are self-contained and we have not attempted to link them together into an overall theme. In the belief that a balanced diet is a varied diet, we have made the topics as diverse as possible and have treated them in different ways. Some are on specialized subjects, others are more general; some deal in detail with experimental evidence for well established theories, others are on controversial issues on which there are no definitive answers. *All* are intended to foster critical thought and we hope that students will read them in that spirit.

Each topic is followed by one or more questions for consideration together with a short list of books and/or articles to which reference can be made for further information. If the book stimulates students to cultivate a critical attitude towards biology and to delve further into those aspects which they find interesting, then we shall have achieved our main aim.

For helping us to make the book as accurate and up to date as possible we are indebted to the following friends and colleagues who read and commented on various parts of the draft: Dr Philip Ashmole, Professor John Tyler Bonner, Dr Vernon Butt, Dr Graham Cairns-Smith, Dr Ken Clarke, Mr Dennis Cremer, Dr Frances Edwards, Dr Nick Evans, Dr Peter Hardcastle, Dr Ron Kille, Drs Patricia and Peter Kohn, Dr Jaleel Miyan, Miss Grace Monger, Dr James Parkyn, Professor David Patterson, Dr Sandy Raeburn, Dr Eric Sidebottom and Dr Peter Wilson. The entire manuscript was read by Mrs Sheila Basford, Dr Neil Ingram and Dr Michael Reiss, all three of them Sixth Form teachers, who gave us valuable advice as to the suitability of the materials for Sixth Form work. We are grateful to them for their frank and helpful comments. One of us (TJK) wrote much of his share of the book while holding a Schoolmaster Visiting Fellowship at Fitzwilliam College, Cambridge, and he would like to express his appreciation to the Master and Fellows of the college for providing hospitality and facilities.

Finally we must thank our publisher, Thomas Nelson and Sons Ltd., particularly Mrs Donna Evans and Mr Patrick McGuire, for their encouragement and efficiency.

G.H. Harper,
T.J. King,
M.B.V. Roberts.

Acknowledgements

The authors and publishers are grateful to the following for permission to use their photographs: (The numbers in italics refer to the page number).

J. Allan Cash Ltd., *125* (28.2); Allelix, *137*; Heather Angel, *67*; Biophoto Associates, *11*, *35*; J.T. Bonner, University of Princeton, *16*; R. Branton, Wye College, *138*; K.B. Buckton, Western General Hospital Edinburgh, *128*; Dr. Jennifer Dee, *17*; Dista Products/Carl Fox, *141*; R. & C. Foord/Edward Arnold, *108*; Z. Glowacinski, *147–8*; Dr. G.H. Harper, *125* (28.1), *152*; David Hosking/D.P. Wilson, *111*; ICI, *142*; Dr. B.E. Juniper, University of Oxford, *123*; National Trust for Scotland, *161*; Dr. D.H. Northcote, University of Cambridge, *117*; Oxford Scientific Films, *110*; Picturepoint, *153*; L.J. Reed, Clayton Foundation Biochemical Institute, *31*; Science Photo Library, *1*, *2*, *4*, *89*.

The publishers have made every effort to trace the relevant copyright holders, but if they have inadvertently omitted any appropriate acknowledgements they will be pleased to make suitable corrections at the soonest opportunity.

Contents

Authors
GHH 1, 2, 3, 21, 22, 23, 25, 26, 28, 29, 33, 34, 35, 36, 37
TJK (2), 4, 5, 6, 7, 9, 10, 12, 17, 24, (26), 27, 30, 31, 32
MBVR 8, 11, 13, 14, 15, 16, 18, 19, 20

1 . Introducing Biology

So that's life

The simplest questions are sometimes the hardest to answer. For instance, what is life? We can usually tell easily enough whether a thing is alive. Suppose you prod it, and it just wobbles or falls over; then this passive response tells you it is probably a jelly, a brick, or some other non-living material. Living things on the other hand have a habit of doing something active, like squawking or biting.

At the other extreme, some organisms may be almost entirely unresponsive even though not dead. Fungus spores, seeds, bacteria, some insect larvae, rotifers, roundworms, and tardigrades (figure 1.1) – all these can enter a state called **cryptobiosis**, meaning 'life hidden'. They become compact in form and sometimes completely desiccated. Chemical reactions inside them may stop altogether. Cryptobiotic rotifers and tardigrades, which are small water or soil-living animals, have been kept dried for more than a century, or heated to 150°C, or cooled to within a hundredth of a degree above absolute zero. In each case they have returned to an active state after being moistened at normal temperature. This suggests that overt responsiveness and internal chemical activity are not necessarily essential features of life.

Figure 1.1 Scanning electron micrograph of a tardigrade (*Macrobiotus richtersi*), a small segmented invertebrate which lives in water and damp places. If the environment becomes dry the body contracts into a 'barrel' which can survive for long periods. (×150)

Most organisms keep themselves in one piece; and so does a stone. They exchange matter and energy with the environment; so does a flame. They can respond in an active manner to stimuli; not unlike a burglar alarm or a bomb. They grow; so do crystals, though in a different way. Some individual organisms reproduce, but not all. It may, after a thorough search, be possible to find some characteristic which is true for all organisms, and only organisms. But there is a much easier way to appreciate the peculiarity of life. Organisms are the only things which combine *all* the characteristics just listed. A fly is living because it can keep itself in one piece, *and* exchange matter and energy, *and* respond to stimuli, *and* grow. Admittedly a cryptobiotic roundworm cannot do any of these things except keep itself in one piece, but it can acquire the other properties when conditions permit.

Biological research has suggested an important idea. All organisms, from humans to bacteria, seem to be made from the same set of chemical compounds. In particular they all contain nucleic acids and proteins. Even the cells of very different organisms may look and behave identically under the microscope. We know enough about cells and their chemical components to see, at least in principle, how they can grow, respond to stimuli, and exhibit all the other characters of life. In that sense, life itself is no longer a mystery.

But there is still plenty to puzzle us. We really have little idea how the wonderfully complex and beautiful structures of some organisms develop from the egg – such as orchid flowers, the flight mechanism of a bluebottle, or a peacock's plumage. The brain too has so far yielded few of its secrets.

Most likely, though, we shall not need to look for any new kinds of basic component; we probably know them already. It is like a toy construction kit. There is only a small variety of blocks, rods, wheels, and so on, but given enough of each kind it is possible to build a structure of any complexity we choose. Similarly, the development of an orchid or the function of the brain may only become clear when we know how the already familiar components are connected together. This idea may also explain cryptobiosis. So long as the roundworm is dried out slowly, the arrangement of its molecules probably remains intact and the structure is not damaged or irreversibly altered. Then it can withstand extremes of temperature, radiation and so on – yet still return to active life after re-absorbing water.

For consideration

1. Try to think of a property which *all* living organisms possess, but which is not present in any non-living object.

2. Is it more sensible to think of all objects and materials being (a) divided into two sets, the living and the non-living, or (b) having some degree of life in a continuum from the completely non-living to the completely living?

Further Reading

J.H. Crowe and A.F. Cooper 'Cryptobiosis' (*Scientific American*, vol. 225, no. 6, 1971)
This article contains some extraordinary facts about cryptobiotic organisms.

The case of the EBV

In the late 1950s, the Irish surgeon Denis Burkitt was working in Uganda. He discovered a new form of cancer, now called **Burkitt's lymphoma**. It is most common in children between six and eight years old. Typically, tumours of lymphoid tissue grow in the jaw bones, and later crop up in other parts of the body.

In Uganda the disease was found in the same areas as some infections carried by mosquitoes. However no micro-organism had been found, and so it was suggested that the cause of the lymphoma was a virus. Three biologists at the Middlesex Hospital in London tried to show that this was true: Anthony Epstein, B.G. Achong, and Yvonne Barr. They had tumour material sent from Uganda, from which they hoped to establish tissue cultures of the cancer cells. This meant growing the cells in nutrient fluid in a glass dish. But even when tumour samples arrived in good condition, the small lumps of tissue they cut from the tumour refused to grow.

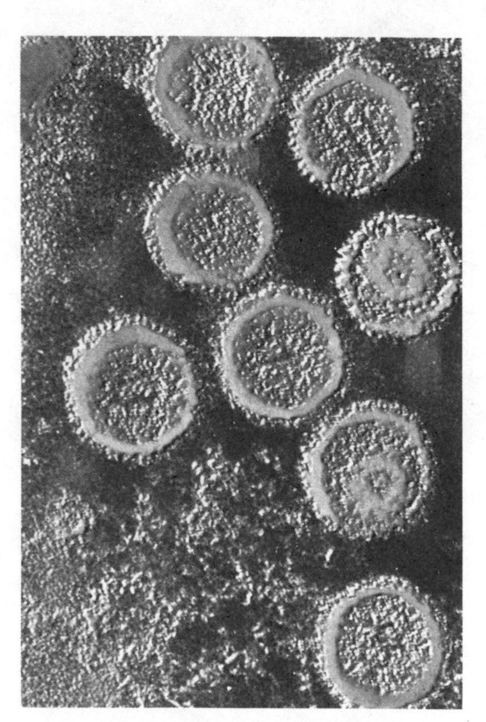

Figure 1.2 Electron micrograph of EBV particles. The outer ring of each particle is the protein coat. In addition some have an inner ring, which is the nucleic acid. (×150 000)

For several years results were negative. Epstein and his colleagues could not find a virus in the tumour cells; nor could they culture them, which would have made it easier to study the cells. Then one day in 1964 a sample arrived, like some others before it, immersed in cloudy liquid. The immediate thought was, 'Oh no, not another batch contaminated with bacteria.' This time, though, the cloudiness was not due to bacteria, but to free-floating cancer cells that had separated from the tumour. When the biologists tried culturing these cells, to their excitement it worked.

One of the first things they now did was to prepare some of the cells for the electron microscope. This was a relatively new technique at the time and previous attempts had failed. But the first attempt with the cultured cells was brilliantly successful. There in front of their eyes were the geometrical shapes of virus particles within the cells (figure 1.2). This virus is now known as the **Epstein–Barr virus** or **EBV**.

BIOLOGICAL RESEARCH

As a biologist you deal with two kinds of knowledge. One is observations, or facts. There ought to be no question about the truth of scientific facts. Ideally any competent biologist should be able to repeat your observations, and even your worst enemy will normally accept them. But collecting facts just for the sake of it is not biology; that would be more like stamp collecting. Most facts are gathered as evidence in order to test the truth of a theory. Unlike facts, theories are not automatically accepted. On the contrary they often cause heated controversy. The theory that a virus causes Burkitt's lymphoma started one such debate.

Originally the theory started off as no more than a hunch. Dreaming up new theories is probably the most disorderly part of scientific research; it doesn't matter how they come or where they come from. But a theory will only be accepted by other biologists if it can be backed up with convincing evidence. It is time-consuming and expensive collecting evidence, so part of a biologist's job is to decide carefully what facts will be most useful. Ideally they should support one theory and show that all the rival theories are false. In most cases this does not happen. For instance, although the virus found by Epstein and his colleagues gave powerful support to the theory about Burkitt's lymphoma, it was still possible that the virus had nothing to do with the disease.

Important though this work was, it involved no real experiments. Burkitt collected facts about the disease in East Africa, and Epstein's team tried all sorts of ways to culture the cells in order to look for the virus. Experiments are just one way of arriving at the truth. They usually involve controlling the environment to see what effect is produced on the thing being measured. For example an animal may be infected with the virus to see whether a cancer appears. (This has been tried, and it works.) Experiments are an excellent way of testing theories, but they are not an essential part of biology. In fact there is no rigid procedure that all biologists must follow in their research. There is plenty of scope for personal choice, enterprise, and luck. This is one thing that makes research so exciting.

DOES EBV CAUSE CANCER?

Evidence has now been collected over more than twenty years. Most of it supports the theory that Burkitt's lymphoma is caused by the virus in conjunction with other factors. One of these may be infection with the malaria parasite. This would explain why most people in the world have the virus but do not contract the disease. Nevertheless there are still a few awkward facts that do not fit the theory. For instance, some people have Burkitt's lymphoma but apparently not the virus.

For consideration

Think up an alternative theory for the facts given in this topic – one not involving EBV as a cause of the disease. How can you reconcile it with the facts, and what evidence would you need to collect to show which theory was preferable?

Further reading

G.H. Harper, *Tools and Techniques* (Nelson, 1984)
The first three chapters describe research methods in biology

W. Henle, G. Henle and E.T. Lennette, 'The Epstein–Barr Virus' (*Scientific American*, vol. 241, no. 1, 1979)
A well-illustrated article.

Viruses

This topic might have been called 'The Discovery of Viruses'. But in a sense that never happened. Instead the modern theory of virus structure and function has been gradually developing ever since viruses were first seen a century ago.

In 1886 the Scottish physician John Brown Buist was studying smallpox. This disease, incidentally, is now believed by the World Health Organisation to have been eliminated everywhere – a notable achievement. Back in the 1880s, Buist was trying to find the microbe he thought caused the disease. Eventually he found a stain which made visible some tiny particles in the fluid from a smallpox pustule. They had a diameter of 0.15 μm, and although Buist thought they were spores of bacteria we now believe they must have been the smallpox virus.

Much of the early work was done on tobacco mosaic disease, so called because the tobacco leaves go blotchy. Two important findings were made in the 1890s. First, the infectious agent could be passed through fine filters, fine enough to catch all known bacteria, and yet still cause the disease. One theory to explain this was that the agent was a toxin produced by bacteria. But that was discounted when the second discovery was made: the disease could be transferred, from one plant to another through an indefinite series, with undiminished strength. If the symptoms were caused by a dissolved chemical, the toxin would

Figure 1.3 Tobacco mosaic viruses as seen in the electron microscope. Each one consists of a protein cylinder with a nucleic acid helix inside. The nucleic acid in this case is ribonucleic acid, RNA. (×70 000)

have been diluted each time it entered the sap of an inoculated plant, and so become weaker with each transfer. This 'filterable virus', on the other hand, could evidently make more of itself in each diseased plant. Tobacco mosaic virus was eventually isolated in pure form in 1935 and this has made it possible to examine it in the electron microscope (figure 1.3).

It is only since 1940 that we have come to know the remarkable properties of these extraordinary particles.

ORIGIN OF VIRUSES

The main reason for studying viruses over the last century has been that they cause disease. We now know how viruses can destroy cells and generally create havoc for the host organism. But this may not be the typical kind of virus-host relationship. It has been suggested that disease is caused only where the host and its virus parasite have not yet managed to achieve a harmonious way of getting on together.

For instance, most people carrying the Epstein–Barr virus described in the last topic – including probably you and I – suffer no ill effects. The virus persists indefinitely in cells called B lymphocytes without doing any harm. In some of these it replicates, then entering the saliva and so spreading to other people. Disease may occur only when this peaceful relationship goes wrong. Much the same applies to the virus which causes acquired immunity deficiency syndrome (AIDS): it is possible to carry the virus without having the disease itself.

Besides the variety of host-virus relationships, there is an amazing diversity in the structure of viruses. This has led to the idea that not all viruses have originated in the same way. The larger kinds may be derived from prokaryotic parasites, such as bacteria, which degenerated into a simpler structure. On the other hand it is difficult to imagine the same origin for the simplest forms. Some viruses are known to carry, from one cell to another, not only their own genes but sometimes genes from the host cell. So perhaps some of these viruses never were organisms. They may have started merely as pieces of genetic material wrapped in a bit of plasma membrane – the gene's method of protecting itself and moving to other cells. The important thing is that such a structure would be capable of replicating itself.

ARE VIRUSES LIVING?

This question was the subject of debate for many years. It is now accepted that there are no chemical reactions going on in a pure virus. Chemical processes and in particular replication only happen when the virus is inside a living cell. Any cell requires a large amount of biochemical machinery if it is to show the properties of being alive (see page 1), and viruses have only some of that machinery. So they are not alive in the generally accepted sense of the word and are not usually included in classifications of organisms. If we imagine a cell as a programmed computer, then viruses are best thought of as extra programmes that can feed themselves into it.

For consideration

1. How would you try to establish whether a particular virus was originally a prokaryotic cell, or instead merely a fragment from a cell's genome?

2. What use could be made – and has been made – of some viruses' ability to transport one cell's genes into another cell?

3. Are you satisfied with the argument that viruses are best considered as non-living?

Further reading

P.J.G. Butler and A. Klug, 'The Assembly of a Virus', (*Scientific American*, vol. 239, no. 5, 1978)
A well illustrated account of how tobacco mosaic virus is constructed.

S.S. Hughes, *The Virus: a History of the Concept* (Heinemann, 1977)
A detailed but readable story.

F. Kingsley Saunders, *Viruses* (Carolina Biology Reader no. 64. 2nd edition, 1981)
This is a good general survey.

2 . Structure and Function in Cells

The cell's skeleton

In days of yore, before the electron microscope was invented, biologists stared down light microscopes to find out what they could about cytoplasm. It seemed to be a gelatinous or viscous fluid, with little structure other than a few organelles.

But many cells look nothing like a blob of jelly. Some have a definite and stable shape. In the case of bacteria, fungi, and plant cells, the shape is maintained by a **cell wall** functioning as an external skeleton. In other cells this is not so, and instead they seem to have an internal skeleton. This is suggested by their ability to change shape rapidly. For instance the Sun Animalcule *Actinosphaerium*, a protist, can extend and retract long straight spikes from its surface. Other cells in tissue culture may adopt a star shape when they are not moving about (figure 2.1).

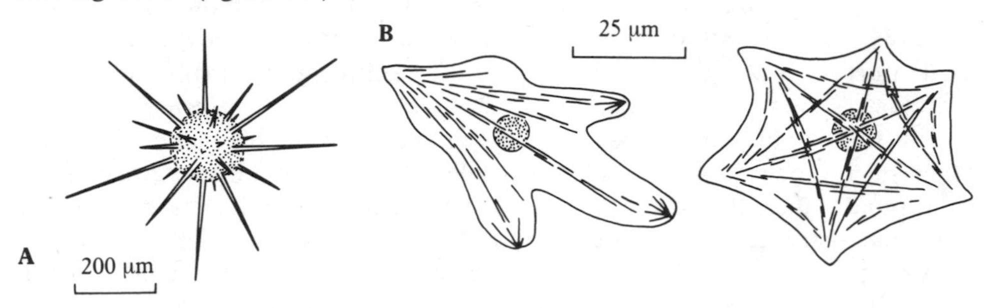

Figure 2.1 **A** The Sun Animalcule *Actinosphaerium*. **B** A cell in tissue culture, at left moving in amoeboid fashion and stationary at right. The long lines in the cell are microfilaments, just under the plasma membrane. In addition to these stress fibres, the moving cell has tufts of short microfilaments in the tips of the pseudopodia.

Several kinds of internal skeleton are now known. The two most important are the **microfilaments** and **microtubules**. A curious feature of these long structures is that they are made, not of stretched-out polypeptide chains, like keratin and collagen, but from *globular* protein molecules – actin and tubulin respectively (figure 2.2). As in haemoglobin and enzymes, their polypeptide chains are folded back on themselves several times, so that the molecule looks something like the inner tube of a bicycle tyre after it has been stuffed into your pocket. The actin molecule has a single polypeptide chain, while the tubulin molecule has two. We shall encounter a similar protein in the next topic – flagellin, of which the bacterial flagellum is made.

These structures are now being investigated using the electron microscope. Another ingenious method is **immunofluorescence**. Certain molecules can be made which will attach themselves only to actin, or to tubulin. Moreover these molecules can fluoresce – that is, give out light – when illuminated with ultra-violet radiation. So when living cells treated with these molecules are illuminated with the ultra-violet, the microfilaments and microtubules fluoresce. It is literally a way of making the cell's skeleton light up in the dark.

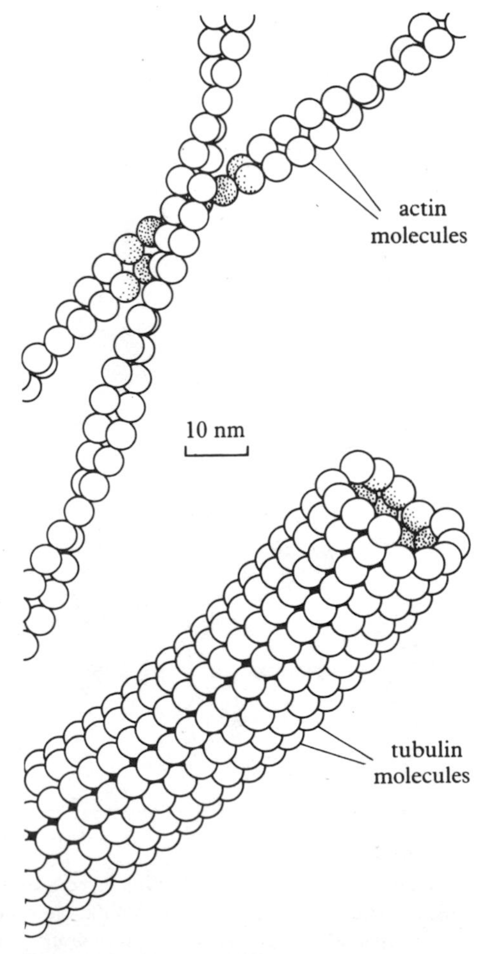

Figure 2.2 Two microfilaments and a microtubule drawn to scale; each is a polymer of a globular protein.

MICROFILAMENTS

Figure 2.2 shows how **actin** monomers (single molecules in solution) are joined together to form a microfilament. The microfilament is like a two-strand rope.

Actin is abundant in some cells, such as amoebae and blood platelets – up to 20–30 per cent of the total protein. It is in the form of soluble monomers and microfilaments. The speed with which one is converted into the other is sometimes incredible. For instance sperm cells of sea urchins and many other animals produce a long spike called an **acrosome**, as a means of penetrating the egg cell; the acrosome lengthens by the growth of microfilaments inside it, and this may happen at 10 $\mu m\ s^{-1}$. In other words over 2000 actin molecules are joining each microfilament every second.

Microfilaments often occur in bundles. In this form they give rigidity to structures like microvilli and acrosomes, and as **stress fibres** maintain the shape of many cells. If on the other hand microfilaments are randomly orientated in the cytoplasm, and joined together at frequent intervals to make a three-dimensional network, they convert the cytoplasm into a gel. Most microfilaments are also attached to the plasma membrane.

In addition to these skeletal functions, molecules of a protein called myosin can in effect 'walk' along a microfilament; so this is a second way in which movement is generated. It is seen in its most developed form in muscle (see page 93). Microfilaments are also involved in amoeboid movement and cell division (see page 103).

MICROTUBULES

The protein molecules used to assemble microtubules are **tubulin**. They spontaneously aggregate into tubes with a diameter of about 25 nm, and consisting of about thirteen rows of monomers (figure 2.2). Although normally only a few micrometres long, microtubules can reach several centimetres in nerve cells.

As with microfilaments, a cell may contain a pool of soluble monomers, from which microtubules can be rapidly assembled. One end of a microtubule grows more rapidly than the other, and under certain conditions it may be growing while the other end is disintegrating at the same rate. So the microtubule remains the same length but all its constituent parts are effectively moving along it – a process called **treadmilling**.

This could be used to generate movement within a cell, if the object to be moved is attached to one section of a microtubule. Another method is to block one end of the microtubule and allow the other end to grow or disintegrate. Many microtubules in the cell arise from **microtubule organising centres**, and it is thought that these function by blocking one end. In this way, microtubules grow out from the centre. Examples will be found in chapter 23, where the mitotic spindle is described.

In a third form of movement, microtubules are once again similar to microfilaments. If the protein dynein is attached to a microtubule, it can cause a second microtubule to slide along the first one. The best known example is eukaryotic cilia, described in the next topic.

Microtubules also perform a purely skeletal function, scattered throughout the cell and helping to keep its shape. A more unusual use is seen in *Actinosphaerium* (figure 2.1A) where each long spike is built around a massive bundle of microtubules, and when an item of prey is pulled in by the retraction of a spike, this shortening occurs through the rapid disintegration of the microtubules.

Some functions of microtubules in plant cells are described in chapter 26.

For consideration

1. Write a list of cell types having a recognisable shape. Red blood cells are one example. In each case decide how you think the shape is maintained. What evidence supports your opinion?

2. What other uses can you think of for immunofluorescence?

3. What advantage might there be in the two strands of a microfilament being twisted around each other as in a rope? (Hint: think about the rope.)

Further reading

N.F. Cooper, 'The Cytoskeleton' (*Biologist*, vol. 33, no. 4, 1986)
A short useful summary.

K. Weber and M. Osborn, 'The Molecules of the Cell Matrix' (*Scientific American*, vol. 253, no. 4, 1985)
A detailed account of the cytoskeleton.

Moving cells

EUKARYOTIC CILIA AND FLAGELLA

The cilia and flagella of eukaryotic cells range in length from 1 μm to 2 mm, and can move in a variety of ways. Despite such a diversity, the **axoneme** – the '9 + 2' arrangement of microtubules – always has the same structure and a diameter of about 0.2 μm. This suggests there is but one kind of 'motor' for generating power.

Where is this motor? It could be either inside the cell, or at the base of the cilium, or the axoneme itself may be the motor. A simple experiment suggests that the last idea is correct. Fine laser beams can be used to cut flagella away from the basal body and cell. These flagella can keep swimming for a considerable time, even if their plasma membrane is removed, so long as they are supplied with ATP and certain ions.

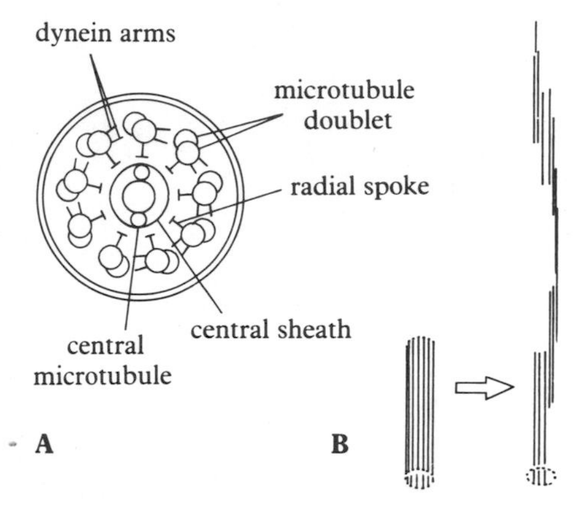

Figure 2.3 A Cross-section of a eukaryotic cilium, showing the radial spokes extending from each microtubule doublet towards the central sheath, and the two dynein arms on each doublet. B An isolated cilium with no radial spokes can increase length up to nine times, presumably by the doublets sliding along each other.

Certainly the axoneme is sufficiently complicated to contain a motor (figure 2.3A). High resolution electron micrographs show arms and radial spokes projecting from the pairs of peripheral microtubules (doublets). The arms are composed of the protein dynein. The axoneme appears to work by a **sliding microtubule mechanism**. If enzymes are used to digest away the radial spokes but not the dynein arms of isolated cilia, the electron microscope shows that each cilium can increase its length up to nine times (figure 2.3B). But if instead only the dynein arms are destroyed, the cilium cannot move. Dynein is an ATPase, and like the myosin in muscle it changes shape in the presence of ATP. It is believed that the dynein arms of one doublet can attach themselves to the next doublet and push it along a short distance before releasing. This repeated action would cause the doublets to slide along each other.

An important feature of the sliding microtubule theory is that the doublets do not change length. Each doublet becomes a singlet in the tip of the cilium, and figure 2.4 shows how a mixture of singlets and doublets can be expected to appear in cross-sections of the tip when the cilium is bent. Electron micrographs showing these patterns have provided powerful support for the theory.

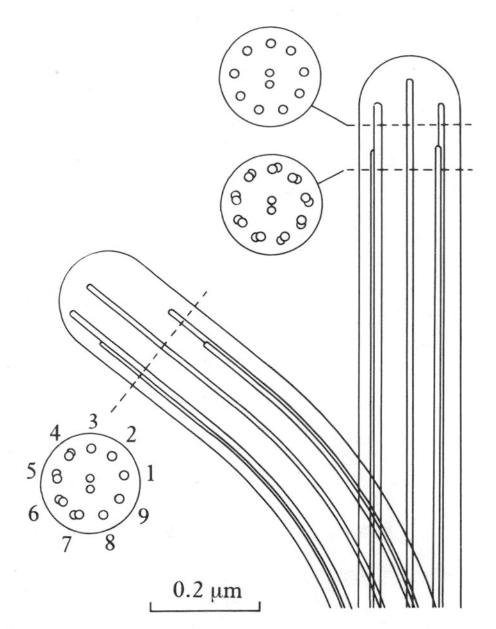

Figure 2.4 According to the 'sliding microtubule' theory the doublets do not change length, so those on the inside of a bend project further into the tip of the cilium. The longitudinal sections show two doublets, and a central microtubule or singlet. The transverse sections show the observed '9 + 2' arrangements looking down the cilium.

Given a doublet's ability to slide along its neighbour, why does the cilium in fact *bend*? Imagine two flexible rods lying beside each other and loosely attached to each other all the way along (figure 2.5A). Suppose now that the left rod is forced upward past the other one; what will happen if they are rigidly anchored at the bottom? The only way they can move past each other, except at the bottom, is by bending into a curve.

This is thought to be what happens in the cilium. In figure 2.4, the only doublets that would need to move past each other are those on the sides of the bends – that is, 2-3-4-5 and 7-8-9.

Figure 2.5B suggests how, according to the sliding microtubule theory, the cilium can throw itself into complex shapes. A local bend is produced in one region, and then an opposite bend will form higher up if the dynein arms there push in the opposite direction. Now imagine that the whole pattern of instructions to the dynein arms moves up the rods: the result would be a wave moving in one plane. This is the way the flagellum normally beats. As a further complication, the pattern of instructions could not only move up the flagellum but also spiral around it, repeatedly passing from one pair of doublets to the

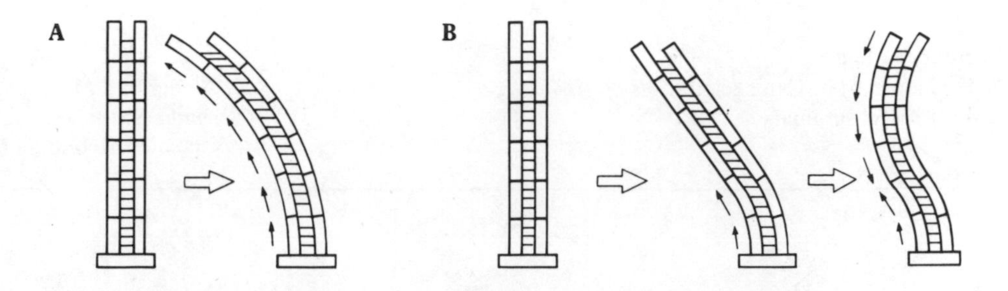

Figure 2.5 **A** A model of the 'sliding microtubule' theory. Two flexible rods loosely attached to each other represent a pair of doublets. If they are anchored at the bottom and the right-hand rod pushes the other one upward, they must curve to the left. **B** The flexible rods can form a wave: pushing the left rod up produces bending to the left, while pushing it down gives a bend to the right.

next. Now the wave would be in three dimensions, looking like a helix passing along the flagellum.

It is not clear how the pattern of instructions passes along the cilium or flagellum. Some electron micrographs show the radial spokes touching the central sheath – but only on the bends, not where the cilium is straight. The function of the spokes is not known; they may merely tie the whole axoneme together, or they may coordinate the behaviour of the cilium.

THE BACTERIAL FLAGELLUM

As just described some eukaryotic flagella can bend themselves into a spiral, and make this spiral wave move towards or away from the cell. The flagellum of a prokaryote cell looks precisely the same when it is moving. However, direct observation under the microscope will not tell you whether the prokaryote flagellum is flexible like the eukaryotic one, or is instead a stiff structure like a pulled out spring.

Compared with an axoneme, the diameter of a prokaryotic flagellum is only about one tenth, or 20 nm. It is a tube made of a globular protein, **flagellin**, in rather the same manner as a single microtubule is made of tubulin (see page 6). This suggests that the filament of the flagellum is not nearly complex enough to contain a 'motor'.

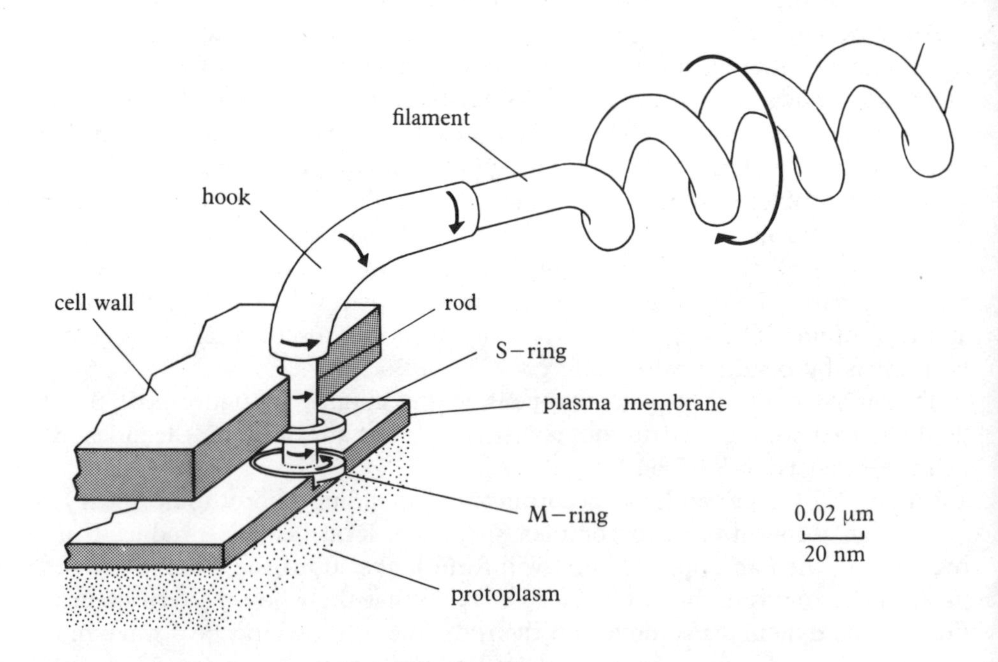

Figure 2.6 A bacterial flagellum, as in *Bacillus subtilis*. The M-ring is thought to spin in the plasma membrane.

Ingenious experiments have shown that the filament is in fact a rigid helix. The whole thing is rotated by a 'motor' at the base of the flagellum, where it is attached to the surface of the cell (figure 2.6). Micrographs of isolated flagella depict two discs at the base. It is believed that the lower one, the **M-ring**, is fixed immovably to the **rod**, and is embedded in the plasma membrane. The **S-ring** is probably attached to the cell wall, and has a hole in it through which the rod passes.

It is thought that the M-ring spins in the plasma membrane, causing the rod to rotate. The most likely energy source is the concentration gradient of hydrogen ions across the plasma membrane. Although the rod always points at right angles to the cell surface, the flexible **hook** acts as a universal coupling; in this way, the rotation is transmitted to the filament no matter what direction it points in. Moreover the 'motor' can go into reverse. So the bacterium clearly has the means to perform complex movements forwards and backwards.

For consideration

Suggest reasons why the microtubule pattern in eukaryotic cilia is always 9 + 2, and not 10 + 2 or 8 + 2.

Further reading

H.C. Berg, 'How bacteria swim' (*Scientific American*, vol. 233, no. 2, 1975)
This details the evidence for a motor.

P. Cappuccinelli, *Motility of Living Cells* (Chapman & Hall, 1980)
A general review covering most kinds of cell locomotion.

P. Satir, 'How Cilia Move' (*Scientific American*, vol. 231, no. 4, 1974)
A well illustrated article.

Did Golgi ever see his body?

The structure of the Golgi body is now generally agreed, but there was great confusion about it before the electron microscope was applied to the problem. One textbook of the 1950s noted that 'As Golgi bodies have rarely been seen in living cells, and as hardly any two workers agree on what they look like when fixed and stained, the elementary student need not be concerned with them.'

Fixing a tissue means adding a chemical to stop the proteins decaying. Staining is adding a dye which colours only certain structures in the cell. The Italian biologist Camillo Golgi studied nerve cells, and in the 1890s he invented a new method of fixing and staining with silver nitrate. By this means he 'discovered' what was later named after him – the Golgi body. Far from being a localized stack of flattened vesicles, as now described by electron microscopists, what he saw was a network of threads extending through all the cytoplasm.

Later a controversy arose about the exact nature of the organelle, and it was still going strong in the 1950s. One of the problems was that when Golgi's method was used on different cells, it gave different results. Tiny objects looking like discs, caps, crescents, rods, rings and blobs were described, and they were all said to be the Golgi body or Golgi apparatus. You will now understand why the textbook writer thought that students need not waste their time on them.

ARTEFACTS

One argument ran as follows. What Golgi saw does not exist in living cells; it is only something created in the process of fixing and staining. In other words the network is an **artefact**.

For instance, the Oxford cytologist J.R. Baker thought that the real structure in the living cell was a group of spherical drops, consisting mostly of fats. He suggested that Golgi's stain merely collected in the cytoplasm between the drops; if they were closely packed, the stain would then look something like a network. Baker put it like this: 'If one calls the intervening substance a net, one must be consistent and call the floury part of a currant bun a net.'

Ideally biologists would like to avoid artefacts altogether. The most extreme artefact corresponds in no way at all to what exists in the living material. This is obviously unhelpful, and merely leads biologists up the garden path.

On the other hand, an intermediate kind of artefact may provide two kinds of information – true and false. For example, a particular stain may collect in certain parts of the cell, and appear as a distinctive shape. The position may tell us something important about the biochemical properties of that region of the cell, while the shape may be purely artificial. Some of the discs, caps, crescents, and so on may have been like this – just funny shapes produced by the fixing and staining, but possibly showing the position of what we now call the Golgi body.

The question is, of course, how do we *know* something is an artefact, and not a true picture of the living structure? One approach is to observe only living material, which has been interfered with as little as possible. Another is to use several methods to investigate a structure; if the structure is shown up by all the methods it is probably not an artefact.

One of the most vociferous exponents of artefacts in microscopy, particularly electron microscopy, is Harold Hillman of the University of Surrey. He and his former colleague Peter Sartory argue that not only the Golgi body but also the endoplasmic reticulum, nuclear pores and mitochondrial cristae are all artefacts resulting from preliminary treatment of the material. Most authorities feel that Hillman and Sartory are wrong but, insofar as they have prompted cell biologists to question their assumptions and look critically at their methods, they have performed a useful service.

For consideration

It has been claimed that nuclear pores and the endoplasmic reticulum, as seen in the transmission electron microscope, are artefacts, and do not exist in the living cell. How would you set about showing that they do?

Further reading

G.M.W. Cook, *The Golgi Apparatus* (Carolina Biology Reader no. 77, 2nd edition 1980)
A well-illustrated introduction.

H. Hillman and P. Sartory, 'A Re-examination of the Fine Structure of the Living Cell and its Implications for Biological Education' (*School Science Review*, vol. 62, no. 219, 1980)
Hillman and Sartory present evidence that certain cell organelles are artefacts.

R.W. Horne and J.R. Harris, 'The Electron Microscope in Biology' (*School Science Review*, vol. 63, no. 222, 1981)
An answer to Hillman and Sartory by two practising electron microscopists.

The Fluid-Mosaic Model of the Cell Membrane

In the mid-1970s the Danielli–Davson model of cell membrane structure was superseded by the Singer–Nicholson fluid-mosaic model. In this topic we shall consider some of the evidence which led to the replacement of the one hypothesis by the other, and briefly discuss the current concept of cell membrane structure.

In Danielli and Davson's hypothesis, formulated in the 1930s, a cell membrane was envisaged as a sandwich. The core of the membrane was a lipid bilayer 3–5 nm wide, with the charged or polar ends of the phospholipid molecules sticking outwards and their long hydrophobic tails directed inwards. Protein molecules were attached to both the inside and the outside of the bilayer, forming the 'bread' of the sandwich. This model was apparently confirmed when electron microscope pictures of fixed membranes showed two dark zones on the outside and a lighter zone between them.

EVIDENCE AGAINST THE PROTEIN-LIPID 'SANDWICH'

The rate at which molecules pass across a cell membrane is strongly related to their lipid solubility (figure 2.7). Charged ions pass through slowly whilst non-polar molecules pass through rapidly. Why was this effect so marked if the molecules first met a layer of protein? The technique of freeze-etching soon provided the answer. When cells were rapidly cooled in liquid nitrogen and their membranes were fractured down the middle of the lipid bilayer, numerous bumps and hollows were seen, which were later shown to represent protein molecules (figure 2.8). Thus it seemed likely that some proteins, at least, extended right through the lipid bilayer. When membranes were treated with solutions of extreme pH, some proteins were separated easily, but others proved almost impossible to isolate. This suggested that parts of them might lurk permanently in the hydrophobic interior of the lipid bilayer.

In the original Danielli–Davson model, the proteins in the membranes were regarded as having fixed positions. Around 1970, evidence began to appear that the proteins in the membrane could move. David Frye and Michael Ededin were working with hybrid cells, produced by fusing human with mouse cells in aqueous suspensions under the influence of Sendai virus (see page 128). The proteins in the human cell membranes and the mouse cell membranes could be 'labelled' in different colours using specific antibodies to which were attached green or orange fluorescent markers. When a human and a mouse cell first fused, one half of the hybrid cell was green and the other half was orange. After forty minutes incubation at 37 °C, the green and orange markers were randomly mixed in the cell membrane. Experiments on the membranes of red blood cells showed also that the spacing of marked proteins changed when the pH of the surrounding solution was changed.

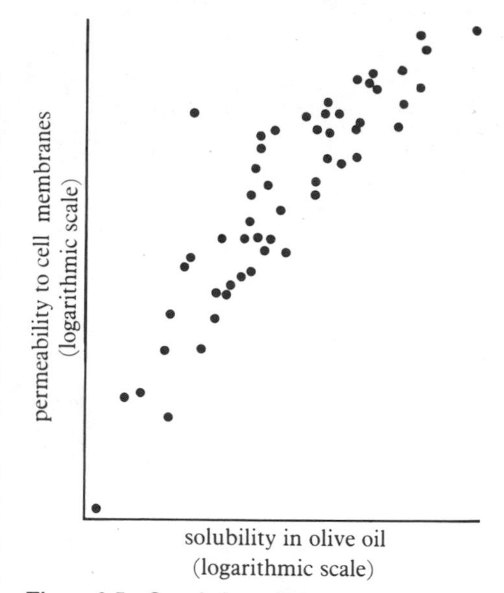

Figure 2.7 Correlation between the permeability of cell membranes of the alga *Nitella mucronata* to uncharged organic molecules and the tendency of the molecules to dissolve in olive oil.

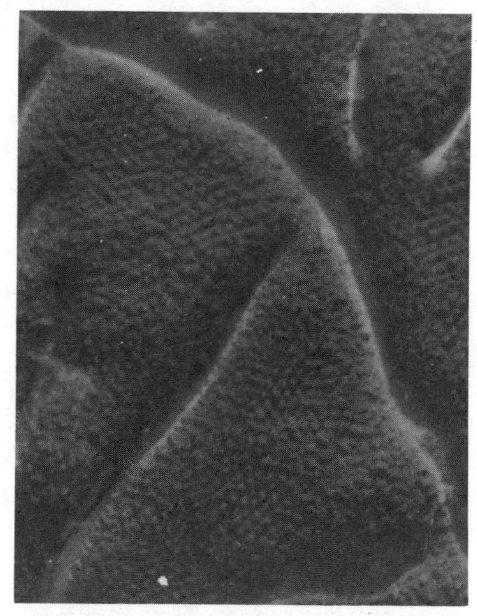

Figure 2.8 Electron micrograph of the inside of a freeze-etched cell membrane. In the intact membrane the bumps and hollows are occupied by protein molecules. (×132 000)

THE SINGER–NICHOLSON MODEL

These observations led to the development of the fluid-mosaic model, formulated by Singer and Nicholson in 1972–5. A thin lipid bilayer with the consistency of oil covers the surface of the cytoplasm. Proteins float in it, some passing through the bilayer, protruding from both surfaces, others concentrated either on the inside or the outside (figure 2.9).

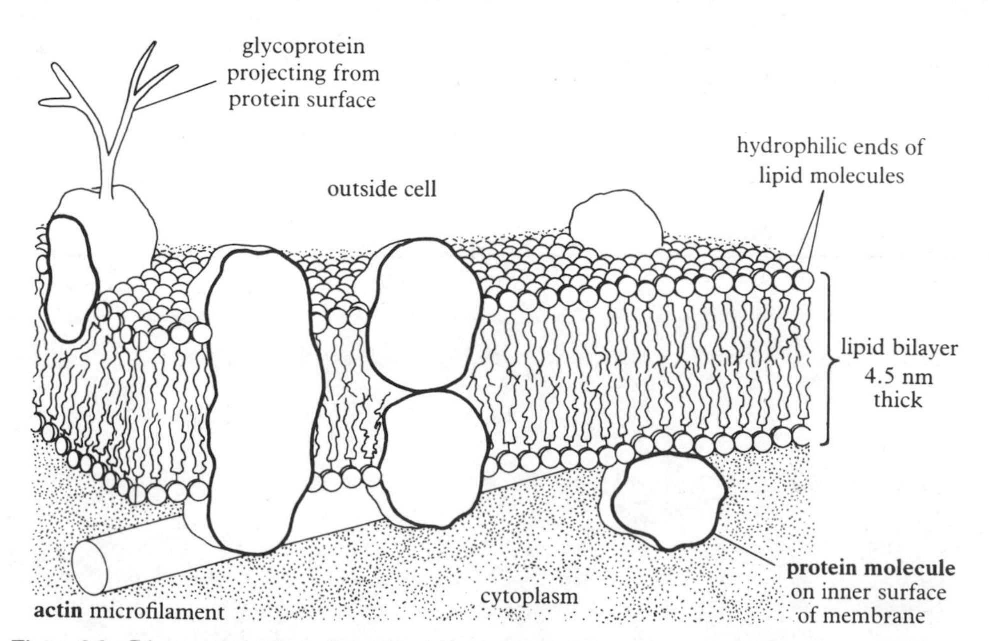

Figure 2.9 Diagrammatic three-dimensional view of the structure of a cell membrane according to the Singer–Nicholson hypothesis. The tail-like features shown in the lipid bilayer represent the hydrophobic fatty acid part of the lipid molecules.

Most of the lipid molecules in the membrane can move. In fact they exchange positions with neighbouring lipid molecules in the same layer at about a million times a second. Many lipids, however, have their charged ends bound to nearby protein molecules, and can only move in the membrane when the proteins move.

Some proteins and glycoproteins, particularly those concerned with cell adhesion and recognition in animal cells, are fibrous and protrude a long way from the outside surface of the membrane. The majority of membrane proteins, however, are globular. Their charged amino acid residues are concentrated on the insides and outsides of the membranes, whilst non-polar residues occur in the water-repellent interior of the lipid bilayer. Some of these protein molecules seem to be anchored in fixed positions in the membranes by microtubules and microfilaments (see page 6).

Membrane proteins have a variety of functions. Some are receptor molecules, with a shape complementary to polypeptide hormones such as insulin. In plant cells, the enzymes for cellulose synthesis float in the membrane. Some proteins form channels through which water molecules pass during osmosis. Others are 'carriers' which carry out the active transport of ions or small organic molecules across the membrane, such as is involved in the sodium–potassium pump (see page 82). Some proteins form 'gated channels', protein lined pores which can be opened or closed and through which ions, and small molecules such as vitamins, can be exchanged between cells or between a cell and its environment. There are also protein systems concerned with electron transport and the pumping of hydrogen ions, systems which create electrical potential differences across the membranes of most cells. Thus many activities of organisms, including nervous conduction, hormone action, ionic composition and cell co-operation during development can be understood in terms of the proteins in cell membranes.

For consideration

1. How do gas molecules such as oxygen, carbon dioxide and ethene pass across a cell membrane, and how is this related to the molecular structure of the membrane?

2. What problems might arise for a cell if its cell/plasma membrane were constructed entirely of proteins?

3. How might cells benefit from the movement of their membrane proteins, as demonstrated so spectacularly in the hybrid cell experiment?

Further reading

A. P. M. Lockwood, *The Membranes of Animal Cells* (Studies in Biology no. 27, 3rd edition, Arnold, 1984)
A sound general introduction.

J. A. Lucy, *The Plasma Membrane* (Carolina Biology Reader no. 81, 1975)
A very clear account of the structure and functions of cell membranes.

R. B. Freeman, 'The Fluid-Mosaic Model' (*School Science Review* vol. 65, no. 233, 1984)
A critical discussion of the evidence for the Singer–Nicholson hypothesis.

3 . Tissues, Organs and Organisation

Cell size and the development of the multicellular state

Nowadays a cell is regarded as a mass of protoplasm surrounded by a plasma membrane composed of phospholipids and proteins. There may be any number of nuclei in a cell – from zero in the case of mammalian red blood cells, to many millions in a myxomycete plasmodium. (The plasmodium is described in the next topic.)

Cells range in size from the minute to the relatively enormous (figure 3.1). Typical green plant cells measure 50–100 μm, much of the space being occupied by a vacuole. Most animal cells are smaller, 10–60 μm. Smaller still are prokaryote cells, usually 2–5 μm long.

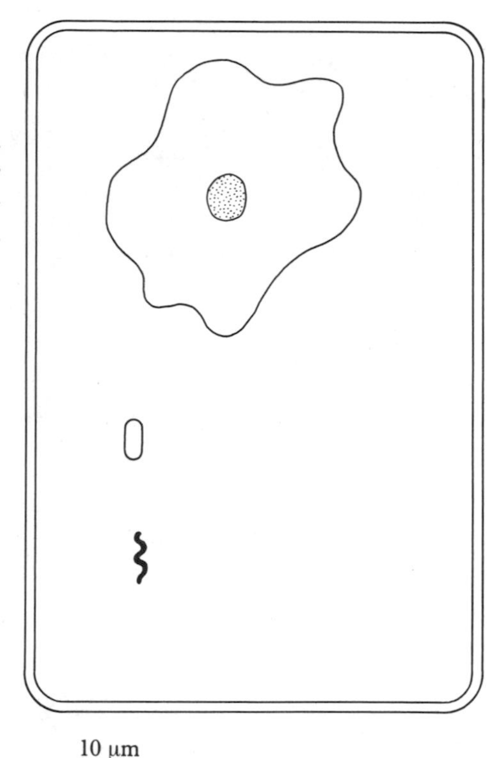

Figure 3.1 The range of normal cell sizes. Inside a plant cell, which is 100 μm or 0.1 mm long, are drawn a leucocyte (30 μm long), a bacterium (3 μm long), and a spiral mycoplasma (0.2 μm wide).

THE SMALLEST CELLS

The smallest cells yet discovered are **mycoplasmas**. Some are spherical, with a diameter varying from 0.1 to 1 μm according to the stage of the life cycle. Others are spirals, 0.1–0.2 μm in diameter but about 5 μm long. Mycoplasmas are prokaryotes, but simpler than typical bacteria, most obviously because there is no cell wall. However, they *are* true cells, because they contain nucleic acids, ribosomes, proteins, and a plasma membrane. Moreover they can grow and reproduce in the absence of other cells. Mycoplasmas were discovered through the diseases they cause. One such disease, bovine pleuropneumonia, was studied by Louis Pasteur in the nineteenth century, although he was unable to see the organism under the microscope. Another name for mycoplasma is PPLO (pleuropneumonia-like organism). Similar to mycoplasmas are the **rickettsias**, all of which are obligate parasites within other cells – in other words, parasites that cannot grow and reproduce except in the host.

Having discovered such tiny organisms, we might wonder whether still more minute cells remain to be found. Is there any lower limit to cell size? It is generally thought that there is: it is simply the smallest volume that can hold the essential chemical machinery for a cell to function. At least a hundred different enzymes would be required to catalyse essential reactions, as well as the genes coding for the proteins, and the machinery for synthesising proteins and nucleic acids. It has been calculated that a minimum viable cell of this kind would be a sphere with a diameter somewhere between 0.04 μm (40 nm) and 0.08 μm (80 nm). The mycoplasmas are rather larger than this, so perhaps smaller cells will be found – but not *much* smaller.

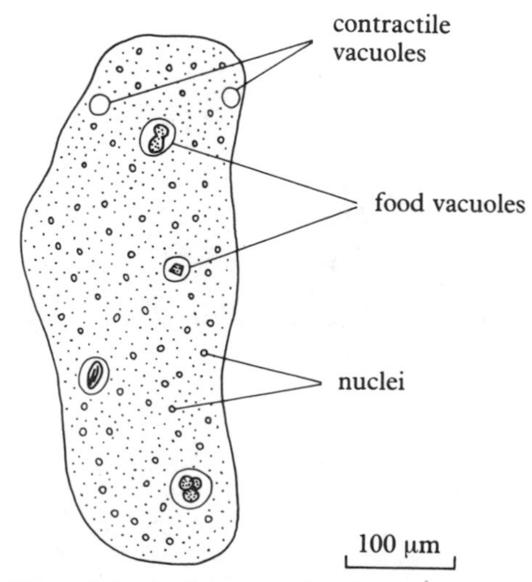

Figure 3.2 A giant amoeba *Chaos chaos*. As well as many vacuoles and other inclusions, there are numerous nuclei.

LARGE CELLS

We have just seen why there is a fixed lower limit to cell size. In contrast, a definite upper limit to cell size probably does not exist. The large amoeba *Chaos chaos* can be 500 μm long (figure 3.2); a squid's giant axon is 1 mm in diameter; the marine alga *Acetabularia* is single-celled and 4 cm high; an ostrich egg is about 20 cm in diameter; a myxomycete may be a metre across; and axons in an elephant's leg are more than 2 m long. Each of these is a single cell, and there seems no real reason why a cell should not be the size of the Eiffel Tower.

It is sometimes said that cells cannot be very large, because of the nucleus: one nucleus can 'control' only a certain volume of cytoplasm. This may be true to some extent, although one way a large cell can solve this problem is simply by having many nuclei, in other words to become a **syncytium**. Examples are striated muscle fibres, myxomycete plasmodia, and *Chaos chaos*. You can probably think of others. But there are other large cells with only one nucleus, such as the neurone in the elephant's leg. Assuming the axon is 10 μm in diameter, you might like to work out the approximate volume of cytoplasm under the 'control' of the single nucleus; mm^3 are convenient units.

A more important limitation on large cells is the surface area. Although plasma membranes are permeable to some molecules, these molecules can pass into or out of the cell only slowly. Imagine a spherical animal cell with a radius of 10 μm. The surface area is about 1200 μm^2, and the volume about 4000 μm^3. Each μm^2 of the surface is supplying oxygen and food molecules to 3.3 μm^3 of cytoplasm.

Now imagine a cell of the same shape, but much larger – say, of radius 50 μm. This might be a cell in a green plant. The surface area is now about 30 000 μm^2, increased $\times 5^2$; and the volume is about 500 000 μm^3, increased $\times 5^3$. If the whole space is filled with cytoplasm, each μm^2 of the plasma membrane would have to supply oxygen and food molecules to 17 μm^3 of cytoplasm. This means passing molecules into the cell, and also CO_2 and waste products out of it, five times as fast as in the small cell. Plasma membranes are just not permeable enough to pass molecules at anything like this rate. So the interior of this large cell would soon become starved, anaerobic, and poisoned by its own waste products.

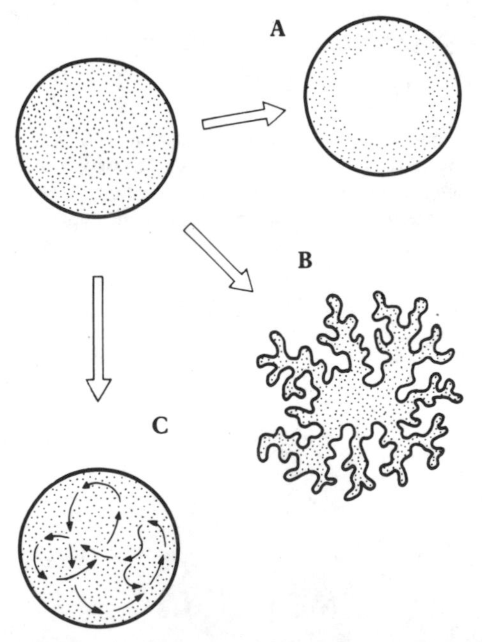

Figure 3.3 Three ways of solving the problems of large cell size. **A** A large vacuole and small amount of cytoplasm; **B** An irregular shape and large amount of plasma membrane; **C** Cytoplasmic streaming.

SOLVING THE LARGE SIZE PROBLEM

There are two main ways in which cells can solve this problem (figure 3.3). First, large cells in green plants are not completely full of cytoplasm. They have relatively little, and this is mostly in the peripheral part of the cell, around the non-living vacuole. So these cells actually have quite a high ratio of plasma membrane to cytoplasm.

Second, instead of reducing the mass of cytoplasm, the amount of plasma membrane can be increased. This means changing the cell's shape. The more unlike a sphere it is, the greater the surface area. Long extensions, such as axons in neurones, enormously increase the surface of the cell, and hence also the number of molecules that can be absorbed or lost each second. The same applies to cells with microvilli or basal infoldings. Some cells are flattened and others, such as myxomycete plasmodia, adopt a network shape. To see the advantage of this, consider how much paint is needed for the Eiffel Tower compared with the same mass of metal melted down to a solid block.

Another feature of many cells is cytoplasmic streaming. Particles in the cytoplasm can be seen moving quite rapidly along well defined routes. This probably functions rather like the blood circulation in a vertebrate. Once oxygen and food molecules have entered the cell through the plasma membrane, it would take a long time for them to reach the interior of a large cell just by diffusion. Active mixing of the cytoplasm speeds up the movement of molecules in the cell, helping to prevent the interior becoming starved, anaerobic, and poisoned. Streaming also removes from the peripheral cytoplasm molecules which have just entered the cell. This can help to maintain a concentration gradient, which is essential for a fast rate of diffusion.

Of course most large organisms are not single cells but multicellular. However, the distinction between unicellular and multicellular organisms is not entirely clear cut. The next topic describes some curious organisms which belong to both camps.

For consideration

1. In fact no organism the size of the Eiffel Tower has yet been found. Is there any reason why such an organism could not exist, either in a unicellular or a multicellular state?

2. Why do nuclei exist?

Further reading

H. J. Morowitz and M. E. Tourtellotte, 'The Smallest Living Cells' (*Scientific American* vol. 206, no. 3, 1962)
This deals with mycoplasmas.

Unicell today, tissue tomorrow: slime moulds

There can be few more unappetising names for an organism than 'slime mould'. Yet these eukaryotes turn out to have some fascinating properties.

In this topic we shall mostly be looking at **cellular slime moulds**, which live in soil. The one most intensively studied is a species of *Dictyostelium*. Another group is the Myxomycetes or **acellular slime moulds**, already mentioned in the previous section.

CELLULAR SLIME MOULDS

The life cycle of *Dictyostelium* is illustrated in figure 3.4. The resting stage is a spore, about 8 μm long. In moist conditions, each spore splits and an amoeba-like cell emerges. This wanders around ingesting bacteria, and grows to a length of about 20 μm. Then it divides by mitosis, and the daughter cells feed and grow. About 1000 bacteria are consumed by each amoeboid cell before it is ready to divide again.

The process continues as long as bacteria are available. Because the amoeboid cells repel each other, they spread over quite a wide area. Eventually they stop feeding and dividing. Now a remarkable change takes place. Instead of spreading further, the amoeboid cells come together. Here and there, a small group of them starts emitting a chemical signal, now known to be cyclic-AMP. This causes all the others in the neighbourhood to travel towards the central group. In the later stages of this **aggregation**, they stream in along several radii.

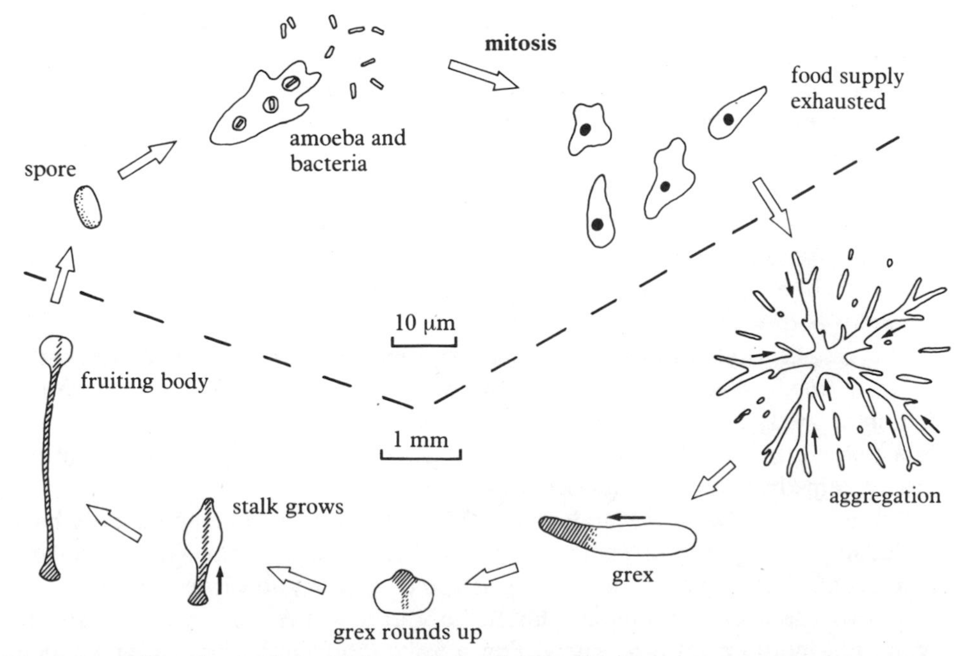

Figure 3.4 Life cycle of *Dictyostelium*, a cellular slime mould. The unicellular stages are drawn highly magnified, and the other stages at a smaller scale. The shaded areas show what happens to the anterior cells of the grex. Small arrows indicate movement.

Figure 3.5 Stages in the formation of a spore body by *Dictyostelium*.

When aggregation is complete, the cells may number hundreds of thousands. They now behave like a multicellular organism. In fact they *are* one, called a **grex**. The grex adopts a slug shape, several millimetres long, and migrates. Cells at the front end may be raised like a head. During this migration the grex moves towards higher intensities of light and temperature. As a result, it leaves the areas where the amoeboid cells were feeding, in the soil or rotting organic material, and comes to the soil surface or into cavities near the surface.

The cells at the front end of the grex are important, and in figure 3.4 they are shaded to show what happens to them. When the grex comes to a halt and rounds up into a compact shape, these anterior cells move on top of the cell mass, and then push down in the middle to the substrate. The same cells then extend upwards, acquiring large vacuoles, strong cellulose walls, and a high turgor pressure. The result is a tall, thin, but strong stalk. As it increases in height, the remaining cells are carried up until they form a yellow ball at the top (figure 3.5). Each cell secretes a tough wall and turns into a single-celled spore. The spores remain viable for years, so enabling the slime mould to survive unfavourable periods and disperse to new sites.

THEIR IMPORTANCE

Cellular slime moulds raise many questions. Are they basically unicellular amoebae, with a brief multicellular stage giving them efficient spore dispersal? Or are they multicellular organisms – fungi perhaps – with an unusual form of larva? On the latter view they can be compared with some insects whose larvae are 'eating machines' and whose adults merely reproduce. Imagine that you had to survive by finding and eating individual bacteria one at a time: you would soon die of starvation or exhaustion. How much more efficient it would be if your cells separated, and they all simultaneously hunted bacteria. In other words, the cellular slime moulds may have arrived at a supremely efficient way of combining an eating and a dispersal phase in the life cycle. This subject is discussed further in chapter 26.

Dictyostelium has become an important organism in biology, particularly because the grex is truly multicellular. Sticky patches develop on the amoeboid cells as they aggregate, causing them to adhere head-to-tail. Later they are more firmly bound together in the grex and spore body. *Dictyostelium* also provides a simple example of cell differentiation. The amoeboid cells eventually turn into just two other cell types – stalk cells and spores. So by studying multicellularity and differentiation in slime moulds, we may also learn much about how more complicated organisms develop, including the growth of our own tissues.

MYXOMYCETES: ACELLULAR SLIME MOULDS

Cellular and acellular slime moulds have several superficial similarities. For this reason they are sometimes classified together as the Mycetozoa or 'fungus-animals'. Both groups for instance have an amoeboid stage, and form a relatively large spore body.

However, the Myxomycetes or **acellular slime moulds** are really very different from *Dictyostelium*. There are numerous species, many forming beautiful and complicated little spore bodies, often yellow, red or pink. Such is their appeal that in the autumn some naturalists get quite obsessional about hunting for them, mostly on rotten wood.

There is no grex in these organisms. After the haploid amoeboid cells have been feeding on a patch of bacteria, two of them fuse to form what is called a **plasmodium**. The diploid cell so formed grows larger and larger, by ingesting all the other amoeboid cells in the neighbourhood. At the same time, the growing plasmodium acquires increasing numbers of nuclei, becoming a syncytium. These arise by mitosis from the original diploid nucleus, not from the ingested cells. So, unlike the cellular slime moulds, there is no aggregation of cells.

While still quite small, the plasmodium may look like figure 3.6. It can move around as it captures amoeboid cells, and the rear part is often a network of cytoplasmic tubes. Vigorous cytoplasmic streaming can be seen in these tubes; frequently the direction of flow stops and goes into reverse (see page 14). There seems no real limit to the size of this syncytium. It will often reach a diameter of several centimetres, and sometimes even a metre. In natural conditions it looks like a patch of slime – often yellow, brown, or white – spread over rotten wood or decaying vegetation.

The plasmodium has several features making it useful in studying the cell cycle, described in chapter 23. For instance, its nuclei divide synchronously. The time from one division to the next may be ten hours, but when one nucleus is seen dividing, all the others, perhaps millions, will do the same within a few minutes.

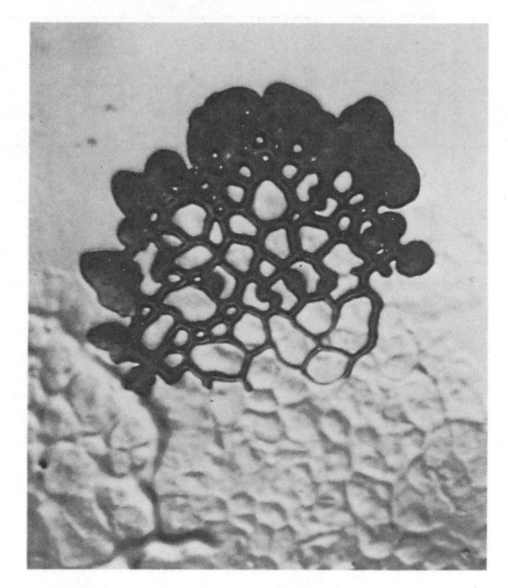

Figure 3.6 A young plasmodium of a myxomycete, seen from above. It is crawling up the page. The cytoplasmic tubes are too small to be seen; they run through the strands of the network of cytoplasm. (×10)

For consideration

1. In both groups of slime mould, separate cells are incorporated into a multicellular organism – but in one case it is by aggregation and in the other by ingestion. What are the advantages and disadvantages of each method?

2. From what kind of organism could the cellular slime moulds have evolved? Suggest why each of the evolutionary changes might have been advantageous to the organism.

3. Are there any other examples of small living units coming together to form a larger 'organism'?

Further reading

J.M. Ashworth and J. Dee, *The Biology of Slime Moulds*. (Studies in Biology no. 56, Arnold, 1975)
A good introduction; it tells you how to culture these organisms, and suggests experiments.

D.W. Mitchell, 'The Bark Myxomycetes – their collection, culture and identification.' (*School Science Review* vol. 58, no. 204, 1977)
This is a useful article for both the naturalist and the biologist.

4 . Movement in and out of cells

The water potential concept

For many years biologists have used the concept of **osmotic pressure** to explain the movement of water between cells and between cells and their watery environments. Now a number of biologists favour moving over to the related concept of **water potential**. This topic explains water potential and its value to biologists.

WHAT IS WATER POTENTIAL?

Water molecules in both liquids and gases are in rapid random motion. The molecules move randomly at speeds which depend on their kinetic energy. If the total energy contents of the water molecules differ in two adjacent regions, there will be a net flow of water molecules from one region to the other. This, of course, is **diffusion**. We can measure the total energy level of the water molecules per unit volume in Joules per cubic metre and call it the water potential of the fluid. Water, like anything else, diffuses from a high potential to a low potential.

Another way to think of water potential is as the pressure exerted by the water molecules in a fluid. Pressure, measured in Pascals (Pa), is dimensionally equivalent to the energy content per unit volume, measured in Joules per cubic metre. As a reference point, the water potential of pure water at 25 °C and atmospheric pressure is taken to be 0 kPa.

OSMOSIS

Imagine a selectively permeable membrane, such as a pig's bladder, separating pure water from a sucrose solution. When sucrose molecules were dissolved in the water to form the sucrose solution, they lowered the water potential of the water molecules in the solution in two ways. Firstly, the sucrose molecules take up space in the solution which would otherwise be occupied by water molecules, thus reducing the number of water molecules per unit volume. Secondly, many of the water molecules become bonded to the solute molecules and in this state are unable to move around freely. This reduces the kinetic energy of the water molecules in the solution.

As a result, the water molecules in pure water exert more pressure on the membrane than those in the sucrose solution on the other side of the membrane. The pure water, by definition, has a water potential of 0 kPa. The sucrose solution must therefore have a water potential below 0 kPa, for instance −3500 kPa for 1 mole of sucrose per litre. Water molecules will pass across the membrane in both directions, but the net flow will be from the higher water potential (pure water, 0 kPa) to the lower water potential (sucrose solution, −3500 kPa).

This net flow of water molecules across the membrane could be halted if pressure were applied to the sucrose solution. If a pressure of +3500 kPa was applied, increasing the energy levels of the water molecules to 0 kPa, a dynamic equilibrium would be set up in which water molecules would pass across the membrane at the same rate in both directions. This is what happens in a turgid plant cell in water. The pressure of the cell wall on the cell membrane counteracts the tendency of water to diffuse into the cell.

EVAPORATION

Imagine the surface of a lake, from which water evaporates into the atmosphere. The water molecules leave the lake surface (water potential, say, −10 kPa) for the air more rapidly than they leave the air (water potential, say, −30 000 kPa) for the lake surface. In the air the water molecules are few and far between. Nevertheless, they are moving much more rapidly in the air than in the liquid, that is, each molecule has much more kinetic energy. Thus the water potential of the atmosphere is very sensitive to changes in temperature and water content. For example, if the relative humidity of the air increased to 100 per cent, its water potential would be 0 kPa, about the same as that of the liquid water, although in the air there are far fewer water molecules per unit volume. In this case a dynamic equilibrium is set up in which there is no net evaporation of water from the lake surface. The same principle applies to water loss from a leaf to the atmosphere in transpiration (see page 54).

THE VALUE OF THE WATER POTENTIAL CONCEPT

The traditional osmotic pressure terminology was devised to explain the movements of solvents such as water across selectively permeable membranes from one liquid to another. Osmotic pressures are unsuitable for describing water flow through the whole soil-plant-atmosphere continuum because the atmosphere does not have an osmotic pressure. Furthermore, water movements are influenced by the bonding of water molecules in solutions to solid objects, and by positive pressures and negative tensions. The water potential terminology can accommodate these forces whereas the osmotic pressure terminology cannot. It focuses attention on the water molecules themselves, not on solutes or the process of osmosis. The water potential concept also links up with chemical thermodynamics. Water potential is simply a special case of chemical potential. It will probably not be long before *all* biologists embrace the concept.

For consideration

Although the water potential concept is used by plant physiologists, it has not gained widespread acceptance by animal and medical physiologists. Choose two cases of water movement in animals which are commonly explained in terms of osmosis and explain them in terms of the water potential concept.

Further reading

J.F. Sutcliffe, *Plants and Water* (Arnold, 2nd edition, 1979)
Transpiration and other aspects of plant physiology are approached through the water potential concept.

C.S. Hutchinson and J.F. Sutcliffe, 'An Approach to the Teaching of Cell Water Relations in Biology at A-level Using the Water Potential Concept' (*Journal of Biological Education*, Vol. 17, No. 2, 1983)
A short, useful summary.

Cell membranes and absorption from the small intestine

Have you ever thought how dependent we are on the epithelial cells which line the villi in the small intestine? These cells absorb food compounds from the lumen and deposit them in blood capillaries. In this topic we shall discuss the part played in absorption by the cell membranes of these columnar epithelial cells. The idea is to relate the general principles underlying the movement of molecules and ions across cell membranes, which you have studied before, to a specific situation.

THE STRUCTURE OF THE ABSORPTIVE CELLS OF THE VILLUS

A typical absorptive cell from a villus is shown in figure 4.1. It has an appearance characteristic of a cell involved in the active uptake of substances. The membrane which is exposed to the lumen has a brush border of microvilli about

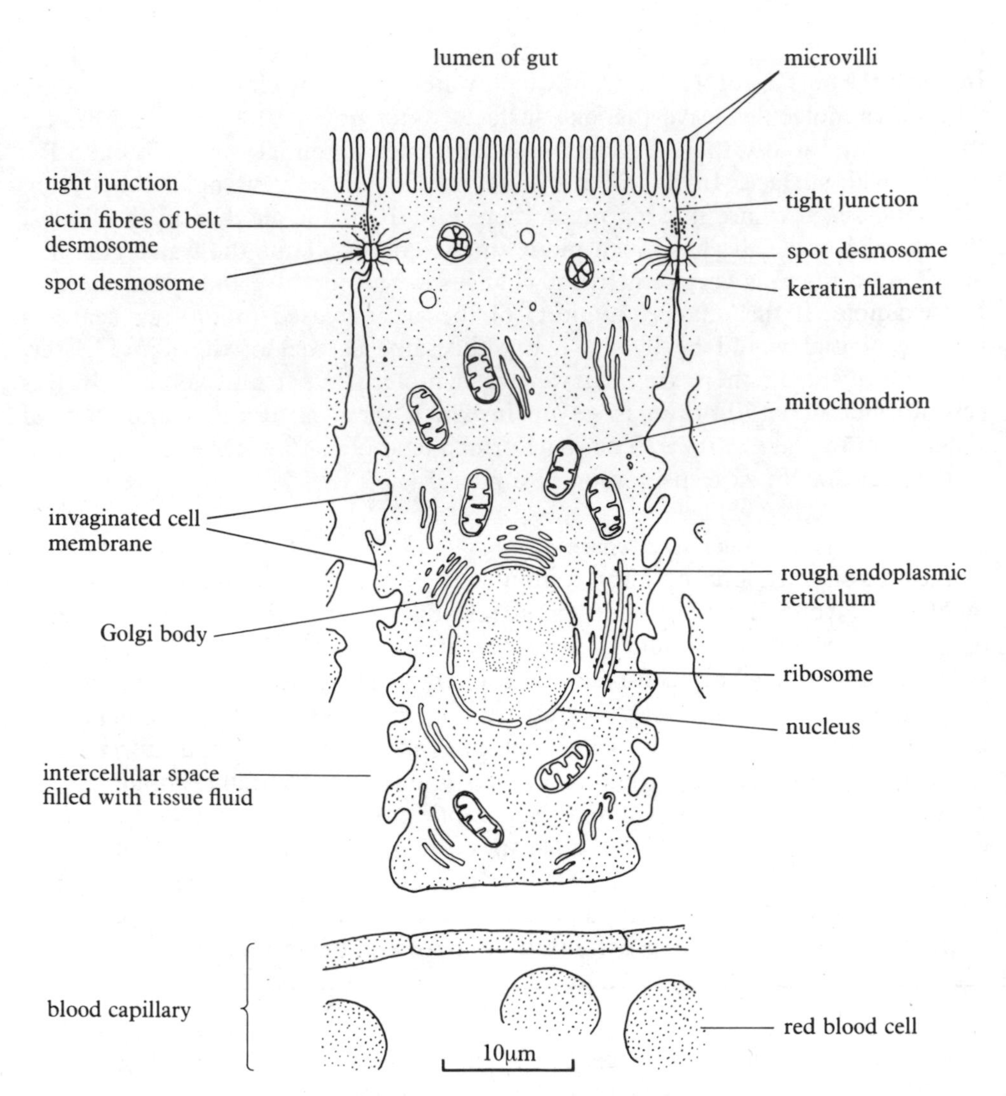

Figure 4.1 Schematic drawing of a cell in the absorptive epithelium of a villus. Notice the microvilli on the surface, the abundant mitochondria, the invaginated cell membrane on the side nearest the blood capillary, and the various types of connections between the cell membranes of adjacent cells. (*After Moog*)

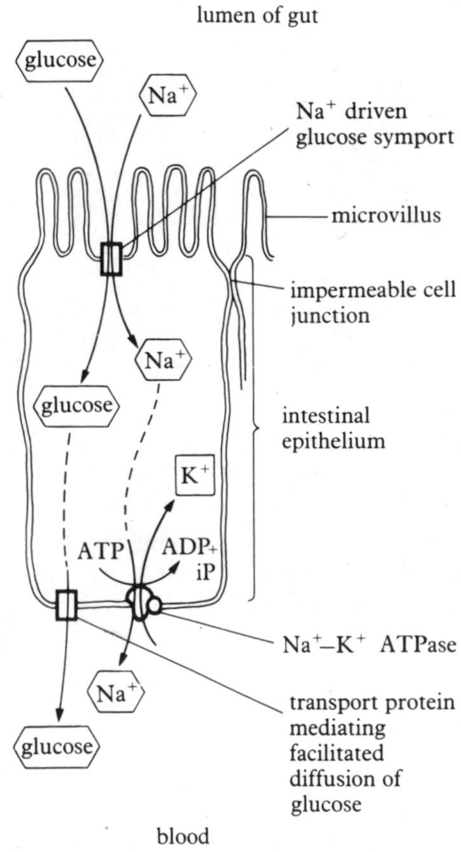

Figure 4.2 Highly diagrammatic view of a villus epithelial cell, showing how glucose is actively transported into the cell by a carrier (symport) which also takes up sodium. The glucose is then transported out of the cell into the blood by facilitated diffusion. The active transport carrier (Na^+K^+ATP-ase) in the basal membrane expends ATP to pump sodium ions from the cell into the blood. This keeps the cellular concentration of sodium ions low. The sodium ions flow into the cell through the symport down their concentration gradient, and this drives the (active) transport of glucose into the cell. The double line which surrounds each cell greatly exaggerates the width of the cell membrane. (*Based on Alberts et al, 1983*)

1 μm high, mitochondria are abundant and the cell membrane near the blood capillary is invaginated, thus presenting a large surface area over which the molecules can be transferred.

The absorptive cells in the epithelium of the villus are held together tightly by compounds attached to the cell membrane. This may prevent them from being rubbed off the villus surface by shearing forces as the food moves down the lumen. The attachments are of two main types:

1. Bands of intercellular material hold together the absorbing cells just below the level of the microvilli. Associated with each band, and running along it on the cytoplasmic side of the cell membrane, is a sheaf of parallel actin fibres. Together they constitute a **belt desmosome**.
2. On the inner surfaces of both the interacting plasma membranes there are hemispherical aggregations of protein molecules. Keratin fibres, which stretch throughout both the cells, run into these aggregations and pass from cell to cell. These act like rivets, holding the cells together. They comprise a **spot desmosome**.

An even more striking feature of absorptive cells in the villi are **tight junctions** (see page 64). Here the cell membranes of the adjacent cells are held together so closely by proteins projecting from the membranes that compounds cannot pass through the gaps between the cells. Tight junctions may be useful for two main reasons. Firstly, they prevent compounds from the lumen of the gut reaching the blood directly, and they prevent tissue fluid from leaking out into the lumen. The compounds which travel from the lumen into the bloodstream can do so only through the absorbing cells, which take up the nutrients selectively. Secondly, the membrane proteins which absorb glucose from the lumen are different from those which transfer glucose from the absorbing cells to the bloodstream, and the tight junctions may prevent these two types of protein from moving to the wrong positions in the membrane.

THE FUNCTIONS OF THE CELL MEMBRANE

Glucose is probably absorbed mainly by **active transport**. In this process a glucose molecule is taken into the cell against the concentration gradient by combining with a protein carrier in the membrane, which then releases it on the other side. An ATP molecule, derived from aerobic respiration, is expended to pump the glucose across the membrane. The ATP may be expended at the site of the carrier, on the microvilli. However, it now seems more likely that the ATP is expended at the other side of the cell, next to the blood capillary. There an enzyme, powered by ATP, expels sodium ions and ultimately powers the uptake of glucose from the lumen (figure 4.2).

An additional mechanism for glucose uptake is **pinocytosis**. In this case the glucose probably combines with a **membrane receptor**. This section of membrane invaginates, and a tiny spherical vesicle of membrane, containing glucose, enters the cell. This mechanism also requires energy. The vesicles then pass to the other side of the cell and the glucose they contain is released by exocytosis into the tissue fluid.

On the blood side of the absorbing cell, however, most glucose molecules probably pass from cell to capillary by **facilitated diffusion**. The glucose concentration is higher in the cell than in the capillary and so the glucose is able to diffuse across the membrane down its own concentration gradient. This is probably accomplished by a protein which projects from one side of the membrane to the other, recognises glucose, and allows it to diffuse out of the cell.

Two other sorts of membrane protein molecule are important in the uptake of compounds by the absorptive cells of the villus. Firstly, many of the enzymes which carry out the final stages of digestion are proteins embedded in the membranes of the microvilli. Among them are disaccharidases, such as maltase and invertase, which split specific 12-carbon sugars into 6-carbon units; and the amino peptidases, which split the final amino acid from short peptide chains. Secondly, cylindrical protein molecules known as **connexons** extend right through the membrane (figure 4.3). The connexons from two adjacent cells form an almost continuous tube. Some areas of the cell membrane, known as the **gap junctions**, possess hundreds of connexons. Presumably these allow adjacent epithelial cells to exchange small molecules and ions.

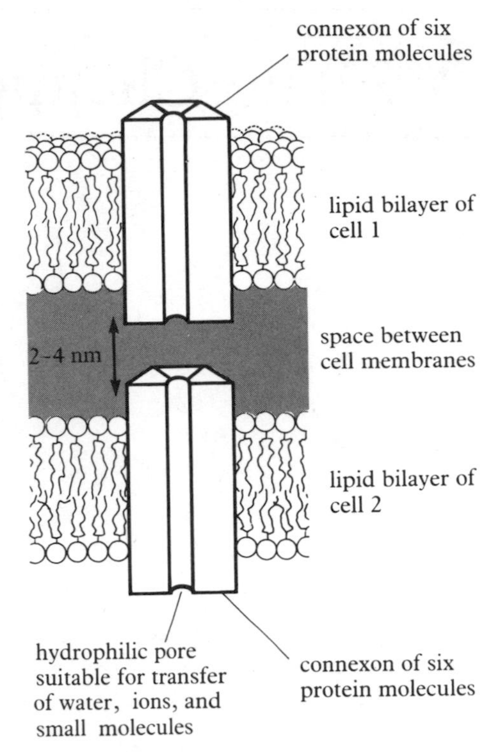

Figure 4.3 Schematic drawing showing a section through two connexons in the membranes of adjacent cells. It is connexons such as these which form the gated channels which are so important in maintaining the resting potential (when closed) and conducting a nerve impulse (when open) in an axon membrane during nervous conduction. Numerous connexons in the same area of cell membrane constitute a gap junction. Connexons seem to allow ions and molecules to pass through the membrane provided that they have a relative molecular mass of less than approximately one thousand.

For consideration

1. Where are the lipids and the protein which appear in the cell membrane synthesised? How are they transferred into the cell membrane?

2. What might be the consequences if (a) the gut lumen contained no sodium ions (b) the tight junctions weren't tight.

3. Why should adjacent epithelial cells exchange small molecules and ions?

Further reading

B. Alberts *et al*, *The Molecular Biology of the Cell* (Garland, 1983)
This contains a clear modern account of cell adhesion and cell junctions.

F. Moog, 'The Lining of the Small Intestine' (*Scientific American*, vol. 245, no. 116, 1981)
This article covers all aspects, from the structure of the gut wall to hypotheses explaining the uptake of molecules.

5 . The Chemicals of Life

Protein sequencing

The primary structure of a protein is the order of its amino acids. The primary structure is important because it alone determines the three-dimensional shape into which the whole protein molecule folds. In turn, its three-dimensional shape is crucial to its activity, because whether fibrous or globular, the shape determines the molecule's ability to interact with other molecules. In this topic we shall describe how a protein's primary structure can be investigated.

HOW CAN WE DETERMINE A PROTEIN'S PRIMARY STRUCTURE?

The classical method of determining the sequence of amino acids in a purified protein is based on that used by Sanger in his original determination of the primary structure of insulin. There are four main steps.

1. The amino acids present in a polypeptide chain, or a fragment of it, are identified by hydrolysing it with concentrated hydrochloric acid. In practice this is carried out in a sealed tube at 105 °C for 24 hours. This allows complete breakdown of the polypeptide into its amino acids under conditions in which none of the amino acids is likely to be oxidised. The mixture of amino acids in acidic solution is then fed into an ion-exchange column containing resinous beads. Positively-charged amino acids become electrostatically bonded to the negatively-charged molecules which have already been attached to the resin. Buffers of gradually increasing pH are fed through the column. Since each amino acid leaves the column at a different known pH, the amino acids always emerge from the base of the column in a predictable order. The relative amount of each amino acid can be determined by treating small samples of the emerging solution with a compound which colours amino acids. The intensity of colour in each sample is then determined quantitatively with a colorimeter. This process has been automated and is now routine.
2. At one end of a polypeptide chain is an amino acid with a free amino ($-NH_2$) group, and at the other end an amino acid with a free carboxylic acid ($-COOH$) group. The amino-terminal (N-terminal) and the carboxy-terminal (C-terminal) amino acids of any protein fragment can be easily determined. For instance, the N-terminal end reacts with the yellow dye 1-fluoro-2,4-dinitrobenzene (FDNB). When the polypeptide chain is subsequently hydrolysed, the identity of the amino acid attached to the yellow marker can be determined by chromatography.
3. Certain protein-digesting enzymes are known to split polypeptide chains next to specific amino acid residues. If the polypeptide chain is digested with one of these enzymes or compounds, the identity of the amino acid on the right-hand (C-terminal) end of each fragment can be ascertained. For example, trypsin cleaves linkages with lysine or arginine on the left, as shown in figure 5.1. On the other hand chymotrypsin splits linkages with phenylalanine, tryptophan or tyrosine on the left (see page 40), and cyanogen bromide splits linkages with methionine on the left.
4. The fragments which result from digestion by enzymes can be separated from one another by chromatography or electrophoresis, and their amino acid composition can be determined by repeating the procedure described above, using a different protein-digesting enzyme.

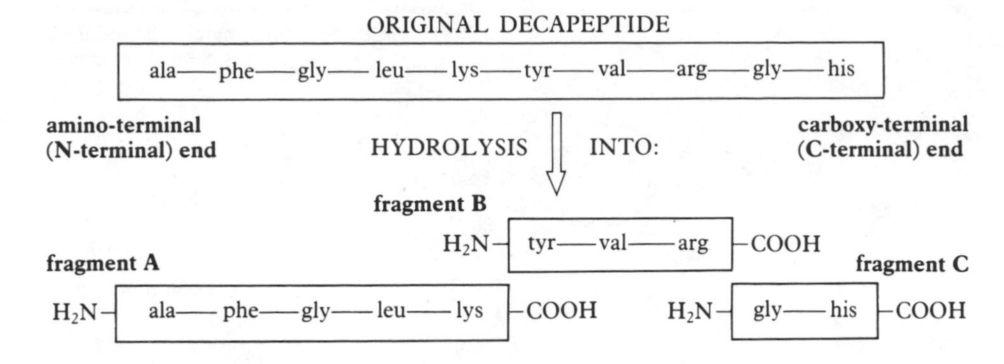

Figure 5.1 In the peptide illustrated here hydrolysis with trypsin produces the three fragments A, B and C. The three fragments can be isolated and the amino acids which they contain can be determined. The N-terminal amino acids in each fragment can be identified with FDNB (see text). The C-terminal amino acids in fragments A and B are known from the specificity of trypsin digestion. The amino acid sequences of fragments B and C are thus established, but the amino acid sequence of A and the order in which the three fragments A, B and C exist in the intact polypeptide chain can only be determined after digestion with a different proteolytic enzyme. (*After Wynn*)

A MORE MODERN METHOD

Nowadays there is a more elegant way to determine the primary structure of a protein, provided that the messenger RNA which codes for it can be isolated. Using a reverse transcriptase enzyme, molecules of DNA with a complementary base sequence are synthesised on the RNA template. The DNA molecules are then fed into an automatic analyser, a 'gene machine' which can determine the base sequence of a gene at the rate of about two bases an hour. Knowing which particular bases stand for particular amino acids in the genetic code, we can work out the sequence of amino acids in the protein.

For consideration

Hydrolysis of a polypeptide chain with chymotrypsin produced three fragments, A, B and C. A contained glycine (N-terminal), tryptophan and alanine; B contained lysine (N-terminal), proline, cysteine and glycine (C-terminal); C contained methionine (N-terminal), tyrosine and arginine.

Hydrolysis of the same polypeptide chain with pepsin produced four fragments, D, E, F and G. D contained lysine on its own, E had cysteine (N-terminal), glycine and proline, F contained methionine (N-terminal) and arginine, and G consisted of tyrosine (N-terminal), lysine, alanine, glycine and tryptophan.

The order of the fragments A–C and D–G in the peptide, and the order of the amino acids in each fragment, are not necessarily in the order given above.

Determine the primary amino acid sequence of this peptide.

Further reading

G.H. Harper, *Tools and Techniques* (Nelson, 1984)
The author describes clearly many of the techniques mentioned in this topic.

C.H. Wynn, *The Structure and Function of Enzymes* (Studies in Biology no. 42, 2nd edition, Arnold, 1979)
The methods for determining protein structure are covered in detail.

Massive molecules

The function of a protein is both determined by, and depends on, its three-dimensional structure. Many enzymes, for example, are globular proteins with active sites which combine precisely with their substrates, as in the case of the serine proteases (see page 40). In this topic we shall examine the structures and functions of two fibrous proteins from the connective tissue of humans – **collagen** and **elastin**.

THE STRUCTURE AND FUNCTION OF COLLAGEN

Collagen makes up a quarter of the dry mass of protein in mammals. It is the main protein in bone, cartilage, tendons and connective tissue. A single functional collagen 'molecule' consists of three polypeptide chains wound around one another like the strands of a rope (figure 5.2).

Each of these polypeptide chains is about a thousand amino acids long, but when formed at the ribosomes within a collagen-secreting fibroblast they are even longer. At both ends they have **extension peptides** which help to establish the triple helix by forming disulphide bonds with one another. Attached to the end of one of these extension peptides is another peptide which helps the

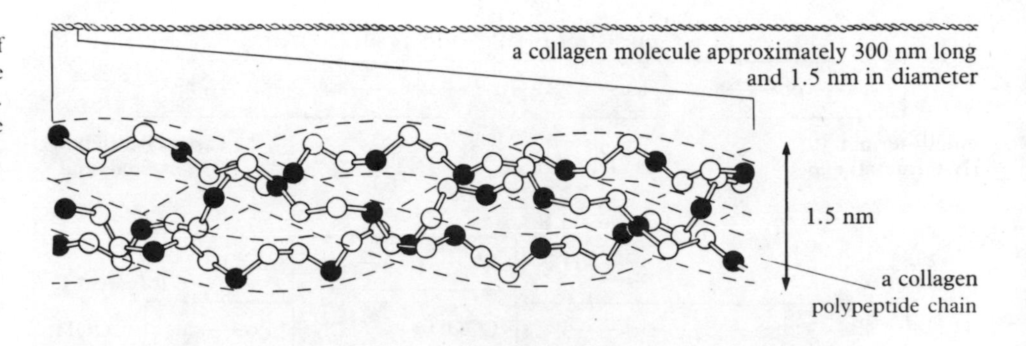

Figure 5.2 A diagrammatic representation of the rope-like collagen molecule. Three separate polypeptide chains are wrapped around each other. Each of these chains is an α-helix with glycine every third residue.

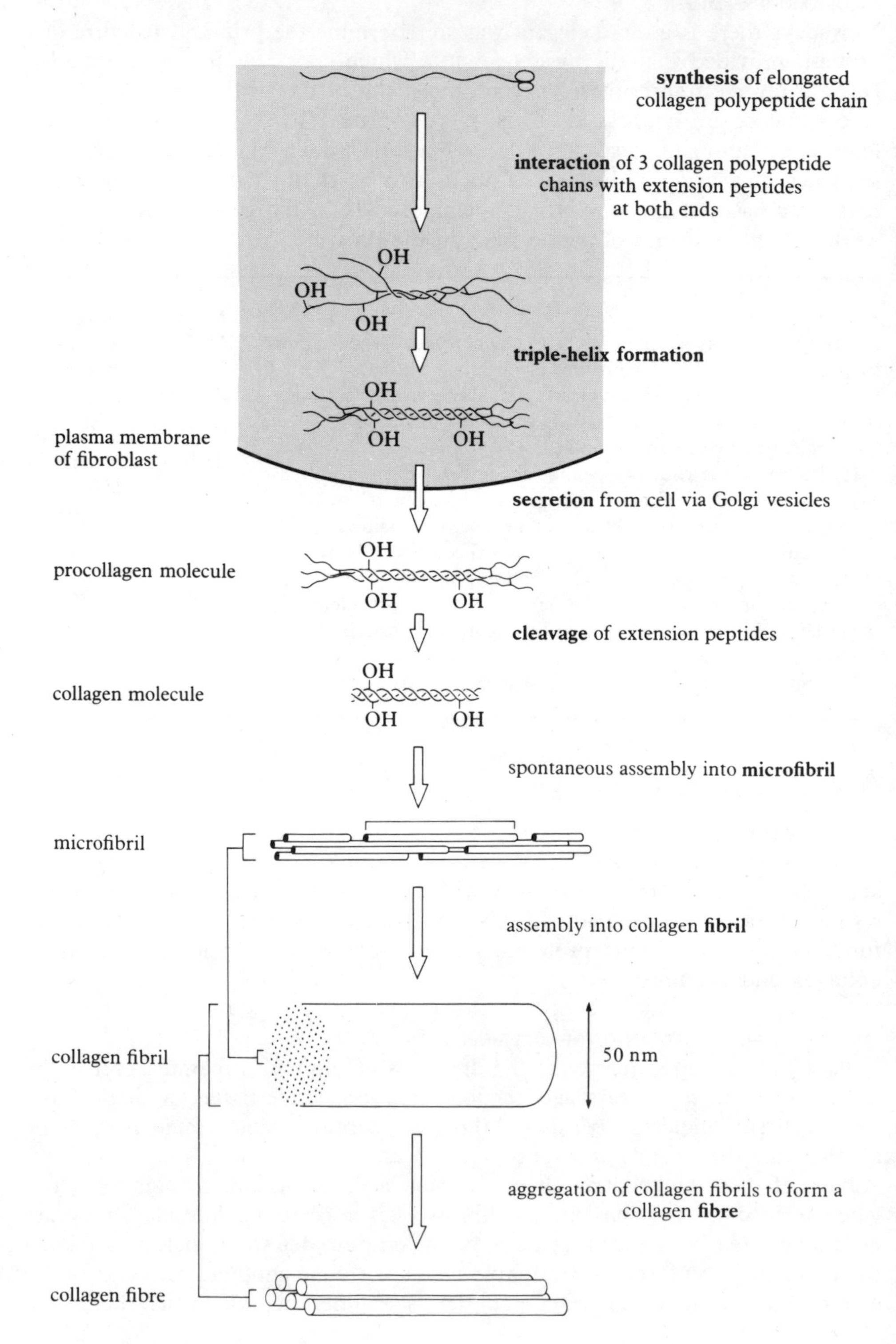

Figure 5.3 The sequence of events involved in the formation of a collagen fibre of a size which is visible under the light microscope. The hydroxyl side groups projecting from the triple helix belong to modified lysine residues which hold adjacent collagen molecules together in the microfibril.

molecule to dissolve in a cell membrane for export via the Golgi body (upper part of figure 5.3). The extension peptides are then cut off the triple helix by an enzyme.

Compared with globular proteins, a polypeptide chain of collagen is unusual in its amino acid composition and in its repeating structure. Each polypeptide chain is an α-helix. It has a great deal of glycine, alanine, proline and hydroxyproline. These all have small R–groups, and glycine, which occurs every third residue, may be the only amino acid with a small enough R– group (a hydrogen atom) to occupy the interior of the triple-stranded functional molecule.

Each of these triple-stranded units is about 300 nm long and 1.5 nm in diameter. However, numerous collagen molecules are associated outside the cell to form collagen fibrils, which are thousands of nm long and up to 300 nm in diameter (lower part of figure 5.3). These fibrils form spontaneously in a test tube when collagen molecules are mixed, and so the information for their formation must reside in the collagen molecules themselves, rather than in enzymes.

Electron micrographs show that the collagen fibrils are striated, with alternating light and dark bands every 67 nm along their length. In the fibres each triple helix is displaced by 67 nm, or 234 amino acids, in relation to its neighbours. This both generates the banding pattern and adds to the strength of the fibril.

From the order of amino acids in each strand of the triple helix the numbers of covalent, electrostatic and hydrophobic bonds which a triple helix can form with other triple helices may be predicted for each point on the molecule. Using a computer, the numbers of bonds which could be formed between two helices has been determined for all possible relative displacements of the two sequences. The main peaks were at displacements of multiples of 234 amino acids, which corresponds to the displacement length in nature. It seems that at these points modified lysine residues project from the triple helices and can form covalent bonds with one another. In tendons numerous parallel fibrils may associate together to form the collagen fibres which impart such tensile strength to the tissue. The strength is increased by extra cross-linking, for example in the Achilles tendon. Collagen is relatively inelastic – it requires considerable force to stretch and break it, and if stretched, it does not, like a rubber band, rebound into its former shape.

A WORD ABOUT ELASTIN

A different protein, elastin, is present in tissues which are elastic, such as the skin, ligaments, arteries and lungs. Why are elastin molecules more elastic than those of collagen? Unlike most proteins, elastin is not organised into a specific three-dimensional structure. Most of its amino acids are uncharged and hydrophobic. It can adopt a variety of 'random coil' structures, in all of which there is no cross-linking between different parts of the same molecule. Numerous molecules are cross-linked together, by links through lysine residues, to form the loose network which exists in an elastic fibre (figure 5.4). When one end of the fibre is pulled, some of the 'slack' is taken up and the potential energy is stored in the fibre, only to be released when the fibre relaxes and the chains return to their natural state of least entropy.

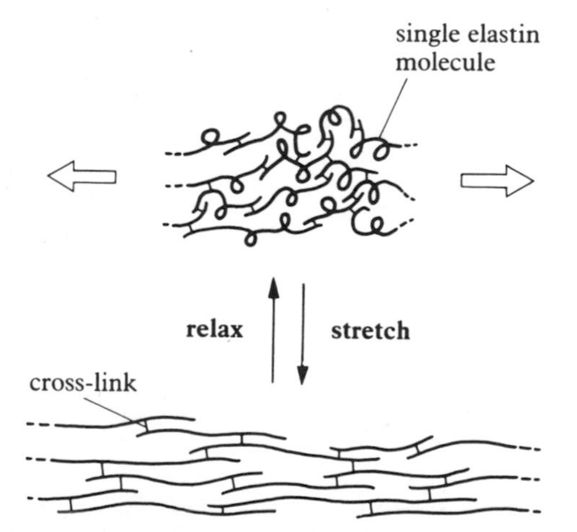

Figure 5.4 Elastin molecules do not have a specific three-dimensional structure. They are cross-linked to form an extensive network.

For consideration

Predict, giving reasons, the molecular structure of (a) keratin, the major protein of skin, horn, nails and claws and (b) resilin, the protein in the legs of fleas which enables them to jump to heights up to 130 times their own height.

Further reading

B. Alberts *et al*, *The Molecular Biology of the Cell* (Garland, 1983)
This deals in detail, but very clearly, with collagen structure and synthesis.

J. Woodhead-Galloway, *Collagen: the anatomy of a protein* (Studies in Biology no. 117, Arnold, 1984)
A fascinating review of the structure and properties of collagen.

6 . Chemical Reactions in Cells

The birth and death of a starch molecule

Each time you swallow a mouthful of mashed potato, a million million million molecules of starch, on average, enter your digestive system. Seventy per cent of an average human diet consists of carbohydrate, and of that, much will be starch. Starch, derived from potatoes, flour and vegetables, therefore provides a high proportion of our energy intake.

Starch is an insoluble but compact energy source. Each molecule consists of over two thousand glucose units covalently bonded end to end. Each glucose unit is potentially capable of yielding energy to synthesise 38 ATP molecules in aerobic respiration. But how has starch obtained its energy?

STARCH AS AN ENERGY STORE

The second law of thermodynamics states that all systems tend towards a state of greater disorganisation (entropy) unless energy is added to the system. From the point of view of starch, its precursors, carbon dioxide and water, are molecules in random motion. If they are trapped in an organised structure, such as the glucose molecules formed in photosynthesis, their entropy decreases. The entropy of the glucose molecules themselves declines still further if they are incorporated into an organised structure such as a starch molecule. It follows that the production of starch from glucose, and the production of glucose from carbon dioxide and water, are both endothermic, requiring an energy source. On the other hand, the breakdown of starch can presumably occur spontaneously. No energy, other than the activation energy required to start the reaction, is required.

Starch in green plants is formed in chloroplasts, for example in the palisade mesophyll of leaves, and in leucoplasts, their colourless cousins, as in potato tubers. The glucose molecules from which starch is formed are made by harnessing solar power in photosynthesis. Glucose will not spontaneously polymerise to form starch. The equilibrium is very much towards starch breakdown, however concentrated the glucose is in the plastid. The organisation of a starch molecule from its glucose precursors requires an energy input from ATP.

One way of providing energy would be for one ATP molecule to be expended each time a glucose unit was added to an existing starch chain, like this:

$$\text{Glucose} + \text{ATP} \rightarrow \text{Glucose-1-phosphate} + \text{ADP}$$

$$\text{Glucose-1-phosphate} + \underset{\textit{Starch}}{(\text{Glucose})_n} \rightleftharpoons \underset{\textit{Longer starch}}{(\text{Glucose})_{n+1}} + \text{inorganic phosphate}$$

Enzymes exist which can catalyze both these reactions. The enzyme **starch phosphorylase**, which performs the second of these two reactions, is particularly abundant in potato tubers. Once extracted, it can be used to polymerize glucose-1-phosphate to starch in a test tube. Glucose-1-phosphate is more reactive than simple glucose.

This reaction, however, does not always go in the direction of starch synthesis. When roughly equal amounts of reactants and products are placed in a test tube with the enzyme, a dynamic equilibrium is soon set up in which, at least at first, the starch is broken down at the same rate as it is synthesized. The reaction can be made to go one way or the other by changing the relative concentrations of reactants and products.

Suppose we place in our test tube the enzyme starch phosphorylase and glucose-1-phosphate solution, thus providing a high concentration of reactant. There is no starch to begin with. The reaction goes towards starch synthesis. In the intact potato, however, the opposite situation exists. Glucose-1-phosphate is scarce and starch is very abundant. The equilibrium moves towards starch breakdown. This is in fact what happens when the potato stem begins to grow out of the tuber and produces a new plant. Stored starch is hydrolysed and starch phosphorylase catalyses the breakdown, not the synthesis, of starch.

FURTHER DISCUSSION

To make the synthesis of starch occur in conditions in which starch already exists, even more energy must be expended on the left-hand side of the equation of produce the organised structure of a starch molecule. In the intact chloroplast starch is made not from glucose-1-phosphate, but from the nucleotide sugar **adenosine diphosphate glucose** (ADPG). This is even more reactive than glucose-1-phosphate, probably because two ATP molecules have been hydrolysed during its manufacture. The equilibrium between ADPG and starch is very much towards starch synthesis:

$$\text{Glucose} + \text{ATP} \rightarrow \text{ADP} + \text{Glucose-1-phosphate}$$

$$\text{Glucose-1-phosphate} + \text{ATP} \rightarrow \text{ADP glucose} + 2\ (\text{inorganic phosphate})$$

$$\text{ADP glucose} + \underset{\textit{Starch}}{(\text{Glucose})_n} \rightarrow \underset{\textit{Longer starch}}{(\text{Glucose})_{n+1}} + \text{ADP}$$

Thus it 'costs' the chloroplast two ATP molecules to add one glucose unit to an existing starch molecule. Why, then, does the cell store starch at all? There are two possible reasons. In the first place, starch is osmotically inactive compared with the glucose from which it is formed. If glucose accumulated within the chloroplast it would lower the water potential so much that water would flow in from outside and the chloroplast would burst. In the second place, the temporary removal of glucose helps the synthesis of glucose to continue. If glucose, as an end product of photosynthesis, accumulated in the stroma, and the reactions which produced it had an equilibrium constant of about one, the rate of the back reaction would steadily increase and the net rate of glucose manufacture would decline.

Starch is only one of several polymers in organisms. The potential chemical energy of ATP is used in the synthesis of all the macromolecules of the cell (see page 33). The synthesis of the other carbohydrate polymers closely resembles that of starch in plants. For example, cellulose in plant cell walls, and glycogen in mammalian liver and muscle, are both made from **uridine diphosphate glucose** (UDPG) which performs essentially the same function as ADPG.

For consideration

Why is it that starch is the main energy storage compound in most higher plants, but fats form the major energy store in mammals and birds?

Further reading

A.W. Galston, P.J. Davies and R.L. Satter, *The Life of the Green Plant* (Prentice-Hall, 3rd edition 1980)
A useful source of information.

A. Lehninger, *Bioenergetics* (Benjamin, 1973)
A clear, elegant introduction to cellular energetics – highly recommended.

The control of metabolic reactions

Have you ever wondered what controls the types of compounds which occur in a cell, and their relative concentrations? In this topic we imagine an individual cell in steady state, and discuss some of the factors which control metabolism.

Of course the reactions which occur within a cell at any time depend on the particular enzymes present. In turn, the production of enzymes is controlled by genes. Gene action is influenced by the age of the cell, its position in the organism, and the concentrations of various ions, hormones and other organic compounds to which the cell is exposed. However, much of the short-term regulation of the levels of compounds in the cell is carried out by the enzymes themselves.

Protein molecules are not only remarkably specific, able in many cases to recognise one or two compounds among thousands, but their molecules are flexible. Many protein molecules are like delicate spiders' webs. If one strand is pulled, the shape of the whole structure changes. The binding of a protein molecule with another molecule frequently alters its ability to bind with another compound at a different active site. Such allosteric effects are involved in many of the mechanisms which regulate the metabolic activity of a cell. We shall look at an example in a moment.

REGULATION AT THE GENETIC LEVEL

In eukaryotes the long chains of DNA in the nuclei of cells are wrapped around spools made from histone proteins to form structures called **nucleosomes**. It seems likely that in this state the genes cannot express themselves, in other words they cannot act as sites for messenger RNA synthesis. However, what determines which genes are inactivated in this way is unknown.

Whether the unwrapped genes are switched on or off at any one time probably depends on their binding with certain small protein molecules. Two types of system seem to exist. When hormones such as ecdysone act on the salivary gland cells of Drosophila larvae, the chromosome puffing observed shows that certain genes have become active. Presumably the ecdysone has combined with repressor protein molecules attached to the DNA, and removed them. A similar situation occurs in the case of the **lac operon** in the gut bacterium *E. coli*. Here the disaccharide lactose is the inducer, removing the repressor protein by the allosteric mechanism shown in figure 6.1.

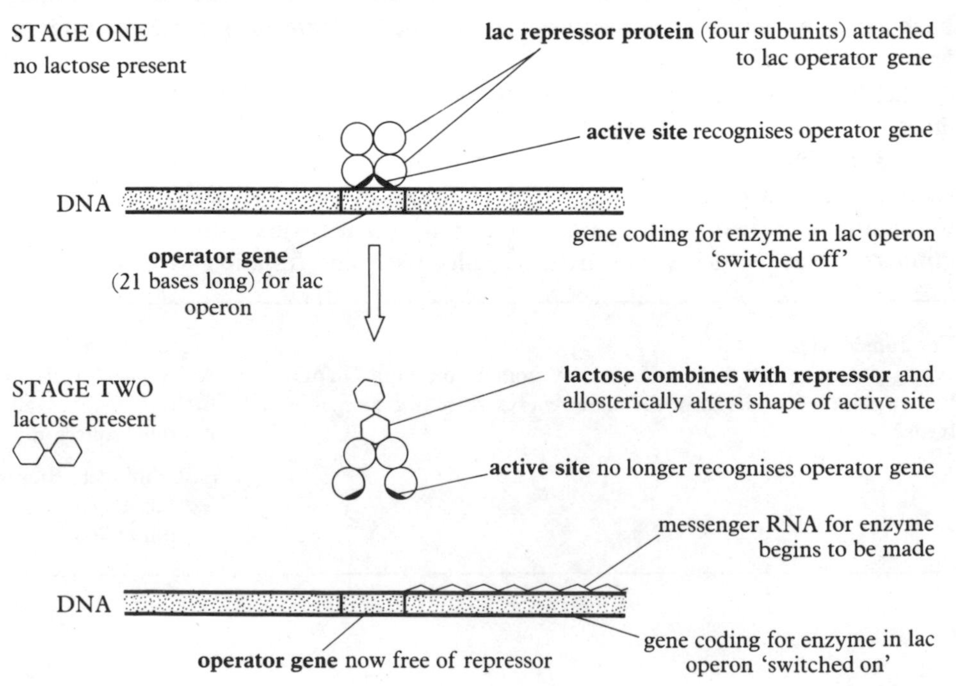

Figure 6.1 An example of an allosteric effect. The lac repressor, a protein molecule, has two active sites. One binds with DNA, the other with lactose. However, the binding with lactose prevents the binding with DNA. This is important because the presence of lactose inside a cell of the bacterium *E. coli* prevents the lac repressor from binding with the DNA. This allows the synthesis of enzymes which promote the uptake and utilisation of lactose from the surrounding medium.

The system works the other way round in the induction of tryptophan synthetase enzyme. When the amino acid tryptophan is present in the cell, it combines with the repressor molecule which is then able to bond to the DNA, switching off the gene for enzyme synthesis. When the amino acid is uncommon in the cell, the repressor is not activated by tryptophan and does not bond to the DNA. Thus the gene is not prevented from coding for an enzyme catalysing tryptophan's synthesis.

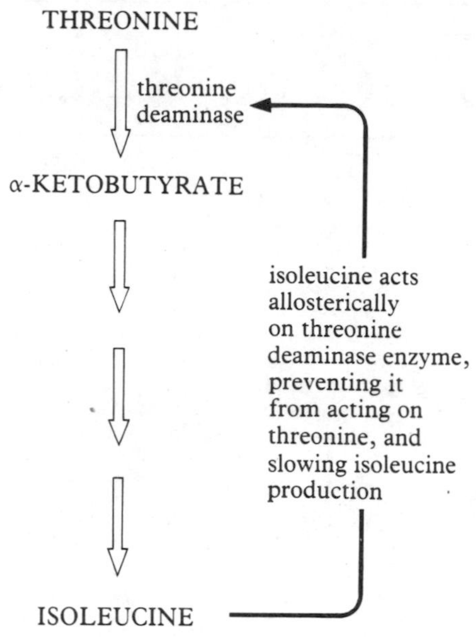

REGULATION AT THE ENZYME LEVEL

Once enzymes have been produced, their activity can be regulated both from outside and inside the cell. Hormones such as glucagon and adrenalin in a mammal, for example, plug into receptor proteins on the cell membranes of liver cells. As a result, the level of cyclic AMP is raised, and this affects the activity of several enzymes concerned with respiration and glycogen synthesis and breakdown (see page 85).

However, intracellular mechanisms also exist. They possibly prevent certain compounds from becoming too abundant, and also make optimal use of the cell's energy resources. One frequent phenomenon is **end-product inhibition**, which occurs when the end product of a metabolic pathway allosterically inhibits the activity of the enzyme which catalyses the first reaction in the pathway. This happens in the synthesis of most of the amino acids. For example, isoleucine inhibits threonine deaminase, which catalyses the first reaction in the metabolic pathway leading to its synthesis (figure 6.2).

Figure 6.2 Feedback inhibition in the synthesis of the amino acid isoleucine in a bacterial cell.

The rate of cellular respiration is of central importance, because it determines the rate at which ATP energy is made available to the cell. In fact the rate of synthesis and other activities which use up ATP may determine the rate of respiration. One of the points at which the rate of respiration is regulated is the reaction in which fructose-6-phosphate is converted to fructose-1, 6-diphosphate near the beginning of glycolysis in the cytoplasm. The enzyme catalysing the reaction, phosphofructokinase (PFK) appears to be allosterically regulated by both the concentration of ATP and the concentration of ADP. When the ratio of ATP to ADP is high the cell needs little additional energy. ATP allosterically inhibits PFK and thus reduces the glucose respired. In this case the fructose-6-phosphate that accumulates can be used in other ways, such as the pentose phosphate pathway (see page 33) or conversion to glucose.

There are also more subtle ways in which metabolism is regulated. At branches in pathways, one enzyme may have a greater affinity for a certain substrate than another enzyme. The rate at which a molecule travels across a membrane into a membrane-bound compartment may limit the speed at which a set of reactions take place. Many metabolic reactions are reversible. All these mechanisms, acting together, ensure the constancy of the intracellular environment as well as the constancy of the composition of the tissue fluid.

For consideration

1. Many metabolic reactions are reversible. What are the consequences as far as their control is concerned?

2. Discuss the various ways in which mutations in DNA could cause some of the control mechanisms mentioned in this topic to fail.

Further reading

C.H. Wynn, *The Structure and Function of Enzymes* (Studies in Biology no. 42, 2nd edition, Arnold, 1979)
Read the concise chapter on the control of metabolic reactions.

L. Stryer, *Biochemistry* (Freeman, 2nd edition, 1983)
The author describes metabolic regulation in detail, aided by some exceptionally clear multicoloured diagrams.

7 . The Release of Energy

The link between glycolysis and the Krebs cycle

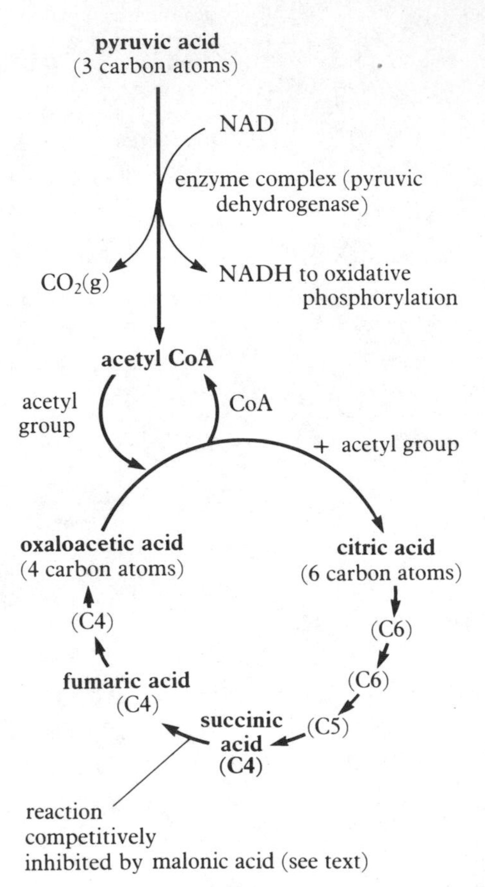

Figure 7.1 A simplified summary of the reactions involved in the transformation of pyruvic acid into citric acid in a eukaryotic cell.

Aerobic respiration takes place in three stages – glycolysis, the Krebs cycle and oxidative phosphorylation. In eukaryotes the first two stages, glycolysis and the Krebs cycle, are linked when **pyruvic acid**, the end product of glycolysis in the cytoplasm, moves into a mitochondrion and, after several chemical modifications, some of it enters the Krebs cycle. In this topic we shall examine the chemical reactions which occur at this major metabolic junction and use them to illustrate several biochemical principles. The chemical reactions involved are summarized in figure 7.1.

PYRUVIC ACID TO ACETYL COENZYME A

In the matrix of the mitochondrion, pyruvic acid meets the **pyruvic dehydrogenase system**, an enzyme complex, and is decarboxylated and dehydrogenated. The remaining acetyl group ($CH_3{-}CO$) combines with the thiol ($-SH$) group on a molecule of coenzyme A to form **acetyl coenzyme A**. The overall equation for this reaction is:

$$CH_3{-}CO{-}COOH + NAD + HS{-}CoA \xrightarrow{\text{pyruvic dehydrogenase}} + CH_3{-}CO{-}S{-}CoA + CO_2 + NADH_2$$

The reduced NAD will be used in oxidative phosphorylation, in which some of the energy in the two electrons which it donates to the cytochrome chain will be used to synthesise three molecules of ATP from ADP and inorganic phosphate.

The pyruvic dehydrogenase system is remarkable for its size – at 23×35 nm it is bigger than a ribosome, and easily visible under the electron microscope (figure 7.2). It contains three different types of enzyme and five types of coenzyme. Coenzymes are large, non-protein molecules essential for particular enzyme-controlled reactions but not normally attached to the enzyme molecules themselves. In the pyruvic dehydrogenase complex, no less than four of the coenzymes contain vitamins. These vitamins are biotin (vitamin H), thiamin (B_1), riboflavin (B_2) which is part of FAD, and nicotinamide (B_3, PP) – part of NAD and NADP. No wonder vitamins are important in the diet.

ACETYL COENZYME A

Coenzyme A is a complex molecule. It is part nucleotide, since it contains adenine, ribose and phosphate, and part vitamin, the vitamin being pantothenic acid (vitamin B_5). If present, the acetyl group is attached to the sulphur atom derived from the thiol ($-SH$) group.

Coenzyme A receives acetyl groups from the breakdown of fatty acids and amino acids as well as from glycolysis. The acetyl groups are the fuel for the Krebs cycle. Acetyl coenzyme A (frequently referred to as acetyl-CoA) is an energised compound which generates on hydrolysis more free energy than an ATP molecule. When it breaks down, the two-carbon acetyl group is released, and the reconstituted coenzyme A can pick up another acetyl group from glycolysis or fatty acid breakdown.

OXALOACETIC ACID TO CITRIC ACID

In the Krebs cycle the acetyl group is decarboxylated and dehydrogenated according to the general equation:

$$CH_3COOH + 2H_2O \longrightarrow 2CO_2 + 8H$$

This happens in a series of separate enzyme-controlled reactions involving acids containing four, five and six carbon atoms (4C, 5C and 6C acids respectively) in the mitochondrial matrix. Oxygen, ATP, ADP and inorganic phosphate are not required for this cycle.

During the first reaction of this cycle, 4C **oxaloacetic acid** combines with the 2C acetyl group released by acetyl-CoA to form 6C **citric acid.** The enzyme **citric synthetase** is the catalyst. Evidence for this reaction began to accumulate in the 1930s when it was found that oxaloacetic acid was oxidised very rapidly on being added to minced muscle. In fact, given a single molecule of oxaloacetic acid – or any other Krebs cycle acid for that matter – the Krebs cycle can continue indefinitely, provided that acetyl groups continue to be fed into the cycle. This is because the oxaloacetic acid is regenerated each time the cycle turns.

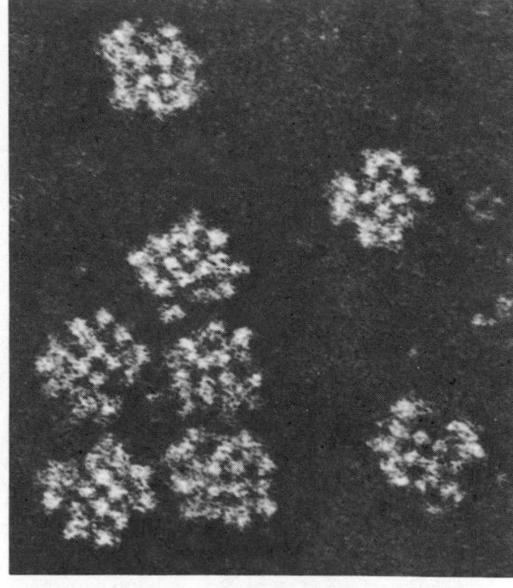

Figure 7.2 An electron micrograph of the pyruvate dehydrogenase complex of the bacterium *E. coli* (×325 000). Notice its large size. How large in comparison would be (a) a prokaryotic ribosome (b) a eukaryotic ribosome?

EVIDENCE FOR THE CYCLICAL REGENERATION OF OXALOACETIC ACID

In 1937 Krebs and his co-workers found that when oxaloacetic acid and pyruvic acid were fed simultaneously to minced muscle in anaerobic conditions, some citric acid was formed. They then blocked the succinic acid to fumaric acid step in the hypothetical cycle with an inhibitor (lower part of figure 7.1). The result was that the level of oxaloacetic acid in the tissue fell to zero. Subsequently each molecule of oxaloacetic acid added was found to cause the disappearance of one molecule of pyruvic acid from the muscle. Similar experimental techniques were used to investigate the other reactions in the cycle.

Of course the oxidation of pyruvic acid and acetyl groups does not proceed with gay abandon without reference to cellular requirements. Both the pyruvic dehydrogenase complex and the citric synthetase enzyme are allosterically inhibited by ATP. Thus, when there is already enough ATP to satisfy cellular demand, the Krebs cycle is effectively switched off and the pyruvic acid in the cytoplasm can be diverted to other purposes.

For consideration

1. Suggest processes apart from the Krebs cycle in which the pyruvic acid formed from glycolysis might participate.

2. When muscle, in a suitable buffer, is first placed in a manometer a high rate of oxygen consumption is observed, but this quickly declines. The addition of a small amount of succinic acid causes the muscle to start taking up oxygen again and this effect continues for more than an hour.

It is found that the oxygen consumed is far more than is needed for the oxidation of the succinic acid which was applied to the muscle.

Explain the implications of this observation.

Further reading

B. Alberts *et al*, *The Molecular Biology of the Cell* (Garland, 1983)
This describes in detail the structure and functions of pyruvic dehydrogenase.

A. Lehninger, *Biochemistry* (Worth, 2nd edition 1979)
This text contains an elegant summary of the evidence for the Krebs cycle.

An A to Z of metabolism

A chart of metabolic pathways is as complex as a street map of London, with hundreds of highways and byways. To help you find your bearings we provide here a map of the metabolic motorways.

In cells in aerobic conditions, **glycolysis** and the **Krebs cycle** play a central role in metabolism. Compounds are fed into aerobic respiration for degradation into carbon dioxide, water and ATP. These breakdown reactions are catabolic. The other side of the coin, however, is that many of the compounds taken into the cell are used to manufacture important molecules which are essential both for the production of new cells and for the maintenance of existing cells. These synthetic reactions are anabolic. In this topic we shall look first at the catabolic reactions and then at the anabolic ones. In reality the two are very much connected and are summarised together in figure 7.3.

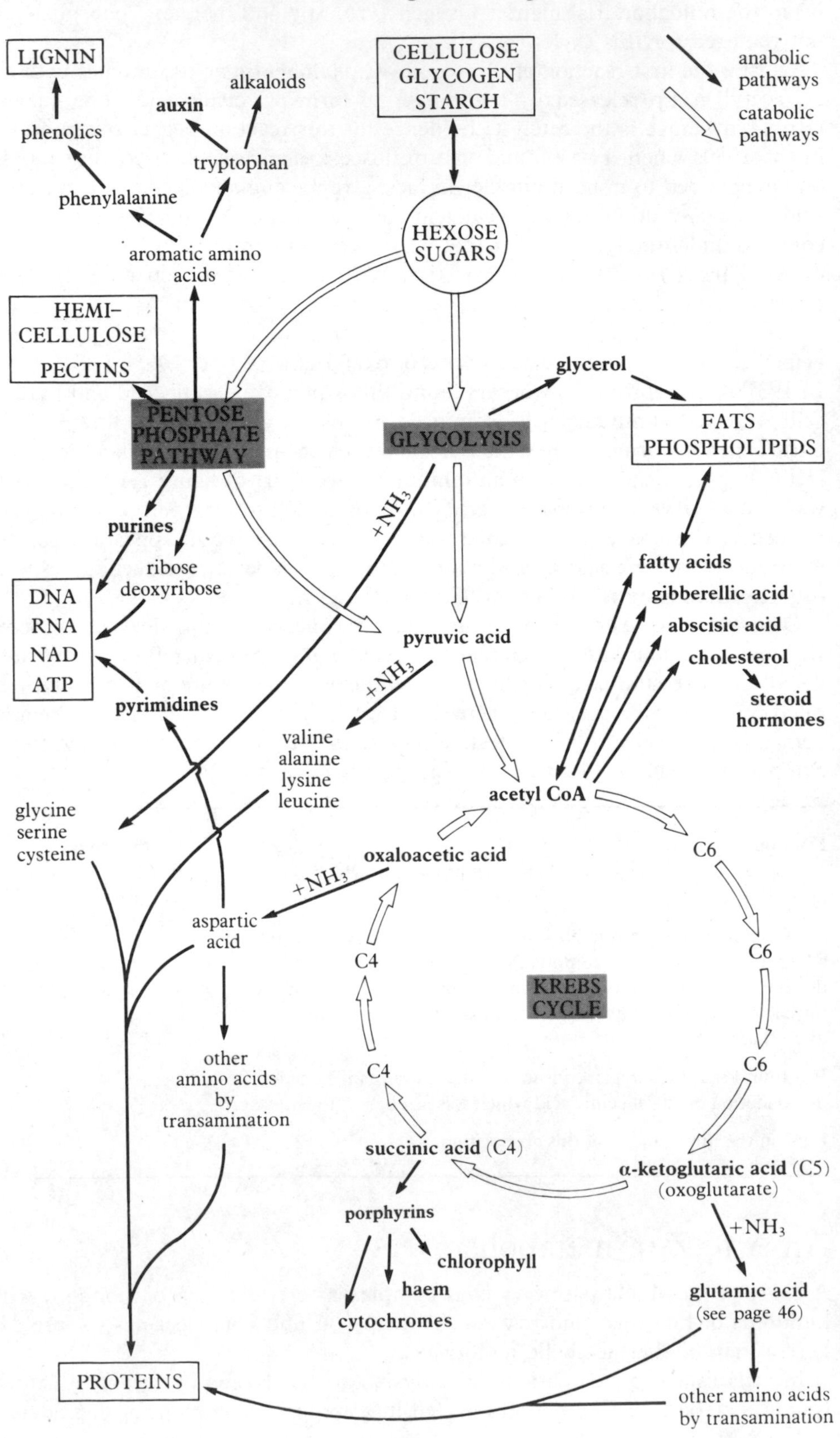

Figure 7.3 The main catabolic and anabolic reactions in a cell. In both catabolism and anabolism glycolysis and the Krebs cycle play a central role. Not all these compounds are synthesised in all cells, but in this diagram we have attempted to show how many of the molecules you have come across are synthesised. For simplicity, we have omitted the reactions of photosynthesis, and the electron transfer (respiratory) chain in respiration. The latter produces the ATP which drives the anabolic reactions of the cell.

CATABOLIC REACTIONS

These are shown by white arrows in figure 7.3. In glycolysis a six-carbon sugar is split into two three-carbon fragments. Each of these fragments is used to make a molecule of pyruvic acid. When this acid enters the Krebs cycle it is first decarboxylated to form a two-carbon fragment, an acetyl group, and then bonded to coenzyme A to form acetyl coenzyme A (see page 30). When the acetyl group is bound to a four carbon acid, a six carbon sugar (in this case a citrate) is formed. Then, in a cyclical series of reactions, it is decarboxylated and dehydrogenated in the Krebs cycle until the same four-carbon acid is regenerated to react again with acetyl coenzyme A.

During this series of transformations, vitamins are required as coenzymes at several different stages. For example NAD, required for dehydrogenation and the transfer of hydrogen atoms to the electron-transfer (respiratory) chain, contains nicotinamide (vitamin B_3). The enzyme succinic dehydrogenase, which converts succinic acid to fumaric acid in the Krebs cycle, needs FAD as a coenzyme – and FAD is a nucleotide containing riboflavin. A deficiency in either of these vitamins reduces the efficiency of the most important metabolic pathway in the cell.

As a result of these reactions and the operation of the respiratory chain ATP is synthesised. This ATP drives a series of anabolic reactions and other energy-requiring processes, such as ion uptake, nervous conduction, cell or muscular movement, and the synthesis of essential polymers. The turnover of ATP is very rapid. A human weighing 75 kg breaks down his own mass in ATP every day. In a human cell ATP has a half-life of only half a second!

ANABOLIC REACTIONS

These are shown by black arrows in figure 7.3. They lead to the formation of a wide range of important compounds in cells, notably nucleic acids, proteins, lipids and various carbohydrates.

Nucleic acids, which indirectly control cell metabolism, are polynucleotides. DNA, carrying the genetic code, is composed of millions of nucleotides, each of which contains a pentose sugar, a nitrogenous base and a phosphate molecule. RNAs are also polynucleotides. The three main types – messenger RNA, transfer RNA and ribosomal RNA – are all formed on DNA templates. Together DNA and RNA determine which proteins, many of them enzymes, the cell will manufacture.

The sugars deoxyribose and ribose, and the nitrogenous bases adenine and guanine (purines) are made from glucose by the **pentose phosphate pathway**, which also produces some reduced NADP ($NADPH_2$) for reductive reactions in the cell. Nitrogenous bases are components of the nucleoside triphosphates ATP (adenosine triphosphate), GTP (guanosine triphosphate), CTP (cytidine triphosphate) and TTP (thymidine triphosphate), from which DNA is synthesised. They are also, with the exception of UTP (uridine triphosphate) substituted for TTP, the building blocks from which the enzyme RNA polymerase constantly makes RNA in all metabolising cells.

Proteins, manufactured at the ribosomes, are synthesised from amino acids. Some amino acids arise from modifications of compounds due to glycolysis in the cytoplasm. Others arise by **transamination** of compounds derived from the Krebs cycle. For example, a Krebs cycle intermediate, oxoglutarate (α-ketoglutarate) can be combined with ammonia to form glutamic acid (see page 46). From glutamic acid several other amino acids can be formed by transamination, in reactions in which vitamin B_{12} (cobalamine) is required as a coenzyme. If destined for protein synthesis, the amino acids are activated by ATP and joined to their appropriate transfer RNAs. Then they are assembled in the right order at the ribosome at the rate of about ten a second.

Glucose not immediately required for metabolism is stored, often as fat. Fatty acids are formed from acetyl coenzyme A. Two-carbon fragments are combined to form acids with even numbers of carbon atoms. In the membranes of the endoplasmic reticulum they are joined to glycerol to form lipids and phospholipids, either contributing via the Golgi vesicles to the cell membrane, or lying as lipid droplets in the cytoplasm.

Glucose is temporarily stored as starch grains in plant cells, and as glycogen in animal cells. The manufacture of starch and glycogen are energy-demanding processes. Most plant cells also continually extrude cell wall polymers. The soluble precursors of cellulose, hemicelluloses and lignin are either glucose, or are synthesised from glucose by the pentose phosphate pathway. They enter the Golgi vesicles, and are transported to the membrane, where they polymerise (see page 117).

For all these syntheses, energy is necessary. This is provided ultimately by ATP and the oxidation of $NADPH_2$. How is the rate of each synthetic reaction controlled? The levels of the compounds in the cell are regulated by gene control and allosteric effects (see page 28).

For consideration

1. Pellagra is a disease which results from the deficiency of nicotinamide (vitamin B_3). Its symptoms are known as the three Ds – dementia, diarrhoea and dermatitis. Explain each of these symptoms on the basis of the vitamin's normal function in the cell.

2. An average human eats each day the equivalent of 700 g of glucose molecules. Assuming that this is all used in aerobic respiration, and that aerobic respiration yields 38 moles of ATP per mole of glucose oxidised, calculate the mass of ATP synthesised by the human each day. (Relative molecular mass of ATP = 507)

Further reading

A.W. Galston, P.J. Davies and R.L. Satter, *The Life of the Green Plant* (Prentice-Hall, 3rd edition, 1980)

This contains a brief but thorough account of anabolic reactions in the cell.

8 . Gas Exchange in Animals

The alveolar barrier

In the ten minutes that it will take you to read this topic, your lungs will have transferred over two litres of oxygen into your bloodstream across a surface the size of a tennis court. This surface forms a barrier between the alveolar air and the blood, and as the layer of tissue across which gas exchange takes place it has received considerable attention from physiologists over the years.

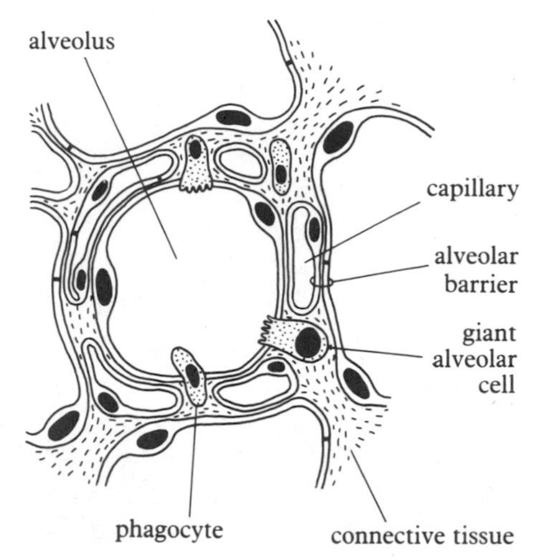

Figure 8.1 Several alveoli and associated capillaries as seen in a section of the mammalian lung.

STRUCTURE OF THE BARRIER

The **alveolar barrier** is extremely thin (figure 8.1) and the electron microscope has been required to resolve it clearly. It consists of two layers of cells, the epithelium surrounding the alveolus and the endothelium of the capillary. Both layers are composed of squamous (pavement) epithelial cells. Each cell has a slightly bulbous nucleated centre surrounded by an extended flattened region; here the plasma membranes of the upper and lower sides of the cell are so close together that the cytoplasm sandwiched between them is almost devoid of organelles (figure 8.2). The total thickness of the alveolar barrier varies from place to place but is generally around 0.3 μm; for comparison an average sized mitochondrion has a width of about 1.0 μm.

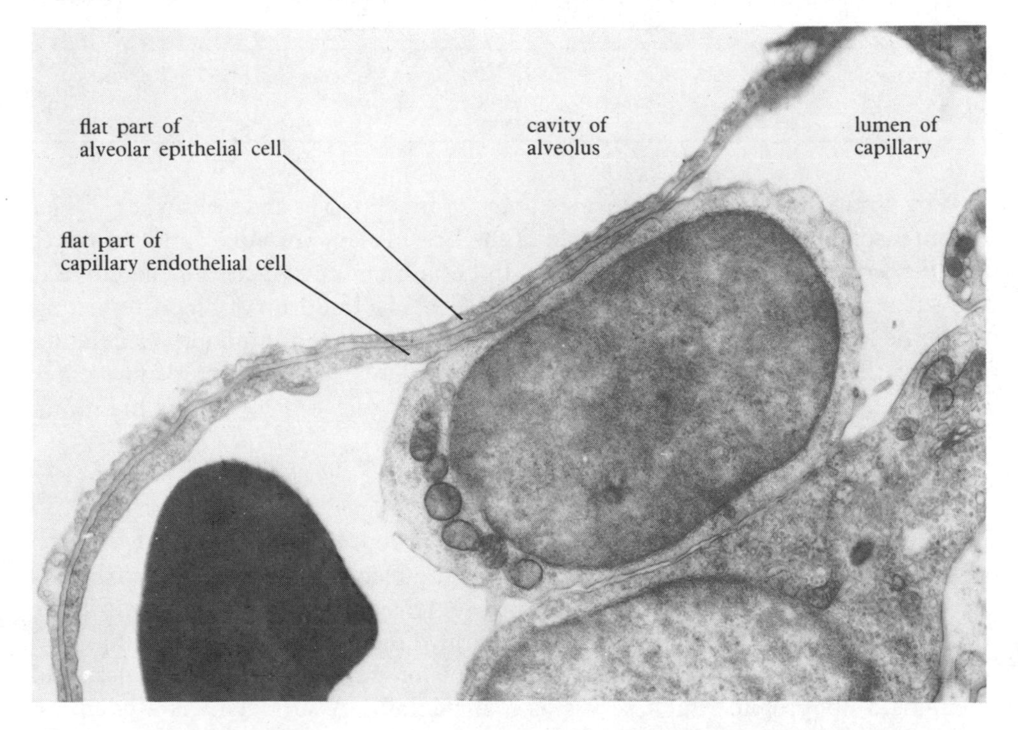

Figure 8.2 Electron micrograph of the alveolar barrier showing the flattened parts of an alveolar epithelial cell and capillary endothelial cell closely applied. What other structures can you identify in this micrograph? (×25 000)

Two other types of cell are associated with the alveolar barrier. Phagocytes, located in the tissue between neighbouring alveoli, may move through the barrier into the alveoli where they ingest small particles that have managed to get into the depths of the lungs. Wedged between adjacent alveolar epithelial cells are large cuboidal cells called **giant alveolar cells**. They contain numerous organelles including a prominent Golgi body and their free surface is covered with microvilli. Their function will be considered shortly.

The alveoli and associated capillaries are bound together by a three-dimensional meshwork of elastic and collagen fibres. These, as well as holding the various structures together, assist in the elastic recoil of the lungs following inspiration.

THE ALVEOLAR FLUID

Although the epithelial lining of the alveoli is moist, the cavity inside is dry – indeed it must be if gas exchange is to take place sufficiently rapidly. Keeping the alveoli dry depends on achieving the right balance between two opposing forces: the hydrostatic pressure of the blood which tends to force fluid out of the capillaries into the alveoli, and the osmotic pressure of the plasma proteins which promotes water movement from the alveoli into the capillaries. In most tissues the hydrostatic pressure at the arterial end of the capillaries exceeds the osmotic pressure (Table 8.1), with the result that fluid leaks out of the capillaries by ultrafiltration. This becomes tissue fluid and any excess either passes back into the bloodstream or is drained into the lymph vessels. Now there are no lymph vessels in the alveolar tissue, and if tissue fluid was formed there in this way the alveoli would fill up with fluid. Gas exchange could no longer take place and in effect we would drown in our own tissue fluid. However, a glance at Table 8.1 will show you that in the lungs the hydrostatic pressure of the blood is lower than the osmotic pressure, with the result that fluid does not leave the capillaries as it does in other organs.

Table 8.1 Comparison of the hydrostatic and osmotic pressures at the arterial end of the capillaries in different tissues.

	Hydrostatic pressure (kPa)	**Osmotic pressure (kPa)**	**Filtration pressure (kPa)**
Most tissues	4.4	3.6	0.8
Alveolar tissue	0.6	3.6	−3.0

Why is the blood pressure in the lungs so much lower than elsewhere? The main reason is that the right ventricle of the heart is less muscular, and contracts less powerfully, than the left ventricle. In certain circumstances, for example if the left ventricle is not working efficiently, there is a build-up of blood returning to the heart from the lungs, and the blood pressure in the pulmonary circulation rises. If the hydrostatic pressure exceeds the osmotic pressure fluid may enter the alveoli, resulting in **pulmonary oedema**. This can lead to serious breathing difficulties.

THE ALVEOLAR SURFACTANT

The inner surface of the alveoli is covered with a very thin layer of fluid through which gases pass as they diffuse to and from the blood. When a liquid surface is curved, as it is in the alveoli, the surface tension creates a resultant force towards the centre. This force would make it difficult to expand the lungs and might even cause them to collapse were it not for the fact that the alveolar fluid contains a lipoprotein which serves as a **surfactant**. A surfactant is a chemical substance which is capable of changing the surface tension of a liquid. The surfactant in our lungs reduces the surface tension of the alveolar fluid to less than a fifth of what it would be otherwise.

The surfactant is thought to be produced by the giant alveolar cells (figure 8.1). It appears in the foetal lung at about the 28th week of pregnancy. Occasionally premature babies are born at or even before this time and they may experience severe breathing difficulties because their pulmonary surfactant has not yet developed.

For consideration

1. Why can't humans breathe under water like fishes?

2. Why should absence of the alveolar surfactant result in breathing difficulties? What steps would you expect a hospital to take to prevent this problem in a premature baby?

Further reading

G.M. Hughes, *The Vertebrate Lung* (Carolina Biology Reader no. 59, 1979)
This includes a good account of the anatomy of the lung.

G.H. Bell, D. Emslie-Smith and C.R. Paterson, *Textbook of Physiology* (Churchill-Livingstone, 10th edition, 1980)
There is a useful summary of the alveolar surfactant.

The control of rhythmical breathing

Breathing, like the beating of the heart, occurs continuously and rhythmically without our having to think about it. The basic pattern can of course be modified by voluntary intervention, but the underlying mechanism is essentially automatic, as is apparent from the fact that it continues when we are asleep or unconscious. The process is controlled by the hindbrain which contains groups of nerve cells known collectively as the **respiratory centre** or more accurately as the **ventilation control centre.**

HOW WAS THE CENTRE DISCOVERED?

To find out which part of the brain controls breathing, experiments were carried out by T. Lumsden in the early 1920s on cats. First the vagus nerves were cut, thus isolating the brain from any afferent connections with the breathing apparatus. The brain was then cut across (transected) at various levels, and the effect on breathing noted. Lumsden found that when he transected the brain just in front of the medulla, rhythmical breathing continued though it was of a slow, gasping type. However, when the brain was transected just behind the medulla, breathing ceased altogether. It was concluded that the medulla contains cells with an intrinsic rhythmical activity which can sustain regular breathing.

Later, in the 1950s and 60s, attempts were made to locate these cells more precisely. Two techniques were used. In the first, different parts of the medulla were stimulated with a weak electrical current and the effect on the depth and frequency of breathing was observed. In the second technique electrodes were placed in contact with nerve cells in different parts of the medulla and impulses were recorded with an oscilloscope at different stages of the breathing cycle.

The results of these experiments showed that the medulla contains two types of nerve cells. Stimulation of the first type causes breathing to stop in the inspiration position with the diaphragm and external intercostal muscles contracted. Stimulation of the second type causes breathing to stop in the expiration position with the diaphragm and external intercostals relaxed. Recording electrical activity from the medulla indicated that some of the nerve cells were active during inspiration, whilst others were active during expiration. By systematically stimulating and recording throughout the length and breadth of the medulla, the exact positions of the different nerve cells were carefully mapped. Though there is considerable overlap, a distinction can be made between what we may call an **inspiratory** and an **expiratory centre**.

How are these two centres related functionally? If both are stimulated simultaneously with currents of equal strength, inspiration occurs, suggesting that the inspiratory centre is the dominant of the two. When the two centres are stimulated alternately, rhythmical breathing occurs at a depth and frequency which is determined by the frequency of the stimuli. The nerve cells of the two centres are profusely interconnected and when a stimulus is applied to one of them, nerve impulses spread to the other. This reciprocal spread of excitation between the two centres provides the basis of its intrinsic rhythmical activity. How do you think it might work?

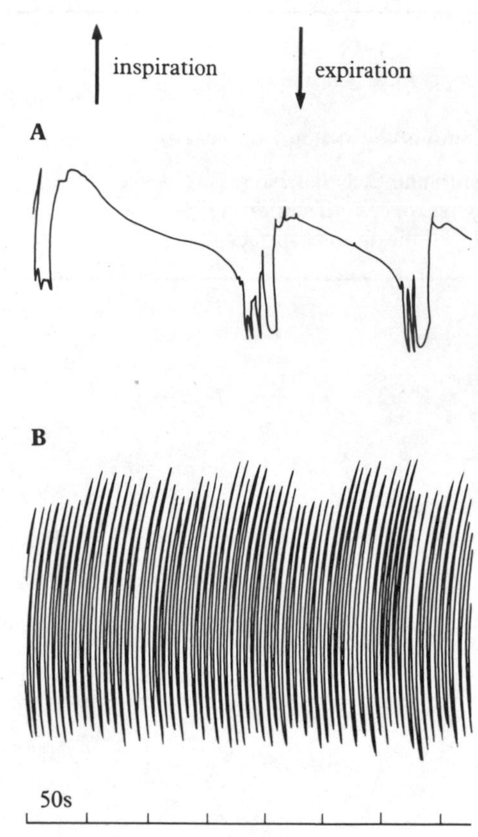

Figure 8.3 Kymograph recording of breathing movements of a cat. (**A**) Apneustic breathing showing prolonged inspirations. (**B**) Normal rhythmical breathing with the pneumotaxic centre intact. Inspiration upwards, expiration downwards. (*After Lumsden, 1923*)

OTHER CENTRES IN THE CONTROL OF BREATHING

The region of the hindbrain in front of the medulla is called the pons. If the posterior part of the pons is stimulated electrically, or if the brain is transected just in front of it and the vagus nerve cut, an extraordinary kind of breathing takes place – it consists of deep, prolonged inspirations separated by brief expiratory gasps (figure 8.3A). This is known as apneustic breathing and the nerve cells in the posterior part of the pons responsible for it constitute the **apneustic centre**.

Now if the anterior part of the pons is stimulated, or if the brain is transected just above it, breathing is normal (figure 8.3B). So it seems that we have here yet another group of nerve cells, in this case promoting normal rhythmical breathing. This is called the **pneumotaxic centre**. Although it promotes normal breathing, it does not have an intrinsic rhythmicity like the medullary centres. It is, if you like, an advisory body rather than the executive.

What part does the pons play in the overall control of breathing? You will recall that the medullary centres on their own produce a slow, gasping type of breathing. The pneumotaxic centre imposes a restraining influence on the inspiratory neurones in the medulla so that breathing is faster and shallower. A similar effect is achieved by reflexes arising from stretch receptors in the bronchial tubes – as the lungs inflate, impulses are discharged via the vagus nerves to the hindbrain where they inhibit inspiration.

The function of the apneustic centre seems to be to promote inspiration by acting on the inspiratory neurones, probably via the apneustic centre.

So we see that breathing is controlled not by one centre but by four. Indeed the term centre is misleading since it implies a clearly defined area with a distinct function. What we have been calling centres are really diffuse collections of neurones extensively interconnected so as to provide for ample 'crosstalk' within the brain. Various internal stimuli, notably carbon dioxide and oxygen, influence the neurones so as to give a highly adaptable and dynamic system (see page 72).

THE PROBLEM OF VIVISECTION

The investigations described in this topic involve experimenting on live animals. The animals were, of course, anaesthetised and unconscious, but nevertheless such experiments are an affront to many people. This would not be the place to rehearse the arguments for and against vivisection, but the problem should not be ignored. If you are reading this book you are probably a committed – or potentially committed – biologist, and as such you should be prepared to form an opinion on this subject if you have not already done so. There is an extensive literature to help you, and one particularly useful reference is given below.

For consideration

1. Why is the term 'ventilation control centre' preferable to 'respiratory centre'?

2. Suggest a possible sequence of events whereby the centres described in this topic control one complete breathing cycle (inspiration followed by expiration).

Further reading

O.J.C. Lippold and F.R. Winton, *Human Physiology* (Churchill Livingstone, 7th edition 1979)
This work contains a particularly clear account of the respiratory centres and their interactions.

R.J. Berry, 'Ethics in Biology and for Biologists' (*Biologist*, vol 31, no. 5, 1984)
This article includes a short discussion on vivisection and has a useful reference list.

9 . Heterotrophic Nutrition

Protein-digesting enzymes at the molecular level

Without digestive enzymes a meal might take fifty years to digest. The breakdown of compounds such as proteins would depend both on random collisions between water molecules and peptide bonds *and* on water molecules with enough activation energy striking the bonds in just the right place.

An enzyme provides an alternative reaction pathway which does not require so much activation energy. It cuts the substrate economically, as if with a pair of scissors or a surgeon's scalpel. In this topic we shall illustrate the mechanism of enzyme action by considering the battery of pancreatic enzymes which digest proteins in humans.

PROTEIN STRUCTURE AND DIGESTION

A protein molecule has a backbone of repeated nitrogen-carbon-carbon atoms with the R– group of each residue projecting from the chain. The pancreatic enzymes which digest proteins in the gut are often lumped together as 'trypsin' and 'peptidases', but in fact no less than four specific enzymes are involved: **trypsin, chymotrypsin, elastase** and **carboxypeptidase A.** As indicated in Figure 9.1, they act at different points on the polypeptide chain. This multiple attack presumably ensures that the polypeptides are split up into small enough units by the time that they reach the ileum to be absorbed by the villi.

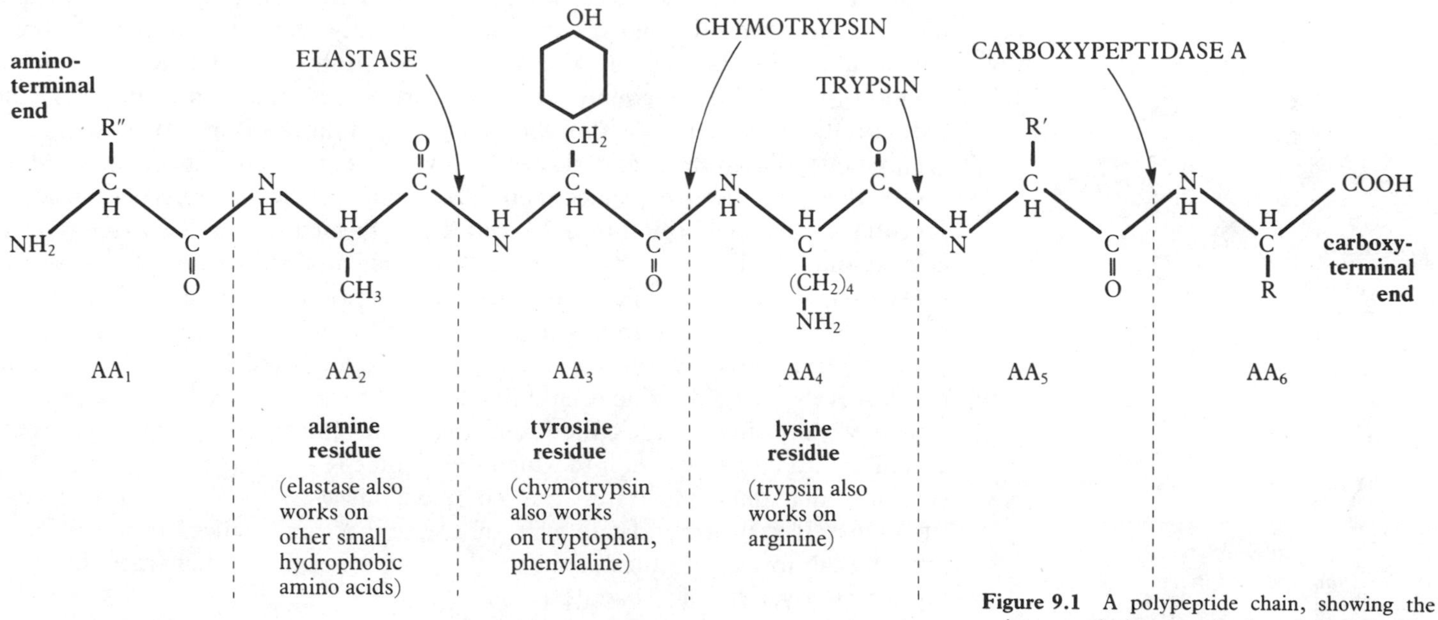

amino-terminal end
ELASTASE
CHYMOTRYPSIN
TRYPSIN
CARBOXYPEPTIDASE A
COOH
carboxy-terminal end
AA_1 AA_2 AA_3 AA_4 AA_5 AA_6
alanine residue (elastase also works on other small hydrophobic amino acids)
tyrosine residue (chymotrypsin also works on tryptophan, phenylaline)
lysine residue (trypsin also works on arginine)

Figure 9.1 A polypeptide chain, showing the points at which it might be attacked by the protein-digesting enzymes mentioned in the text. Carboxypeptidase A is an exopeptidase, since it splits off only the terminal amino acid residue from a polypeptide chain. The others enzymes are endopeptidases, since they can attack polypeptide chains in the middle. The segments AA_1, AA_2 etc indicate the residues of the various amino acids which make up the polypeptide chain.

The four enzymes – trypsin, chymotrypsin, elastase and carboxpeptidase A – are manufactured by the pancreas in the form of inactive precursors. They collect in the Golgi bodies and are stored in the cytoplasm as densely staining zymogen granules, which originated as Golgi vesicles. When the cell receives a nervous or hormonal signal, some of the zymogen granules fuse with the cell membrane and release their contents into the 'pancreatic juice'. The enzyme precursors flow down the pancreatic duct into the duodenum.

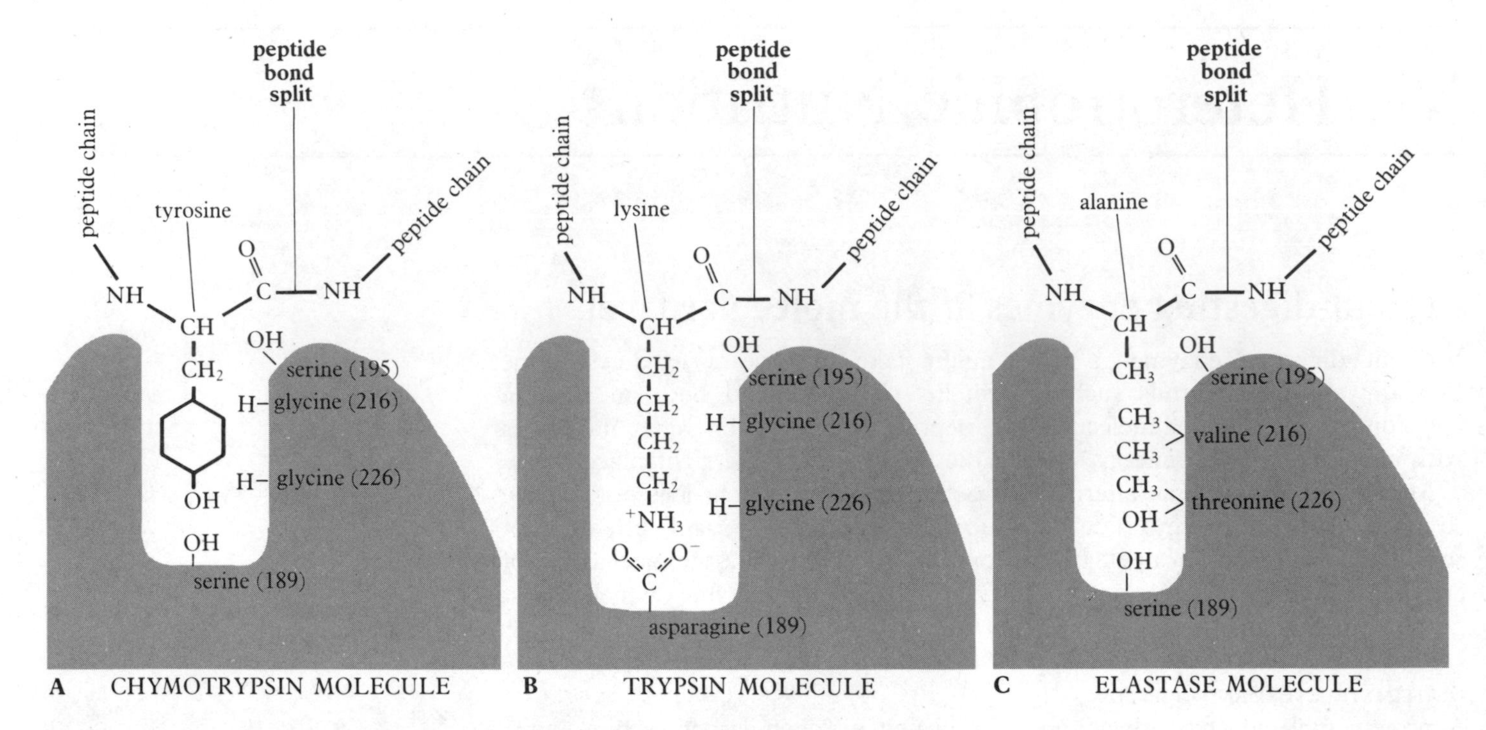

Figure 9.2 The active sites of (**A**) chymotrypsin, (**B**) trypsin and (**C**) elastase are shown as rectangular gaps at the top of each globular enzyme molecule. Into these gaps fits the R–group of a single amino acid in the polypeptide chain which is being digested. Four other R–groups are shown projecting *from* the active site of each enzyme. The most critical of those shown is serine, which breaks the peptide bond opposed to it and so splits the polypeptide in half. (*After Nixon*)

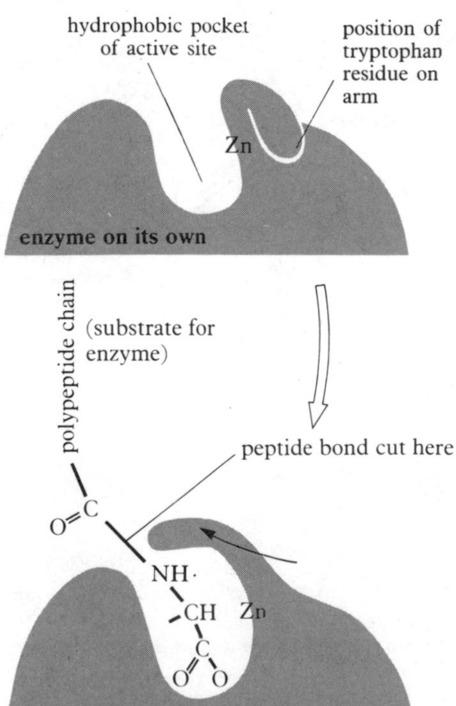

Figure 9.3 The diagram shows the 'induced fit' of the enzyme carboxypeptidase A with its substrate, the amino acid at the carboxylic acid end of a peptide chain.

HOW THE ENZYMES SPLIT THEIR SUBSTRATES

The chymotrypsinogen of humans is a globular protein of 245 amino acids. When it enters the duodenum it is attacked by trypsin, which snips off the first fifteen amino acids (by attacking the peptide bond next to arginine, see figure 9.1) and converts the enzyme from inactive chymotrypsinogen into active chymotrypsin.

The active enzyme begins to snip polypeptide chains at any large hydrophobic R– groups which have rings in them, such as tyrosine, phenylalanine and tryptophan. In the active site of the enzyme there is a large water-repellent pocket, into which the side chain fits (figure 9.2a). Next to the peptide bond which is to be broken in the protein substrate lies the R– group of chymotrypsin's 195th amino acid, a serine residue. This induces the redistribution of electrons within the substrate and digests it into two fragments.

Chymotrypsin, trypsin and elastase are very similar. Their molecules are all about the same length, shape and size, they are all secreted as inactive precursors, their pH optimum is 7.8 and they all have the reactive serine residue at position 195. For this reason these three enzymes are known as the **serine proteases**. However, as might be expected, their active sites differ. This allows them to act on different amino acids in the substrate chain. In trypsin (figure 9.2b) the negatively-charged asparagine (189) résidue is ideally placed to bind to the positively-charged lysine or arginine residue on the substrate. In elastase, on the other hand (figure 9.2c), the projecting valine and threonine residues create a small hydrophobic pocket into which the alanine R– group can neatly fit. This situation not only illustrates the complementary shapes of enzymes and substrates. It may also indicate why, of all the macromolecules in the cell, only proteins can form molecules of such a diversity of shapes and charges that they can attack a variety of substrates.

The other pancreatic protease enzyme, carboxypeptidase A, alters its shape when it acts on its substrate, an example of **induced fit**. Carboxypeptidase A belongs to a different family of enzymes than the serine proteases. It has 307 amino acids, and contains a zinc atom which is necessary for its activity. It breaks the carboxy-terminal amino acid from a protein chain, whatever its R– group. The binding of positively-charged arginine 145 to the negatively-charged carboxylic acid group triggers off movements in the enzyme backbone. These culminate in the movement of an arm of amino acid residues across the enzyme (figure 9.3). The arm has a tryptophan residue at the end. This residue not only clasps the substrate in a hydrophobic pocket, but also adds H^+ to it, converting the $-NH$ group to an amide ($-NH_2$) group and breaking the peptide bond.

For consideration

1. Bearing in mind the similarity between the three serine protease enzymes, how do you think this family of enzymes arose by natural selection?

2. Some seeds, for example those of soya bean, produce compounds which inhibit the serine proteases. Why might such compounds have evolved?

Further reading

J.E. Nixon, 'Aspects of Protein Chemistry – Part 1' (*School Science Review*, vol. 58, no. 203, 1976)
This includes some recent insights into enzyme specificity.

L. Stryer, *Biochemistry* (Freeman, 3rd edn, 1981)
The author provides a clear and detailed account of the action of carboxypeptidase A.

Cellulose digestion

Cellulose in plant cell walls makes up about a third of the mass of all the carbon compounds on Earth. It surrounds the protoplasts of plant cells and makes plant material difficult to digest. Cellulose occurs in the diets of most animals and decomposers, but as far as we know the ability to *digest* cellulose is almost confined to bacteria, heterotrophic protists and fungi. Only they possess the necessary enzyme – **cellulase** – for doing the job. It is difficult to establish whether an animal produces its own cellulase, because its gut almost always contains traces of cellulose-digesting micro-organisms which release the enzyme. The only animal which is known for certain to produce its own cellulase is the silverfish (*Ctenolepisma ciliata*), the tiny primitive wingless insect which occurs in leaf litter and houses in Europe.

Many herbivores, however, can make use of some of the products of cellulose digestion because of a mutualistic association with micro-organisms in their guts. Here we shall discuss the various associations between mammals and micro-organisms which allow mammals to derive maximum benefit from a diet which is rich in cellulose.

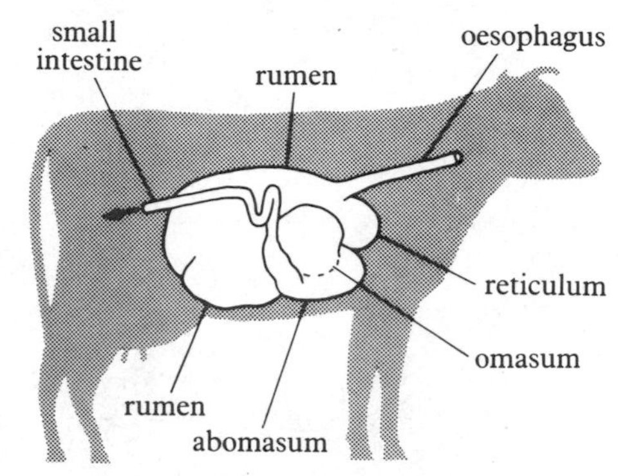

Figure 9.4 The gut of a cow. The rumen makes up 70% of the volume and is the fermenting chamber. Its walls are rich in absorptive cells and blood capillaries. Grass enters the rumen first and begins to ferment. Boluses of food form in the reticulum, and pass up the oesophagus to the mouth by reverse peristalsis. They are mixed with saliva and the cell walls are broken down by the cow's molars. Then the bolus is swallowed, returning it to the rumen. Some of the rumen contents will be digested by the cow itself. They pass into the omasum, which has parallel cartilaginous plates (increasing the surface area). The abomasum secretes pepsin and begins to digest proteins, protists and bacteria. Digestion then proceeds in the normal manner. The rumen, reticulum and omasum are pouches of the oesophagus. The true stomach is the abomasum. (*After Schmidt-Neilson*)

THE RUMEN

Some of the mammals of greatest economic importance – such as camels, cattle, sheep and goats – harbour, in specialised pouches in their alimentary canals, various cellulose-digesting bacteria and protists. The cow's system is shown in figure 9.4. The compounds produced in anaerobic conditions in the fermenting chamber of a cow's oesophagus, **the rumen**, provide cows with 70 per cent of their energy intake. Most of the cellulose is absorbed by the walls of the rumen in the form of ethanoic, butanoic and propanoic acids, and carbon dioxide and methane gases are belched into the atmosphere from both ends of the digestive system (figure 9.5).

The statistics of this process in a cow are impressive. A cow of 500 kg may have rumen contents weighing 100 kg, of which 1 kg may be protists and another 1 kg may be bacteria. These micro-organisms produce about 5 kg of volatile fatty acids a day. They also release about 200 litres of methane gas daily, which amounts to a tenth of the total energy ingested by the cow and is sufficient to light a small house continuously. The total production of saliva each day is about 150 kg, a third of the body mass.

The organisms in the contents of the rumen are also important to the ruminant in the production of protein and vitamins. Many of the bacteria hydrolyse amino acids to ammonia and organic acids, and all the bacteria use the ammonia to make most of their amino acids (see page 47). The protists obtain their protein by ingesting the bacteria. In many ruminants, urea ($CO(NH_2)_2$) is secreted in the saliva or from the rumen wall, and the ammonia produced from it is converted into microbial protein. When the microbes are digested in the

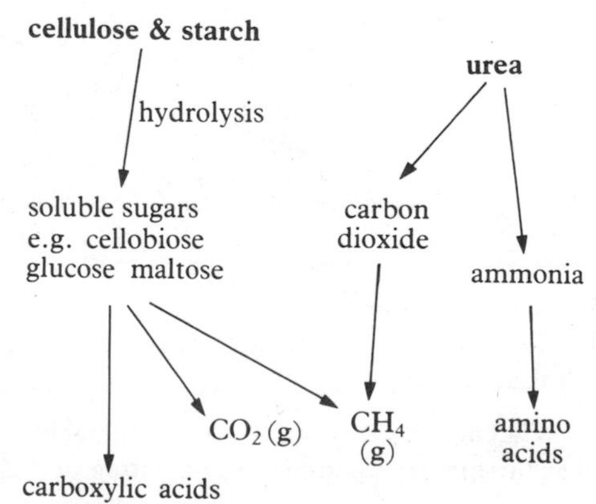

Figure 9.5 Diagrammatic representation of the main conversions of compounds by micro-organisms in the rumen of a ruminant mammal. (*After Wolin*)

abomasum, as about 70 per cent of them are each day, they contribute to the host about 100 g per day of protein. This is why farmers supplement the cows' nutrient-poor feedstuffs with cheap urea.

B vitamins secreted by the bacteria may also be valuable to the ruminant. In particular, cows obtain their supplies of vitamin B_{12} from their mutualistic microbes.

THE CAECUM

Although kangaroos, wallabies and sloths are not strictly ruminants they have digestive systems which work on a remarkably similar principle. They all have large fermentation chambers situated before the stomach. Many other mammalian herbivores, such as the donkeys, horses, rodents and lagomorphs, have a special site for cellulose digestion by micro-organisms in the **caecum**, a blind-ending tube attached to the alimentary canal near the junction of the small and large intestines.

The caecum may not be as efficient as the rumen at cellulose digestion, as you can see by comparing the textures of horse dung and cowpats. When cellulose is attacked in the caecum the food cannot be continually regurgitated and masticated. Furthermore, the products of fermentation cannot easily be shunted forward to be exposed to the villi for absorption in the duodenum and ileum.

In some mammals with caecal fermentation there are additional mechanisms which increase the rate of cellulose digestion. Horses exhibit reverse peristalsis in the large intestine. This enables some of the incompletely digested food to be pushed into the caecum. Rabbits and rats go to even greater lengths: they eat some of their own faeces, a phenomenon known as **coprophagy**. Indeed, when they are prevented from eating their own faeces, they suffer nutritionally.

Rabbits produce two types of faeces, hard and soft. The soft faeces are produced in the caecum. As they emerge from the anus, they are eaten again. They lodge in the fundus of the stomach (figure 9.6) where, separated from the food which is ingested, they become covered with mucus and ferment for hours. By this means a rabbit obtains vicariously some of the advantages of a rumen.

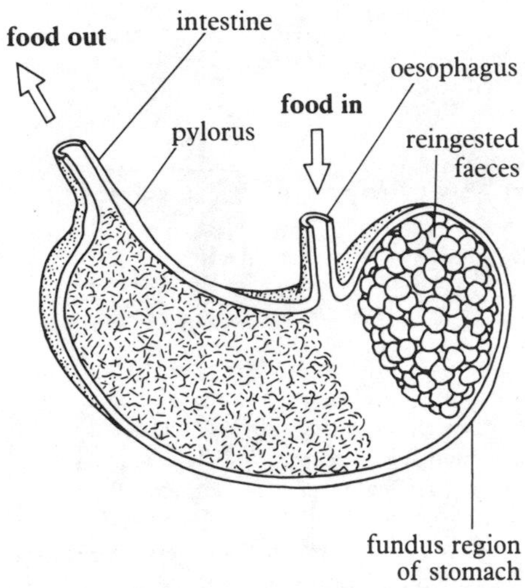

Figure 9.6 The digestion of soft faeces in the stomach of a rabbit. The faecal pellets in the fundus ferment without interfering with normal digestion in the pyloric region. (*After Schmidt-Neilson*)

HUMANS

What about ourselves? Calculations suggest that every day each one of us egests approximately 1.2 million million bacterial cells. In human societies with refined diets low in cellulose (fibre) there has been a marked increase in bowel cancer, colitis and diverticulitis of the colon. This is why it is important to understand the fermentations which go on in our own large intestines.

The chemical reactions which occur in the human colon resemble those in the rumen of a cow, but only a third of humans produce significant quantities of methane and in humans bacteria are much more frequent than protists. Most of the hydrogen and methane produced by the colon bacteria is absorbed by the blood and exhaled from the lungs. In fact, significant quantities of volatile fatty acids and some vitamins enter the human bloodstream from the colon and are metabolised by the host. The action of these bacteria on a natural high-fibre diet is important in maintaining the health of this part of the gut.

For consideration

1. How could you show that the silverfish can produce its own cellulase enzyme?

2. What are the advantages to a mammal of microbial fermentation in a rumen over microbial fermentation in a caecum?

3. If coprophagy in rats is prevented by removing the faeces as soon as they are produced, the rats need additional vitamin K and biotin (vitamin H) in their diets. Even when these vitamins are provided, their growth rates are reduced by 15–25 per cent compared with rats allowed to eat their own faeces. Suggest some reasons for these observations.

Further reading

Clegg, *Biology of the Mammal* (Heinemann, 2nd edition, 1978)

K. Schmidt-Nielsen, *Animal Physiology: Adaptation and Environment* (Cambridge University Press, 3rd edition, 1983)

10 . Autotrophic Nutrition

Chloroplasts and mitochondria

Life depends on chloroplasts and mitochondria. Chloroplasts, in the photosynthetic cells of plants, make use of solar energy to manufacture organic compounds from simple inorganic precursors. Mitochondria oxidise some of these organic products to inorganic molecules, and use some of the energy released to manufacture ATP. In this topic we shall compare the reactions in chloroplasts and mitochondria, concentrating in particular on the way in which they generate ATP.

SIMILARITIES AND DIFFERENCES BETWEEN CHLOROPLASTS AND MITOCHONDRIA

At first sight the chemical reactions carried out by chloroplasts seem to be the reverse of those performed by mitochondria. On a gross scale, the chloroplast takes in carbon dioxide, water, and light energy, and produces organic carbon compounds. The mitochondrion does almost the opposite. It takes in organic carbon compounds, and makes carbon dioxide and water, releasing ATP in which some energy is conserved.

Despite these differences, chloroplasts and mitochondria have many structural similarities. Both have double membranes, and are about the size of a large bacterium. Both have their own circular DNA double helix, their own ribosomes, smaller than those in the surrounding cytoplasm, and their own protein-synthesizing machinery. Totalled over all the organelles in an individual plant or animal, the surface area of the inner membranes is vast. For instance, the total area of cristae in your mitochondria is about the same as a football pitch. The inner membrane of the mitochondrion and the thylakoid membrane of the chloroplast carry stalked particles where ATP is made. In the chloroplasts these stalked particles face outwards, but in mitochondria, they face inwards (figure 10.1).

In fact the similarities and differences between these organelles become clearer if you regard a chloroplast as to some extent a mitochondrion which is inside-out. Let us now pursue this idea with respect to the chemical reactions that take place in these two organelles, and the way ATP is generated.

From the chemical point of view, light energy is utilised by a pigment system to split water inside the chloroplast (figure 10.2A). In the light stage of photosynthesis, the hydrogen atoms reach NADP, and ATP is generated. These compounds are then used in the Calvin cycle to reduce carbon dioxide to sugar molecules. In the mitochondrion, however, the reverse occurs (figure 10.2B). Organic compounds, such as pyruvic or fatty acids, are broken down to carbon dioxide and hydrogen (which is added to NAD). The $NADH_2$ is then used by a pigment system (the cytochromes) to generate ATP, and the hydrogen atoms are used to form water.

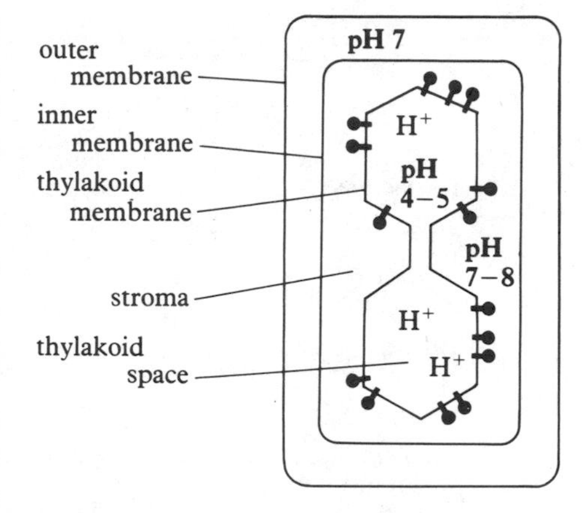

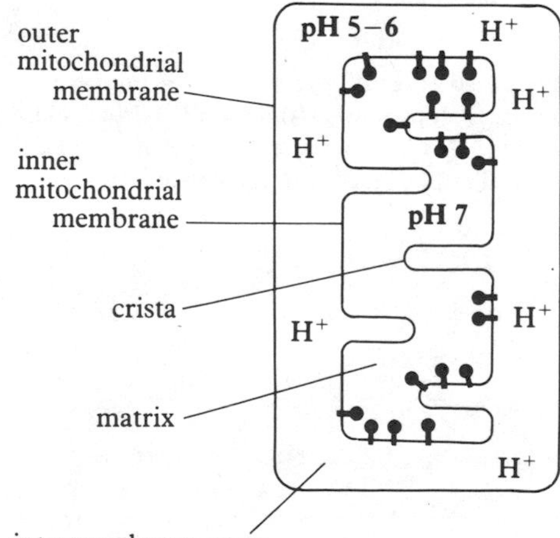

Figure 10.1 A highly diagrammatic representation of the main compartments within (**A**) a chloroplast and (**B**) a mitochondrion, showing the orientation of the stalked particles in which lie the ATP synthetase enzymes (ATPase). The spaces between the inner and the outer membranes have been deliberately exaggerated.

HOW IS ATP GENERATED?

Figure 10.1 shows that, in the chloroplast, the light reaction occurs in the thylakoids. The stalked particles which generate the ATP project into the stroma, where the Calvin cycle and the synthesis of glucose take place. The thylakoid space has a pH of about four whereas the stroma has a pH of about seven. Because the pH scale is logarithmic, the hydrogen ion concentration in the thylakoid is a thousand times that in the stroma. In the mitochondrion, on the

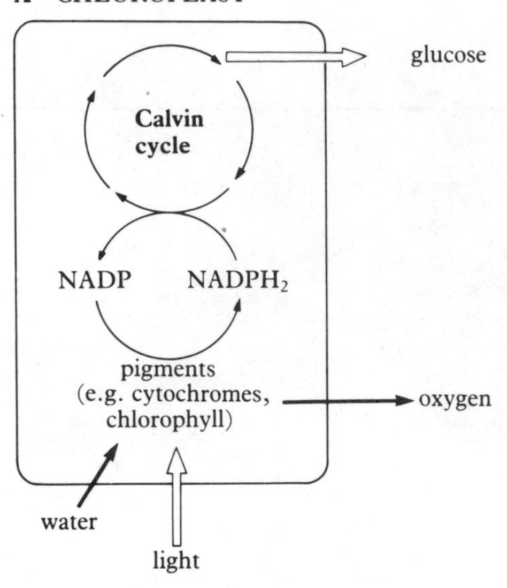

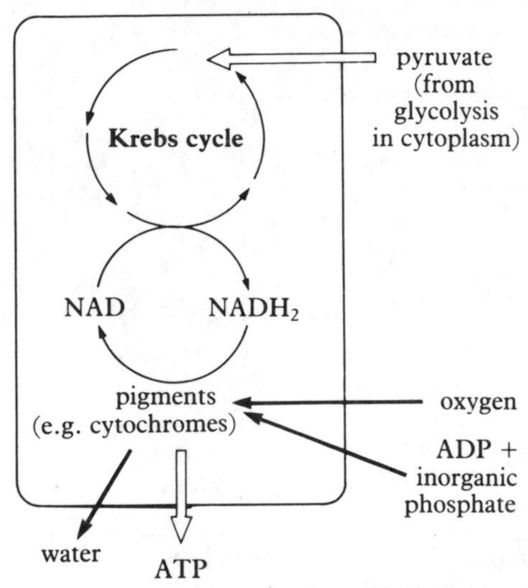

Figure 10.2 Bird's-eye view of reactions within (**A**) a chloroplast and (**B**) a mitochondrion, illustrating that many reactions occur in both organelles but proceed in the opposite direction.

other hand, the stalked particles project from the inner membrane towards the centre of the mitochondrion. Here it is the space between the two membranes, the intermembrane space, which has the low pH – about five or six. The matrix of the mitochondrion, inside the inner membrane, has a pH of about seven. How does this system generate ATP?

In both chloroplasts and mitochondria, much of the energy is delegated to electrons. In cyclic and non-cyclic photophosphorylation in the chloroplast, and in oxidative phosphorylation in the mitochondrion, electrons move between cytochrome molecules which are situated in the membranes in which the stalked particles are embedded. In the chloroplast, this energy has come ultimately from light, and in the mitochondrion, from the oxidation of glucose. In both organelles the energy released as the electrons move about in the membrane is used to pump hydrogen ions across the membrane from one side to the other.

In the case of the chloroplast the hydrogen ions move from stroma to thylakoid space, and in the mitochondrion from the matrix to the intermembrane space. This creates across the membrane a pH difference and a difference in electrical charge of about 200 millivolts. The hydrogen ions are funnelled back through the stalked particles, which contain at ATPase protein, and as they move, the pent-up energy is used to make ATP from ADP and inorganic phosphate. This hypothesis, proposed by Phillip Mitchell in 1961, is known as the **chemiosmotic theory**, and it has revolutionized previous ideas about ATP synthesis.

EVIDENCE FOR CHEMIOSMOSIS

There is plenty of evidence consistent with this idea. For example, in a classical experiment by Jagendorf, isolated chloroplasts were stimulated by a pH gradient to make ATP in the dark. The experimenters stirred spinach chloroplasts for a minute in a solution at pH 4.0 which contained ADP, radioactive inorganic phosphate, and magnesium ions. The aim was to create a pH of 4.0 in both the thylakoid and stroma compartments. When the chloroplasts were transferred for fifteen seconds to a solution of pH 8.0, they synthesized ATP in the dark. They would normally need light energy to make ATP, but presumably the pH gradient between 4.0 in the thylakoid space and 8.0 in the stroma was sufficient to generate an ATP-synthesizing potential.

Any compound which destroys the electrical or pH gradients across the inner membranes prevents ATP synthesis in both chloroplasts and mitochondria. The compound 2,4-dinitrophenol, for example, once used as a slimming agent but now banned, is known to uncouple cellular respiration from ATP synthesis in animals. In mitochondria it combines with hydrogen ions in the inter-membrane space, where the pH is low, and then dissolves in the inner membrane, releasing the hydrogen ions on the other side. This destroys the pH gradient and prevents ATP synthesis. The consequence is that the energy released by the oxidation of glucose is lost as heat and is not conserved in ATP for use in cellular activities.

In fact some cells in the mammalian body have a compound in them which is the natural equivalent of dinitrophenol. These are the **brown fat cells**, important in heat production and the regulation of body mass. The role of brown fat cells in oxidizing glucose and releasing most of the energy in heat depends on a polypeptide. This acts as the uncoupling factor and allows hydrogen ions to diffuse across the membrane, by-passing the stalked particles which synthesise ATP. This is probably why brown fat has a maximum energy output of 500 W kg^{-1} whereas skeletal muscle has an output of only 60 W kg^{-1}.

For consideration

1. (a) Why do you think *chemiosmosis* is so called?
(b) To what extent is the chemiosmotic theory consistent with modern ideas about the structure of biological membranes?

2. 'Photosynthesis is the opposite of respiration'. Discuss.

Further reading
B. Alberts *et al*, *The Molecular Biology of the Cell* (Garland, 1983) Chapter 9 contains a clear, well illustrated account of chemiosmosis in mitochondria and chloroplasts.

D.G. Nicholls, 'Chemiosmotic ATP Synthesis' (*Biologist*, vol. 30, no. 1, 1983)
A brief, simple and elegant account of the process.

J. Hannay, 'Fluid-Mosaic Membrane and the Light Reactions of Photosynthesis' (*Journal of Biological Education*, vol. 19, no. 3, 1985)
The author relates the events occurring in the thylakoid membranes to their fluid-mosaic structure.

How are amino acids made?

Proteins are essential components of all organisms. They are composed of amino acids. Animals obtain their nitrogen compounds, such as amino acids, ultimately from plants. Yet no green plants have been been shown to take up amino acids directly from the soil. Where do plants get them from?

Green plants are adept at converting inorganic compounds into organic ones and nitrogen metabolism is no exception. Plants take up both nitrate and ammonium ions from the soil, and convert them to amino acids. The mechanism is as follows. Nitrate is first converted to nitrite. The nitrite is then reduced to ammonium ions, which are combined with various organic compounds to yield amino acids.

We shall first consider the processing of nitrate and then the conversion of ammonium ions to amino acids.

NITRATE TO AMMONIUM

Most plants have to absorb nitrate ions and reduce them to ammonium ions before amino acids can be made. This reduction is an energy-requiring process. Eight electrons are needed to reduce nitrate to ammonia and the energy and electrons have to come from respiration or photosynthesis.

The nitrate ions are absorbed across the cell membranes of root cells by active transport (see page 56). In some plants the site of reduction is the roots, in some it is the stem and in others it occurs in both places. The enzyme involved is **nitrate reductase**, a large protein incorporating the vitamin riboflavin. The enzyme seems to act in the cytoplasm of, for example, the cortex cells of the root.

The higher the concentration of nitrate ions in the solution which bathes the cell membranes of the cortex cells of the root, the more nitrate reductase is produced by the cells. Some plant species such as nettles, which often grow on sites disturbed by humans, seem to be able to produce nitrate reductase rapidly on demand. Other species from nutrient-poor sites have only limited ability to make it.

The reduction of nitrate to nitrite requires $NADH_2$:

$$NO_3^- \xrightarrow[\text{nitrate reductase}]{NADH_2 \curvearrowright NAD} NO_2^-$$

This reduction does not occur in mitochondria. Where then does the necessary $NADH_2$ come from? It seems that malate ions leak from the mitochondria and are converted in the cytosol to oxaloacetate, as in the Krebs cycle. This reaction generates $NADH_2$. The oxaloacetate goes back into the mitochondrion and the $NADH_2$ reduces the nitrate to nitrite.

In contrast, the reduction of nitrite to ammonium, which requires six electrons, occurs in the chloroplasts in the leaves. Here we find the enzyme **nitrite reductase**, probably associated with the thylakoid membranes on the side next to the stroma. This enzyme is thought to take electrons from ferredoxin, one of the electron carriers in non-cyclic photophosphorylation in the light reaction of photosynthesis. It transfers the electrons to nitrite in three steps:

$$NO_2^- \xrightarrow[+2e^-]{\text{nitrite reductase}} NOH \xrightarrow[+2e^-]{\text{nitrite reductase}} NH_2OH \xrightarrow[+2e^-]{\text{nitrite reductase}} NH_4^+$$

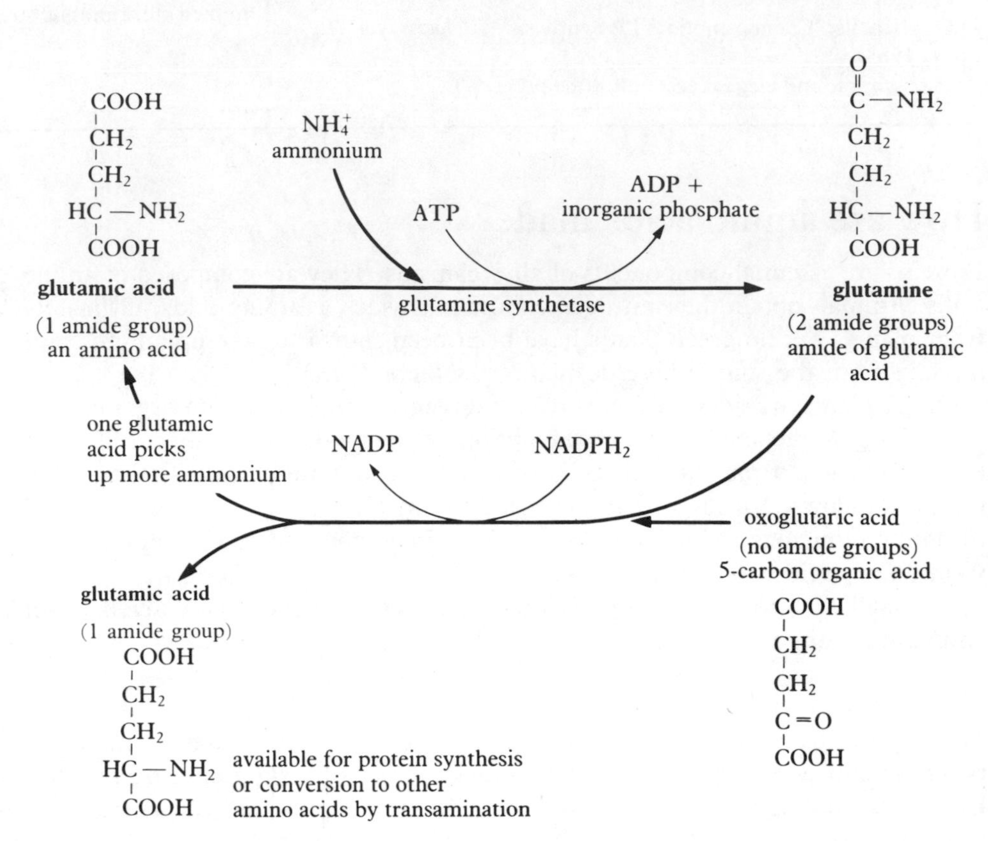

Figure 10.3 The reactions which result in the incorporation of ammonium in glutamic acid in plant cells.

AMMONIUM IONS TO AMINO ACIDS

Ammonium ions are toxic to the chloroplast in high concentrations. Figure 10.3 shows what happens to them. In the chloroplast is the enzyme **glutamine synthetase**. This rapidly combines ammonia with the amino acid glutamic acid to produce the organic amide glutamine. This reaction requires ATP. The nitrogen in the glutamine is in two amino groups ($-NH_2$). The glutamine then reacts with a molecule of oxoglutaric acid to produce two molecules of glutamic acid, each with one amide group. One of the glutamic acid molecules is able to react with more ammonia, and the other is available as a starting point for the manufacture of some other amino acids, and ultimately of proteins.

The metabolic pathways in which amino acids are made are complex (see page 32). Several amino acids, however, can be formed from glutamic acid by **transamination**. For example, when an amide group is transferred from glutamic acid to pyruvic acid, the latter is converted into alanine (figure 10.4).

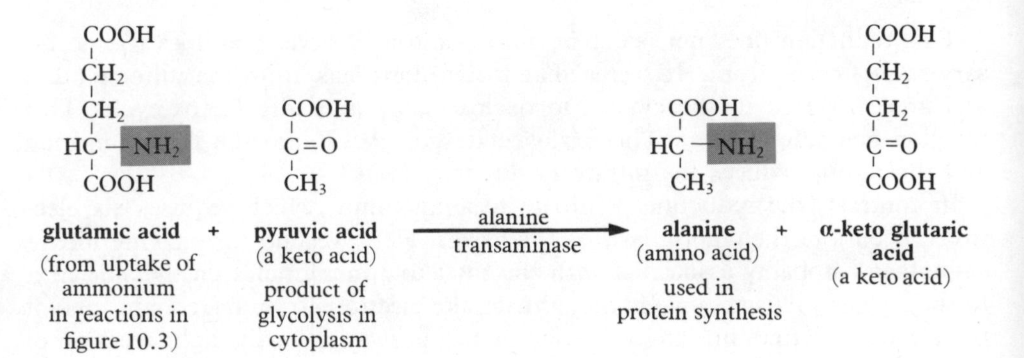

Figure 10.4 An example of a transamination reaction, in which an amide group is transferred to a keto-acid from glutamic acid to produce a new amino acid, in this case alanine.

In acidic soils in particular, the ammonium ion is more abundant than the nitrate ion and is taken up directly by plant roots. Ammonium ions do not need to be oxidised or reduced to produce amino acids. Ammonium ions are acted on by the glutamine synthetase in the cortex cells of the root and converted to glutamine as already described. Glutamine is then transported around the plant in the phloem to the sites of amino acid manufacture.

For consideration

1. Which compounds in organisms contain nitrogen?

2. Suggest why ammonium ions are more abundant than nitrate ions in acidic soils.

3. Some plants, such as Groundsel (*Senecio vulgaris*) on seabird nest sites, seem to have evolved the ability to take up urea. Carnivorous plants can absorb amino acids into their leaves. Yet no green plants have been shown to take up amino acids directly from the soil. Can you suggest why?

Further reading

W.J. Davies and P.G. Ayres (editor), *Biology in the 80's – Plant Physiology* (University of Lancaster, 1982)
The relevant section is 'Uptake of Nitrogen by Plants' by K. Jones.

O.A.M. Lewis, '*Plants and Nitrogen*' (Studies in Biology no. 166, Arnold, 1986)
This recent book provides useful detail.

Nitrogen fixation

In order to reduce atmospheric nitrogen to ammonia in the Haber-Bosch industrial process, a massive input of energy is required. To produce ammonia in reasonable quantities temperatures of 400 °C and pressures of several hundred times atmospheric pressure are needed. Yet many prokaryotes have the ability to perform this same reduction at normal temperatures and pressures. These are the **nitrogen-fixers**. They fix each year, in natural ecosystems, five times the nitrogen gas fixed by human industries. In this topic we shall explain how this remarkable feat is accomplished.

Nitrogen fixation occurs in a wide range of prokaryotes. We will concentrate on the association of the bacterium *Rhizobium* with the root nodules of legumes.

BEANS, BACTERIA AND NODULES

Nodules in field bean (*Vicia faba*) only form in soils with low concentrations of nitrates and ammonium ions. Each nodule is a conglomeration of cells with its own apical meristem and vascular supply. Sucrose and other organic compounds are supplied to the nodule via the phloem. The nodule exports the amino acid glutamine and other nitrogenous compounds, which are picked up by transfer cells next to the xylem, pumped into the xylem and carried up the plant in the transpiration stream.

The fixation of nitrogen gas is accomplished by the bacteria, which are packed in large numbers into most of the cells in the nodule. The bacterial cells absorb the glucose supplied to the nodule by the host plant and respire it to produce ATP. They use this ATP, and electrons from an unknown source, to reduce the nitrogen gas to ammonia. In fact for each nitrogen molecule reduced to ammonia, eight electrons and sixteen ATP molecules are required. A molecule of hydrogen gas is produced as a by-product in this endothermic reaction:

$$N_2 + 10\ H^+ + 16\ ATP + 8\ e^- \xrightarrow{\text{nitrogenase}} 2\ NH_4^+ + H_2 + 16\ ADP + 16\ iP$$

(iP = inorganic phosphate)

NITROGENASE AND THE FATE OF AMMONIUM IONS

The enzyme which carries out this conversion inside each bacterium is called **nitrogenase**; it has two parts with different functions (figure 10.5). The first part, known as **Fe–protein**, contains iron. It picks up electrons from ferredoxin and hydrolyses ATP. The second, called **MoFe–protein**, contains molybdenum as well as iron, and has on its surface a nitrogen-binding site. The reaction is

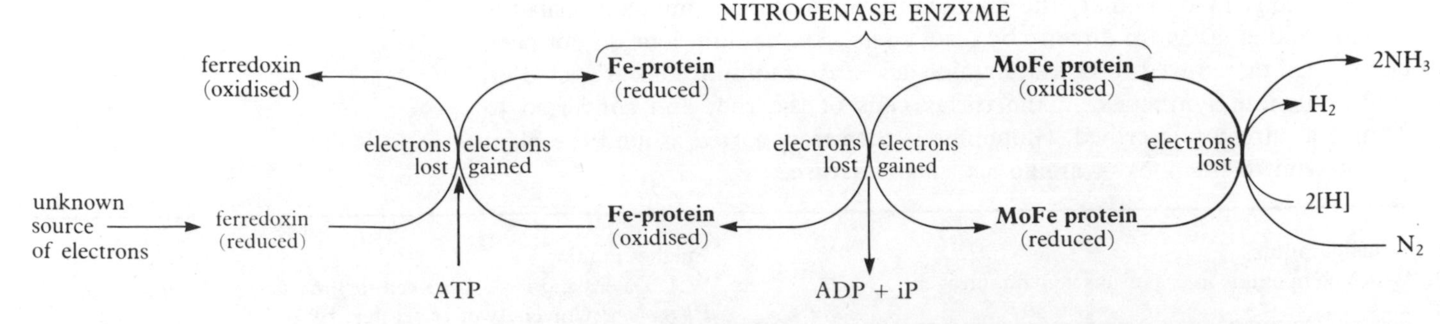

Figure 10.5 The chain of reactions involved in nitrogen fixation by the two parts of the nitrogenase enzyme in bacteria of the genus *Rhizobium* in field bean (*Vicia faba*). The Fe-protein part of the nitrogenase enzyme contains iron and sulphur. The MoFe–protein contains two molybdenum atoms, thirty-three iron atoms and some sulphur. For nitrogen fixation by their mutualistic bacteria, legumes therefore require both molybdenum and iron. Iron is also an essential constituent of leghaemoglobin. Some bacteria can produce a hydrogenase enzyme which allow them to use some of the hydrogen evolved to make ATP.

very slow, and it takes about 1.25 seconds to reduce each nitrogen atom. This seems to be because the two parts of the enzyme separate whenever electrons are transferred between them.

One remarkable property of the nitrogenases in all prokaryotes is that they are destroyed by exposure to oxygen. In nitrogen-fixers such as *Rhizobium*, which need oxygen gas for normal cell respiration, this presents a problem. In the field bean and most other legumes the nodules are bright red because of a compound called **leghaemoglobin** in the cytoplasm of the host cells. Its structure resembles that of mammalian haemoglobin, but the haem group is produced in the bacterium and the proteins are coded for by the host cell DNA. Possibly this leghaemoglobin provides a high current of oxygen to the bacterial cells, but at low concentration. In that way aerobic respiration can proceed and nitrogenase is not damaged.

The ammonium ions which are formed diffuse into the cytoplasm of the surrounding bean cells to be incorporated into organic nitrogen compounds. These ammonium ions cannot be processed by the bacterium's own glutamine synthetase enzyme (see page 46) because the host cell produces a compound which diffuses into the bacterium and inhibits it. Instead, the ammonia is picked up by the host cell's glutamine synthetase and ultimately converted to glutamine for export.

WHY DO BACTERIA FIX NITROGEN?

Nitrogen fixation is an energy-demanding process and it has been estimated that the need to feed the bacteria substantially reduces the potential growth rate of plants with nodules. If nitrogen fixation requires so much ATP energy, why has it evolved? Possibly nitrogen fixation has been advantageous in environments, and in periods of evolutionary history, when ammonium and nitrate ions have been scarce.

In view of the energy demand of nitrogen fixation, it is not surprising that nitrogen-fixing prokaryotes have formed mutualistic associations with angiosperms. For in nature nitrogen fixation is often limited by lack of organic compounds. Add these compounds to a habitat and the rate of nitrogen fixation increases by up to 300-fold. The rate of bacterial growth and nitrogen fixation by *Rhizobium* in bean roots is probably not limited in this way by lack of energy.

For consideration

1. How and why did haemoglobin come to be present in legumes?

2. How might the genes responsible for nitrogen fixation be transferred from bacteria to wheat or rice so as to produce a nitrogen-fixing cereal crop? What are the main problems which might be encountered in producing a viable nitrogen-fixing cereal and what would be the advantages of successfully achieving this?

Further reading

W.J. Brill, 'Biological nitrogen fixation' (*Scientific American*, vol. 236, no. 3, 1977)
This article summarises the agricultural importance of nitrogen fixation.

J. Postgate, *Nitrogen fixation* (Studies in Biology no. 92, Arnold, 1978)
An excellent elementary account.

J. Postgate, 'Nitrogenase' (*Biologist*, vol. 32, no. 1, 1985)
Rather an advanced account of the enzyme and how it works.

11 . Transport in animals

How does haemoglobin work?

Haemoglobin is the oxygen-carrying pigment found in the blood of all vertebrate animals and many invertebrates. Its functional importance can be appreciated by considering what would happen if the haemoglobin was suddenly removed from a person's blood and replaced by a simple aqueous solution. To deliver oxygen to the tissues at the same rate as before, the blood would have to be pumped round the body approximately thirty times faster.

What makes haemoglobin so good at carrying oxygen? To answer this we must first look at its molecular structure.

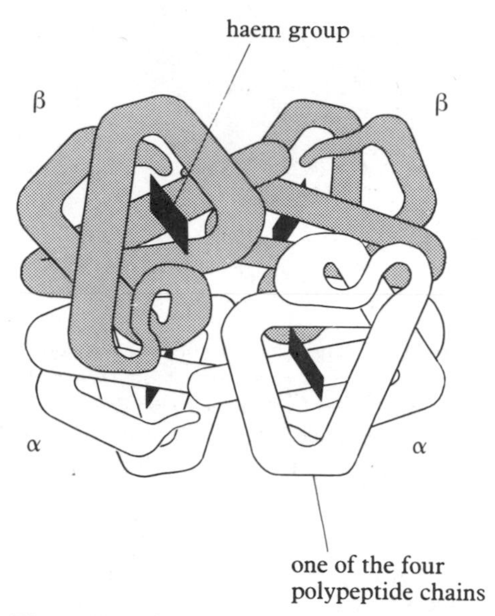

Figure 11.1 A haemoglobin molecule with its four constituent polypeptide chains in their correct positions.

THE MOLECULAR STRUCTURE OF HAEMOGLOBIN

The haemoglobin molecule is illustrated in figure 11.1. The molecule is made up of four sub-units (monomers) grouped together. The whole structure is called a tetramer, derived from the Greek word for four. Each of the four monomers consists of a molecule of the porphyrin **haem** which is attached to a polypeptide chain (**globin**). The four globin chains are not identical; two are called **α chains** and, in adult humans, the other two are **β chains**. Each type of polypeptide chain contains about 140 amino acids, but the precise number and sequence of amino acids differ. Each alpha chain is firmly linked to a corresponding beta chain by hydrogen bonds, and the two alpha-beta pairs (dimers) are linked with each other by weak ionic bonds. The bonding determines the overall shape of the haemoglobin molecule which, we shall see later, is intimately bound up with the way oxygen is carried.

The haem component of each monomer consists of four **pyrrole rings** arranged around a central iron (II) ion (figure 11.2). The iron is linked to each of the four pyrrole rings and to a histidine residue in the associated globin chain. When oxygen is being carried, an oxygen molecule combines with the iron (II) ion. It is important to appreciate that the iron is permanently in the divalent (iron II) state and when it binds with an oxygen molecule there is no change in valency. In fact if haemoglobin is treated with an oxidising agent which converts it to the trivalent (iron III) state, it is incapable of taking up oxygen.

How then is the oxygen taken up? The answer is related to the way the iron is attached to the nitrogen atoms of the pyrrole rings. The iron atom can exist in two different states. In one state it fits neatly into the space between the four nitrogen atoms. In the other state it slips slightly out of position and binds with an oxygen molecule. For this to happen it is necessary for the haem to be attached to a histidine residue in the associated globin chain; it will not happen if the haem is on its own. So it is clear that the globin has an important function in the oxygen-carrying process.

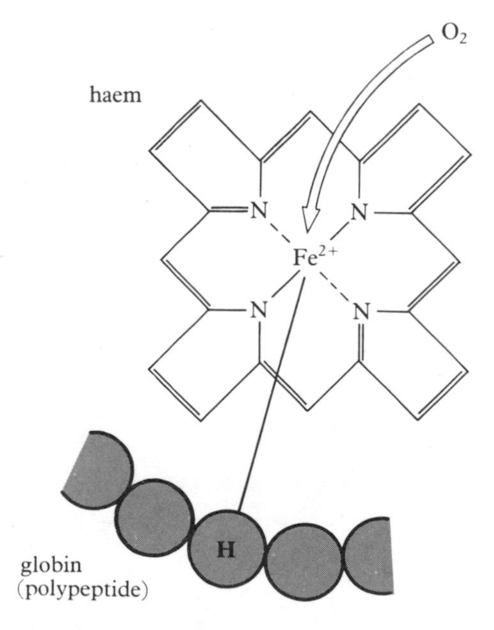

Figure 11.2 Diagram showing how a haem group is bonded to its associated popypeptide chain. H stands for histidine, the amino acid to which the iron (II) ion in the haem group is attached. Oxygen is carried by the iron atom as shown. Only a proportion of the amino acids in the polypeptide chain are included in this diagram.

THE PART PLAYED BY GLOBIN

It was suggested by physiologists several decades ago that haemoglobin's attraction for oxygen depends on the amount of oxygen with which it has already combined; and conversely the ease with which haemoglobin (Hb) gives up oxygen depends on the amount of oxygen which it has already released. We can therefore depict oxygen carriage taking place in four steps like this:

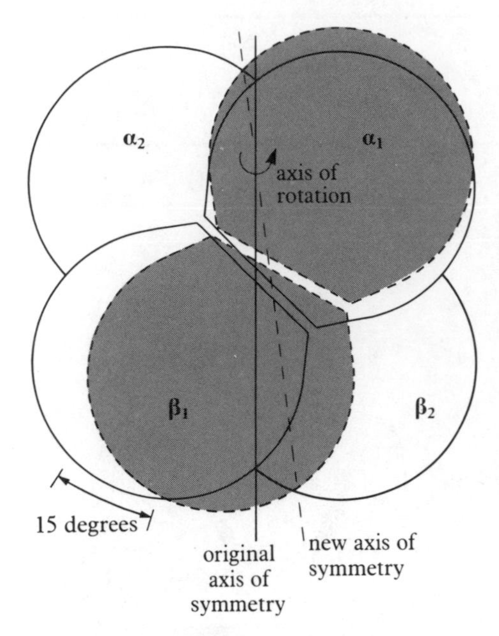

Figure 11.3 The two states of the haemoglobin molecule, tense (T) and relaxed (R). When the molecule changes from the T to the R form, one pair of polypeptides (α_1 and β_1 in the diagram) rotates as a single unit relative to the other pair. The rotation occurs through an angle of 15 degrees about an off-centre axis and there is a slight shift along it – what Perutz describes as a 'rock and roll' action. The position of the polypeptides after rotation is shown by broken lines.

(1) $Hb + O_2 \rightarrow HbO_2$
(2) $HbO_2 + O_2 \rightarrow HbO_4$
(3) $HbO_4 + O_2 \rightarrow HbO_6$
(4) $HbO_6 + O_2 \rightarrow HbO_8$

Note how each step influences the next step. With the discovery by John Kendrew and Max Perutz of the molecular structure of haemoglobin by X-ray crystallography, the way was clear for the formulation of a theory of oxygen uptake by haemoglobin.

The theory, in its current form, proposes that when one of the haem groups takes up an oxygen molecule, the shape of the haemoglobin molecule changes from a tense state to a relaxed state. This makes it easier for the remaining haem groups to bind with oxygen (figure 11.3). The change in shape of the molecule is miniscule, involving a movement of a mere fraction of a nanometre, but it greatly improves its affinity for oxygen.

When the haemoglobin molecule gives up its oxygen the process works in reverse. The release of oxygen from one of the haem groups makes it easier for the other haem groups to release their oxygen.

If this theory is correct haemoglobin is functioning rather like an **allosteric enzyme**: this is an enzyme whose combination with one molecule induces a change in shape which alters its ability to combine with another molecule. Indeed haemoglobin has many features in common with enzymes and shares the same properties of sensitivity to temperature, pH and inhibitors. The mechanism of oxygen uptake and release also helps to explain the sigmoid shape of haemoglobin's oxygen-dissociation curve.

More generally, the carriage of oxygen by haemoglobin demonstrates the importance of a protein's quaternary structure in determining its overall functioning. Some implications of this are discussed on pages 134–6.

For consideration

1. Haemoglobin has much in common with enzymes. How is this similarity related to its molecular structure and the way it works?

2. How does the mechanism of oxygen carriage by haemoglobin help to explain the sigmoid shape of its oxygen-dissociation curve?

Further reading

M.F. Perutz, 'The Haemoglobin Molecule' (*Scientific American*, vol. 211, no. 5, 1964)
Perutz describes some pioneering investigations into the molecular structure of haemoglobin and the way it might work.

N. Maclean, *Haemoglobin* (Studies in Biology no. 93, Arnold, 1978)
The first two chapters cover the structure and physiology of haemoglobin.

M.F. Perutz, 'Haemoglobin Structure and Respiratory Transport' (*Scientific American*, vol. 239, no. 6, 1978)
This article contains recent ideas about the functioning of haemoglobin.

The biological basis of coronary heart disease

Of all the deaths that occur in Britain each year, over a quarter are caused by **coronary thrombosis** or, in everyday language, heart attacks. This is slightly more than the deaths resulting from all kinds of cancer, and twenty times more than those resulting from road accidents.

A heart attack is caused by blood clotting inside one of the coronary arteries serving the muscle tissue of the heart wall. In the blood there are two opposing mechanisms, one tending to make the blood clot and the other tending to keep the blood fluid. To understand what causes a heart attack it is necessary to appreciate how these two opposing mechanisms work. Let us look at each in turn.

BLOOD CLOTTING

When blood clots, the soluble plasma protein **fibrinogen** is converted into a meshwork of fibres called **fibrin**. The conversion is triggered by the enzyme **thrombin** and the fibrin is strengthened by a further enzyme called **fibrin stabilising factor** in the presence of calcium ions. For thrombin to be formed, a series of linked reactions involving numerous enzymes and chemical factors is required. This 'cascade' of reactions is initiated either by substances released from damaged tissues, or by the blood platelets coming into contact with an unfamiliar surface.

BLOOD FLUIDITY

The fluidity of the blood is maintained by processes which oppose clotting. Some of these processes depend partly on chemical agents which inhibit thrombin formation and thus prevent the conversion of fibrinogen to fibrin. One of these anti-thrombin agents is enhanced by **heparin**, secreted by the mast cells, but quite how important heparin is in the natural maintenance of blood fluidity is uncertain.

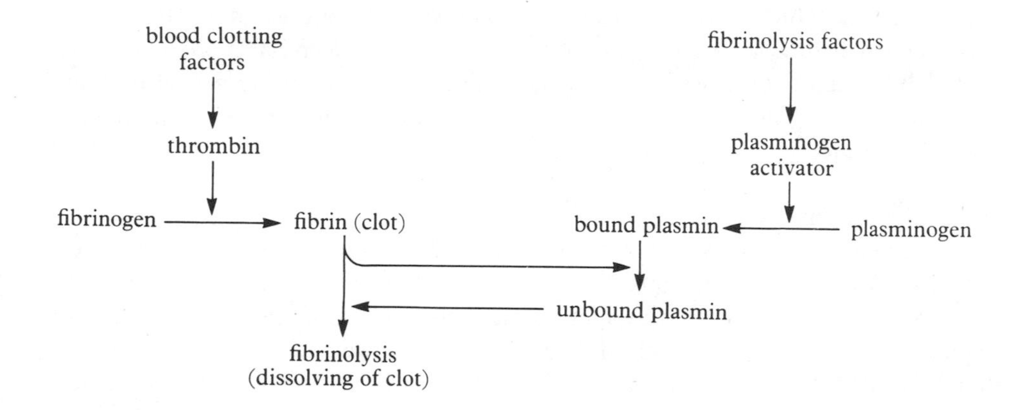

Figure 11.4 Simplified scheme summarising the relationship between blood clotting and fibrinolysis.

Much more important in maintaining fluidity is **fibrinolysis**, the dissolving of fibrin if and when it is formed in a blood vessel. Like clotting, fibrinolysis is the culmination of a cascade of reactions controlled by numerous chemical factors (figure 11.4). The process involves the formation of a protease enzyme called **plasmin** from an inactive precursor, **plasminogen**. Plasmin, once formed, is bound to a plasma protein which keeps it in an inactive state. However, fibrin causes this complex to dissociate, thereby releasing the plasmin which then proceeds to dissolve the fibrin. Thus we have a homeostatic mechanism in which fibrin triggers its own destruction – a good way of making sure that none is present.

The conversion of plasminogen to plasmin is initiated by a **plasminogen activator**. This has been found in low concentration in the blood and in much higher concentration in the tissues, particularly the endothelium of the blood vessels serving vital organs such as the heart.

It is thought that within the blood vessels clotting and fibrinolysis go on side by side all the time – as soon as any fibrin is formed it is removed by fibrinolysis. These two opposing processes are therefore normally in balance. However, if the rate of clotting exceeds the rate of fibrinolysis, fibrin builds up in the blood and a thrombosis may occur.

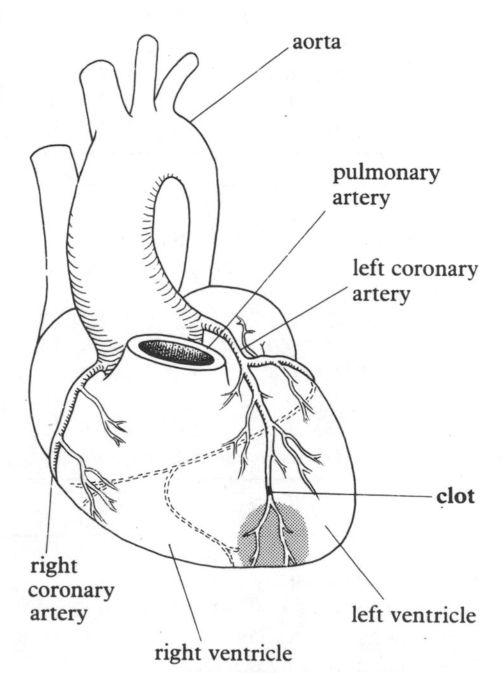

Figure 11.5 Ventral view of the heart showing the principal coronary arteries. A clot in the left coronary artery at the point indicated results in the shaded part of the left ventricle failing to receive oxygen.

THROMBOSIS

Thrombosis is the blocking of a blood vessel by a blood clot. If this happens in one of the coronary arteries, part of the heart muscle normally served by that artery is deprived of oxygen and dies (figure 11.5) – that is what happens in a

Table 11.1 Probability (per cent) of a 45 year-old man and a 45 year-old woman developing coronary heart disease within six years in relation to systolic blood pressure, blood cholesterol level and cigarette smoking. The systolic blood pressure is expressed in millimetres of mercury as is customary in medical circles; the cholesterol level in milligrams per decilitre. The data are based on the Framlingham Heart Study which was carried out on 6000 American volunteers from 1950 onwards.

45-Year-Old Man Non-Smoker	Systolic Blood Pressure					
	105	**120**	**135**	**150**	**165**	**180**
Blood Cholesterol Level **185**	1.5	1.8	2.1	2.5	3.1	3.7
210	1.9	2.2	2.7	3.2	3.9	4.6
235	2.4	2.9	3.4	4.1	4.9	5.9
260	3.0	3.6	4.3	5.2	6.2	7.4
285	3.8	4.6	5.5	6.5	7.8	9.2
310	4.9	5.8	6.9	8.2	9.8	11.5

45-Year-Old Man Smoker	Systolic Blood Pressure					
	105	**120**	**135**	**150**	**165**	**180**
Blood Cholesterol Level **185**	2.3	2.7	3.3	3.9	4.7	5.6
210	2.9	3.5	4.2	5.0	5.9	7.1
235	3.7	4.4	5.3	6.3	7.5	8.9
260	4.7	5.6	6.6	7.9	9.4	11.1
285	5.9	7.0	8.3	9.9	11.7	13.7
310	7.4	8.8	10.4	12.3	14.5	16.9

45-Year-Old Woman Non-Smoker	Systolic Blood Pressure					
	105	**120**	**135**	**150**	**165**	**180**
Blood Cholesterol Level **185**	0.4	0.6	0.7	0.9	1.1	1.4
210	0.5	0.7	0.9	1.1	1.4	1.7
235	0.7	0.8	1.1	1.3	1.7	2.1
260	0.8	1.0	1.3	1.6	2.0	2.6
285	1.0	1.3	1.6	2.0	2.5	3.1
310	1.2	1.6	2.0	2.4	3.1	3.8

45-Year-Old Woman Smoker	Systolic Blood Pressure					
	105	**120**	**135**	**150**	**165**	**180**
Blood Cholesterol Level **185**	0.5	0.6	0.7	0.9	1.2	1.5
210	0.6	0.7	0.9	1.1	1.4	1.8
235	0.7	0.9	1.1	1.4	1.8	2.2
260	0.9	1.1	1.4	1.7	2.1	2.7
285	1.1	1.3	1.7	2.1	2.6	3.3
310	1.3	1.6	2.0	2.6	3.2	4.0

heart attack. Sometimes a clot may block an artery in the brain. The nervous tissue served by the artery dies, resulting in a stroke; the part of the body normally controlled by the affected area of the brain becomes paralysed. A stroke can also be caused by rupture of the brain artery as a result of high blood pressure.

ATHEROSCLEROSIS

Clotting inside a blood vessel is caused by **atherosclerosis**. This is the formation of a localised thickening, called a **plaque**, in the wall of the vessel. Usually a plaque starts as a deposit of fatty substances and cholesterol, capped by fibrous tissue, but later it gets invaded by collagen and may become calcified. This makes the plaque brittle and liable to crack. In small arteries such as the coronaries, the plaque projects into the lumen and impedes the flow of blood. If the plaque ruptures, or gets torn at the edge, platelets congregate on the damaged surface and a local blood clot forms. The clotting process may be triggered by chemical substances released from the ruptured plaque or by contact of the blood with its surface. Either way, the delicate balance between clotting and fibrinolysis is upset and the clotting process predominates.

Substances which activate plasminogen have been found in bacteria, fungi and certain plants, and one such activator from *Streptomyces* is used to induce fibrinolysis in patients with thrombosis. Thrombosis patients are also treated with anti-clotting agents such as heparin and warfarin to help keep their blood in a fluid state.

WHAT CASUES ATHEROSCLEROSIS?

There can be few issues as contentious as the question of what causes atherosclerosis. Most of the data consists of statistical correlations between the incidence of coronary heart disease and factors such as one's sex, age, family history, blood pressure, blood cholesterol level, cigarette smoking, use of the contraceptive pill, hardness of the drinking water, and diet. A sample of such data is shown in Table 11.1.

One thing which emerges from Table 11.1 is that coronary heart disease is associated with a high level of cholesterol in the blood. Less certain is the role of the diet in determining the cholesterol level. On the basis of population surveys, a correlation has long been claimed between coronary heart disease and a diet high in cholesterol, and high cholesterol levels in the blood are associated with a high intake of saturated (i.e. animal) fats. But there are numerous anomalies at the individual level. For example, two people of the same age and sex, both on exactly the same diet, may have completely different blood cholesterol levels. There are anomalies too at the population level. For example, the Masai people of East Africa have a very high intake of saturated fat and yet the incidence of coronary heart disease is exceptionally low. These observations suggest that many other factors are involved, and they may be inter-related in a complex way. However, in matters of this sort it may be best to err on the side of caution and the author of this topic will be starting his low-fat diet tomorrow.

For consideration

1. Present the data in Table 11.1 graphically so that any possible correlations between coronary heart disease and the parameters investigated show up clearly.

2. What conclusions would you draw from the data in Table 11.1? How could you rule out the possibility that they are due to chance?

Further reading

M. DeBakey and A. Gotto, *The Living Heart* (Grosset and Dunlop, 1977)
A popular book by two eminent American heart specialists.

J.R. Hampton, *Cardiovascular Disease* (Heinemann, 1983)
This is written primarily for clinical students, but makes fascinating reading.

12 . Uptake and Transport in Plants

The sensitivity of stomata

'The guard cells . . . constitute miniature sense organs which in an animal would be said to provide a capacity for sight, smell, touch and taste.' (T.A. Mansfield, 1982)

Guard cells are to be found lining the stomata of plants. Since the stomata not only take up scarce carbon dioxide, but also lose water vapour, the reaction of the guard cells to environmental stimuli is crucial to a plant's survival. We shall consider it in this topic.

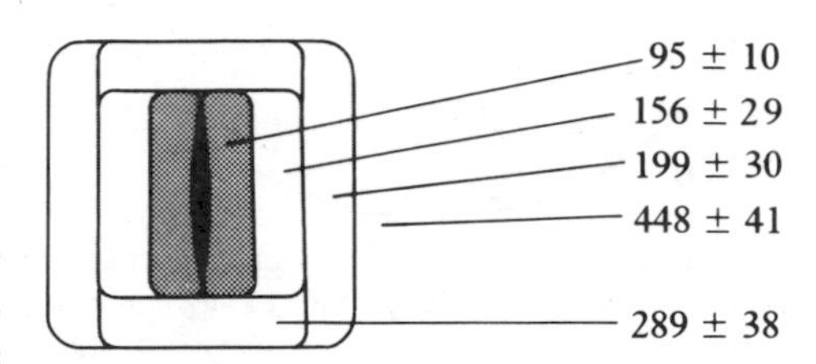

Figure 12.1 Concentrations of potassium ions in the vacuoles of cells in the stomatal complex of *Commelina communis* when the stoma is closed. In this species the guard cells (grey) are surrounded by six subsidiary cells (white). The concentrations were measured in moles per m^3 with an ion-sensitive electrode. (*Based on Penny and Bowling*, Planta, *Vol. 119, 1984*)

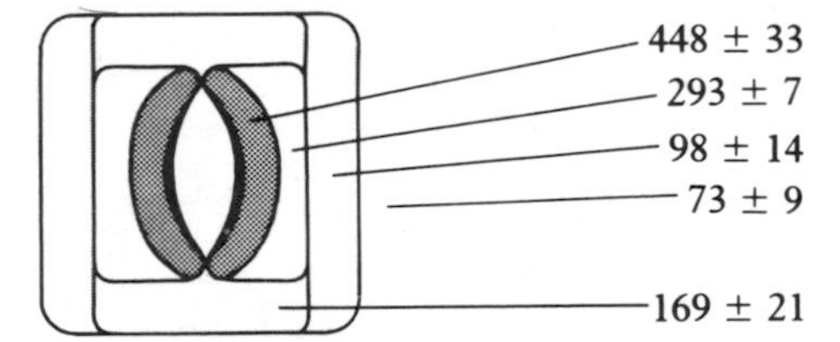

Figure 12.2 The potassium ions are stored in epidermal cells near the guard cells when the stoma is closed, but migrate into the guard cells when the stoma opens. In the light, some of the starch stored temporarily in the guard cell chloroplasts is converted to the four-carbon malate ion. This provides a negatively-charged ion to balance the positively-charged potassium ions which flow into the guard cell cytoplasm.

HOW DOES A STOMA OPEN AND CLOSE?

Many theories have been put forward to explain the opening and closing of stomata. The current hypothesis is based on the movement of potassium ions into and out of the guard cells. This hypothesis runs as follows. Consider a guard cell in the dark when the cells are flaccid, lacking water. At dawn, light stimulates photosynthesis in the leaf. The leaf cells take up carbon dioxide from the intercellular spaces, and by an unknown mechanism the drop in carbon dioxide concentration in the intercellular spaces is sensed by the guard cells. Evidence such as that presented in figures 12.1 and 12.2 suggests that the guard cells begin to take up potassium ions by active transport across their cell membranes. The energy for this uptake is believed to be derived from ATP synthesised by their mitochondria and chloroplasts.

The influx of potassium ions, and the production of some malic acid in the light from the breakdown of starch, lowers the water potential of the solution within the guard cells. So water flows into the guard cells down the water potential gradient, causing the cells to swell. This opens the pore between them, and gas exchange begins. Exactly the reverse happens at dusk, except that the potassium ions flow out passively by diffusion: gated channels are believed to open up in the cell membranes similar to sodium channels in nerves (see page 82).

THE DIFFUSION OF GASES

When all the stomata of a mesophytic leaf are fully open, they are remarkably efficient at gas exchange. The stomata cover only about 1.5 per cent of the leaf surface, yet a leaf hung up to dry loses water at half the rate of a piece of soaked filter paper cut to the same size. How is this possible?

The efficiency of stomata in gas exchange seems to be a property of small holes in general. Simple experiments have shown that the rate at which carbon dioxide travels through small holes is proportional to the circumference of the holes, not their areas. This probably means that a high proportion of carbon dioxide molecules are moving around the edges of the pores rather than through their centres. In fact many small holes can conduct just as much carbon dioxide as a large hole of much greater area (figure 12.3).

STOMATAL APERTURE

The experiments on gas exchange described above obscure the fact that stomata are not simply open or closed. Their apertures differ from minute to minute. Some guard cells, for instance, have tiny holes in their cuticles, and if they lose

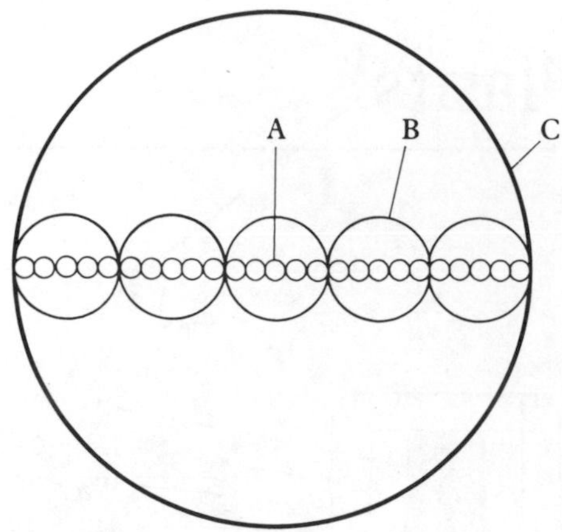

Figure 12.3 Many small holes can conduct gases as rapidly as a much larger hole. The combined diffusion rate through (**A**) the twenty-five small holes should be the same as the combined diffusion rate through (**B**) the five medium holes and the diffusion rate through (**C**) the large hole. This is because diffusion rate is proportional to pore circumference = $2\pi r$, and therefore proportional to radius. If the diameter of each of the smallest holes is 1 unit their combined circumference is 25n; there is also true of the five medium holes and the single large one. (*After Newman*, 1976)

too much water through them, the stomatal pores will close slightly. Let us illustrate the sensitivity of guard cells, however, by examining their responses to wind speed.

Between a mesophyll cell in a leaf, and the atmosphere outside, there are resistances to gaseous diffusion arranged in series. A molecule of water diffusing out of the leaf encounters first the resistance of the intercellular space, then the stomatal resistance, and finally the resistance of the boundary layer of humid air. In still air the boundary layer is thick. It puts up such a resistance to diffusion that water vapour escapes from the leaf at the same rate whatever the stomatal aperture. In windy conditions, however, the boundary layer is thinner and the resistance to diffusion across it markedly decreases. Under these circumstances the main resistance which an escaping water molecule encounters is the stomatal resistance, which varies with stomatal aperture. In windy conditions, then, the pore size influences water loss.

When a molecule of carbon dioxide enters the leaf it encounters the same resistances, and one more. This is the resistance of its diffusion pathway from the wet cell wall of a spongy mesophyll cell to a chloroplast, which is considerable since the rate of diffusion of carbon dioxide in water is 10 000 times slower than its diffusion in air. In windy conditions, when water loss increases, the rate at which carbon dioxide enters the leaf also increases because the resistance of the boundary layer declines. This will be sensed by the inner surface of the guard cells and it will cause a slight stomatal closure. This closure reduces water loss more than carbon dioxide uptake because the major resistance to carbon dioxide uptake is the resistance to its transport to the chloroplasts in the aqueous phase. We can imagine the pore size changing continuously as the wind speed fluctuates, and you can see why Mansfield described the guard cells as miniature sense organs.

If the epidermal cells lose water faster than they can gain it and their water potential declines, the stomatal pores may close rapidly. This often happens in the middle of a hot day, and is known as **midday closure**. This lowers carbon dioxide uptake and is one reason why lack of soil water reduces the rate of growth of crops.

When the hormone abscisic acid is applied to a leaf the stomatal pores rapidly close (figure 12.4). This is accompanied by an efflux of potassium ions across the cell membrane. These observations suggest that the cells of water-stressed plants secrete abscisic acid hormone and that this hormone causes midday closure.

For consideration

1. Suggest another reason, besides its effect on stomatal aperture, why lack of soil water may reduce crop growth.

2. Suggest a mechanism to explain why the stomata of many species of plants begin to close when a leaf is handled or breathed upon.

3. Suppose you sprayed an 'anti-transpirant' onto a leaf and that the compound halved the diameters of the stomatal pores. Would this have the same effect on the rate of water loss as on the rate of carbon dioxide uptake? Explain your answer.

Further reading

E.S. Martin, M.E. Donkin and R.A. Stevens, 'The Stomatal Mechanism – some recent developments' (*School Science Review*, vol. 63, no. 224, 1982)

E.S. Martin, M.E. Donkin and R.A. Stevens, *Stomata* (Studies in Biology no. 155, Arnold, 1983)

A.W. Galston, P.J. Davies and R.L. Satter, *The Life of the Green Plant* (Prentice-Hall, 3rd edition, 1980)

J.F. Sutcliffe, *Plants and Water* (Studies in Biology no. 14, 2nd edition Arnold, 1979)

Water potential and transpiration

When a plant transpires it acts as a wick between the soil and the air. Water is thus moving down the water potential gradient (see page 18). The water potential is a measure of the average energy of water molecules per unit volume of fluid at any point. The water potential of pure water is zero. In most places on

Earth the atmosphere has a low water potential, that is, a strongly negative water potential. On a warm day in Britain the water potential of moist soil might be −10 kPa, and of the atmosphere −30 000 kPa. No wonder that water moves from soil to air through the plant!

STOMATA AND WATER LOSS

In considering the effect of the water potential of the atmosphere on the transpiration rate there are three important points to bear in mind.

Firstly, fluctuations in the relative humidity and temperature have a marked influence on the water potential of the atmosphere. At 20 °C, a decrease from 99 per cent relative humidity to 50 per cent relative humidity lowers the water potential from −1360 kPa to −93 400 kPa. When saturated air at 19 °C (water potential = 0 kPa) is warmed to 20 °C, its water potential falls to −8400 kPa. Plant stems and leaves have water potentials of 0 to −6000 kPa.

Secondly, the rate at which water moves through a plant, or any part of it, follows a physical analogy, that of Ohm's Law. For electricity, $I = V/R$, where I is the current, V is the potential difference between two points and R is the resistance. In plant terms, I is the water flow per unit time, V is the difference in water potential between two points and R is the resistance of a stage in the pathway to water movement.

Thirdly, using this analogy, the largest potential difference in the whole pathway of water movement from soil to air is between the wet cell walls of the spongy mesophyll and the air. Stomata are variable resistances set at this crucial point and that is why their apertures have such a major influence on transpiration rate (see page 53).

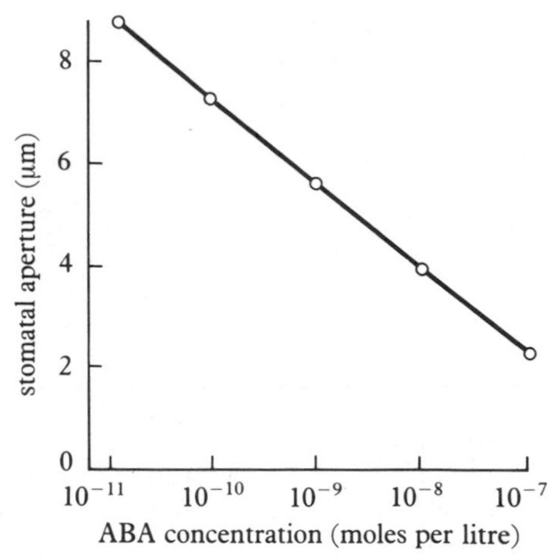

Figure 12.4 The effect of abscisic acid on stomatal aperture in isolated epidermal strips of *Commelina communis*. (*Based on Mansfield, 1982*)

Let us consider the effect of stomatal aperture on the water potential gradient in a transpiring plant over twenty four hours. At dusk, when the stomata close rapidly, there is a large difference in water potential between the top and bottom of the xylem. The cohesive column of water is under tension. The relatively low water potential in the xylem of the roots causes water uptake to continue all night. At dawn the stomata open, markedly reducing the main resistance in the pathway. Transpiration increases tenfold. The evaporation of the most energetic water molecules lowers the potential energy content per unit volume of the remainder. In other words, it lowers the water potential of the water molecules in the wet cell walls of the leaf mesophyll cells, thus steepening the water potential gradient and speeding up xylem transport. However, the resistance of the xylem vessels to water movement prevents the rapid replacement of water. A water deficit develops. This places the water in xylem vessels under so much tension that many tree trunks are marginally narrower during the day than at night (figure 12.5).

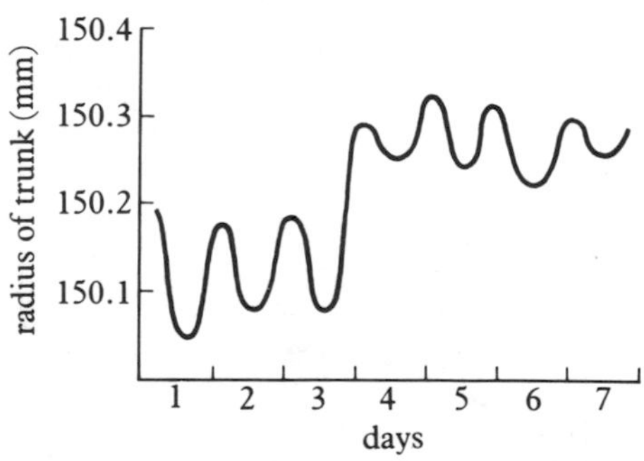

Figure 12.5 Shrinkage and expansion of the trunk of the pine tree *Pinus resinosa* over seven days. (Kozlowski, Water Metabolism in Plants *Harper and Row, 1984*)

SOIL WATER POTENTIAL

We have assumed so far that the soil water potential remains constant. In fact water which lies freely in the pores of the soil, that is **gravitational water**, will be removed first, because it has the highest water potential. The roots then begin to extract **capillary water**, which has a lower water potential because its molecules are bound to soil particles. As water is removed from smaller and smaller pores, its resistance to movement increases and its water potential decreases. Eventually, the continuous films of water rupture.

As water uptake begins to lag further behind water loss from the leaves, the water potential at the base of the xylem will fall. This is because the water molecules at the base of the cohesive column of water will be placed under increasing tension from above.

The soil water potential, however, declines more rapidly still and ultimately equals the root water potential. At this point, about −1500 kPa for most mesophytes, the plant begins to **wilt**. Its leaves and stem flop over and lose turgor, the stomata close and the plant prays for rain to increase the water potential of the soil.

Some species, however, have roots continually exposed to much lower water potentials than −1500 kPa. Desert perennials such as the creosote bush, and halophytes such as sea lavender, usually grow in soil with water potentials of −2500 kPa or less. They solve this problem by making organic solutes like sorbitol and accumulating ions from the soil, thus bringing their root water potentials low enough (−2500 to −5000 kPa) to take up water from such soil.

For consideration

1. Explain how the presence in a cell vacuole of organic solutes with numerous hydroxyl groups (−OH) might lower the water potential of the solution.

2. Make a list of the xeromorphic features of desert shrubs which enable them to conserve water despite the massive difference in water potential between their tissues and the air.

3. The average water potential of the shoots of individual shrubs of the creosote bush, *Larrea tridentata*, in an American desert was measured as −3800 kPa. When adjacent shrubs of the same species were removed, the water potential of these shoots increased to −3200 kPa. What does this suggest?

Further reading

J. Sutcliffe, *Plants and Water* (Studies in Biology no. 14, 2nd edition, Arnold, 1979)
A useful exposition of all aspects of water potential and the flow of water through the plant.

The uptake of nutrient ions by roots

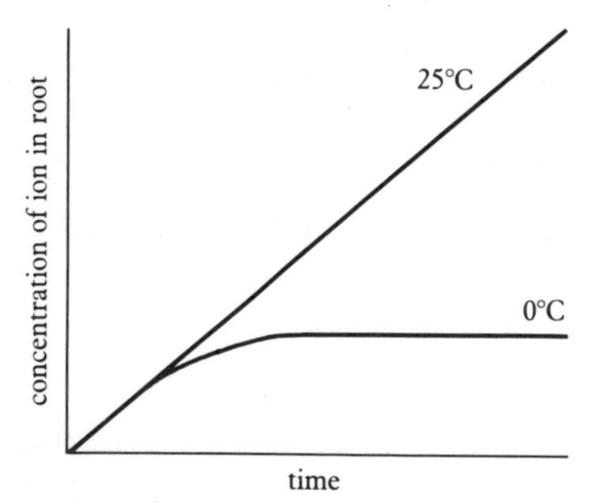

Figure 12.6 Uptake of radioactive rubidium ions by washed barley roots at two different temperatures, 0°C and 25°C. At first ion uptake is independent of temperature. At this stage the ions can be washed out of the roots with distilled water. The concentration of ions which can be removed in this way suggests that they occupied about twenty per cent of the root volume. It seems likely that at this stage the solution has diffused into the wet cell walls of the root hairs and root cortex until its concentration of radioactive rubidium is in equilibrium with the solution surrounding the roots.

Later, ion uptake depends on temperature. This probably represents the phase of active transport of ions across cell membranes, which, being an enzyme-controlled process, is strongly temperature-dependent. As ions are transported across the membranes into the cells, more ions diffuse from the solution surrounding the roots into the cell walls to replace those that have been removed.

Plants depend on a constant supply of essential elements to the meristems. Most of these elements, including N, P, K, S, Mg, Ca, Fe and various micronutrients, are absorbed by the roots as ions. In this topic we shall discuss how this absorption takes place.

In most terrestrial plant communities, roots are abundant in the upper layers of the soil. In a temperate forest such as that at Hubbard Brook, New Hampshire, for example, each square centimetre of soil has beneath it about 2.7 metres of fine roots with a total surface area of 24 cm^2. The root area is twenty-four times the ground area. Indeed, this estimate ignores the massive surface area of the mycorrhizal fungal hyphae associated with roots, which absorb nitrates and other ions in return for a glucose supply.

The ions absorbed come from the decomposition of mineral particles, rainfall, ions washed out of living leaves, and the dead parts and egested matter of plants and animals. The positively-charged ions are stored on negatively-charged soil compounds, both in clay and the organic acids which make up the humus. As ions are absorbed from the soil water by roots, others dissolve into the soil water from the particles.

MOVEMENT INTO THE ROOTS

The roots concentrate many of the ions. The roots expend metabolic energy to take up ions against the concentration gradient. The natural tendency of the ions would be to diffuse from the cells to the soil, so how do the roots concentrate ions?

One clue comes from the experiment outlined in figure 12.6. Thoroughly washed roots were placed in solutions of radioactive ions, and the uptake of the ions was measured at different temperatures. The uptake appears to occur in two stages. At first, in a process unaffected by temperature, the ions diffuse into the wet cell walls of the root. Then, in a process which is strongly temperature dependent, the ions are actively transported across the cell membrane into the cytoplasm of a root cell.

In most plants about half the ions are taken up in the root hair zone and the other half on either side of it. The soil water, containing ions, saturates not only the cell walls of the root hairs, but also the cell walls of the root cortex cells. The root cortex can be regarded as a tissue which exposes a massive surface area of cell membrane to the soil for nutrient uptake.

The membranes pump the ions into the cells. This process requires energy from cellular respiration. The evidence for this is that ion uptake stops if the roots are cooled, deprived of oxygen, or bathed in cyanide solution. Cyanide acts on the terminal cytochrome in the respiratory chain in mitochondria and prevents much of the ATP from being formed.

Just as ATP is used to maintain the resting potential in a nerve cell membrane, so it is used to maintain a similar potential difference across the cell membranes of the root hair and cortex cells. However, in this case, hydrogen ions are pumped across the membrane from the cytoplasm to the cell wall. This creates a potential difference of about -120 mV, the inside being negative with respect to the outside. Potassium ions and other cations can flow into the cell down the electrical gradient thus created.

By analogy with other cells ion uptake in cortex cells is probably regulated by specific protein carriers embedded in the lipid bilayer. They have not yet been isolated, but there is evidence that they exist. For example, some pairs of ions compete for uptake: calcium with strontium, sulphate with selenate, bromide with chloride, and potassium with rubidium. In each case the first ion mentioned in the pair is common, and the second is rare. The ions in each pair are similar in chemical activity, charge and ionic radius. It seems likely that a separate carrier protein exists in the membranes for each ion to which the cortical cells are normally exposed. Occasionally they make mistakes, and pick up the less common ion.

ACROSS THE ROOT TO THE XYLEM

Once inside the cells, the ions can move from cytoplasm to cytoplasm via the plasmodesmata en route for the xylem. The endodermal cells, with their suberised Casparian bands in the cell walls, hardly resist ion transport because their cytoplasm is connected to the cells on either side by plasmodesmata.

What stimulates the ions to move from root hair to xylem? The mechanism is obscure, although there is some evidence for active accumulation of ions in the pericycle cells nearest the xylem. The ions could move from these cells across the membrane into the xylem either by diffusion or active transport.

Not all the ions accumulated by the root cells are valuable to the plant and indeed some inhibit enzyme activity. Many species, however, have enzyme systems which can withstand high concentrations of toxic ions. For instance, some plants normally found growing on alkaline soils in Britain would be killed by 1 part per million of aluminium in the soil water. However, the tea plant, which grows in the tropics and subtropics, not only has up to 30 000 parts per million of aluminium in its leaves, but dies of aluminium deficiency in normal soils. So one plant's meat is another plant's poison.

For consideration

1. There is evidence for the accumulation of ions in the pericycle cells close to the xylem in the root. What are the implications of this in (a) ion uptake and (b) water uptake?

Further reading

D.A. Baker, 'The Transport of Ions across Plant Cell Membranes' (*Journal of Biological Education*, vol. 15, no. 2, 1981)
A detailed account of the membrane proton pump.

E. Epstein, 'Roots' (*Scientific American*, vol. 228, no. 5, 1973)
An excellent introduction to roots and mechanism of ion uptake.

A.W. Galston, P.J. Davies, and R.L. Satter, *The Life of the Green Plant* (Prentice-Hall, 1980)
Chapter 7 is particularly relevant.

13 . Homeostasis

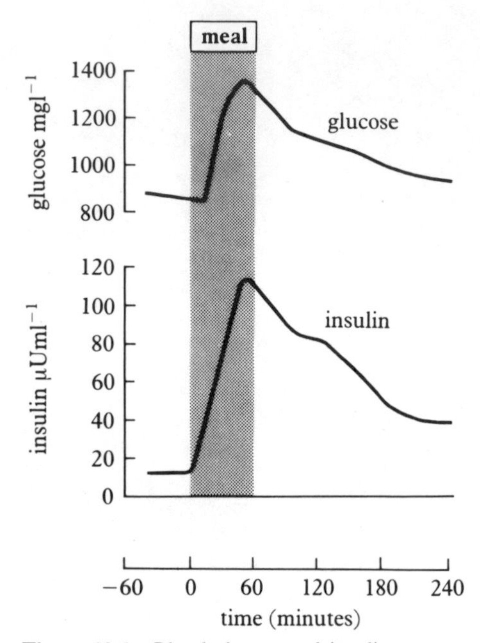

Figure 13.1 Blood glucose and insulin concentrations in the blood of a normal human subject in response to a large carbohydrate meal. (*Based on R. H. Unger*, New England Journal of Medicine, *Vol. 283, 1970*)

What makes an efficient homeostatic system?

If you recorded your body temperature every five minutes or so with a sensitive thermometer, you would probably find that it fluctuated continuously by a fraction of a degree. All homeostatic systems involve fluctuations about a norm or set point – they must do. An efficient system is one which, through negative feedback, keeps the fluctuations small enough to allow the organism to function properly. How is this achieved and what other features contribute to the efficiency of homeostatic systems?

ANTICIPATING A CHANGE

The effectiveness of a homeostatic response can be increased if the system anticipates a deviation from the set point before it actually happens. Take the regulation of blood sugar, for example. Figure 13.1 shows the effect of a large carbohydrate meal on the glucose and insulin concentration in the blood of a normal human subject. The remarkable feature of the graph is that the rise in the insulin concentration precedes the rise in the blood glucose concentration. This is thought to be due to the fact that when glucose is present in the stomach and small intestine it causes the release of hormones which stimulate insulin secretion by the pancreas. Not much is known about the precise mechanism but insulin secretion is certainly stimulated by gastrin and other gut hormones, and also by impulses in the vagus nerve. By this means insulin secretion gets underway *before* the glucose is absorbed into the bloodstream.

ERROR-ACTUATED CONTROL

Consider the artificial homeostatic system illustrated in figure 13.2. The idea is that you should keep the temperature of the aluminium block at a more or less constant 60 °C by noting the temperature and adjusting the power supply accordingly. Suppose the heater is on and the temperature of the block is rising. When the temperature reaches 60 °C, the set point, you switch the heater off. When subsequently the temperature falls below 60 °C you switch the heater on again – and so on. With a simple on-off system like this the temperature of the block would fluctuate wildly on either side of the set point.

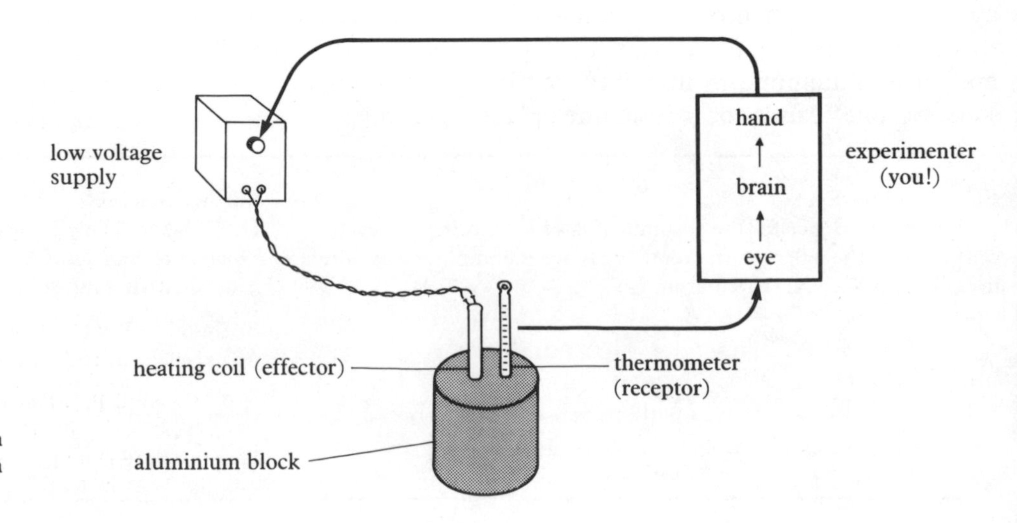

Figure 13.2 An artificial homeostatic system in which you act as the control centre. (Explanation in text)

How could you reduce the fluctuations? One way would be to switch the heater on or off before the set point is reached – the equivalent of anticipating the deviation before it actually happens. Even better would be to adjust the power supply as the temperature rises and falls so that the overshoot is minimised. Thus, as the temperature rises towards 60 °C you gradually reduce the power, and as the temperature falls towards 60 °C you gradually increase it. Table 13.1 shows four different programmes which you might follow.

Table 13.1 Four possible programmes designed to reduce fluctuations in the homeostatic system illustrated in Figure 13.2. To see how the table works consider programme 1. First you turn the power off. When the temperature is below 58.5°C you set the power supply at 16V. When the temperature reaches 58.5°C you reduce the power to 12V, and when the temperature reaches 59°C you reduce the power to 8V, and so on. In other words for every one degree rise in temperature you reduce the power by 8V, giving a sensitivity of 8V deg^{-1}.

Programmme	Sensitivity	Temperature (°C)		Supply Setting (V)
1	8 V deg^{-1}	Below	58.5	16
		At	58.5	12
			59.0	8
			59.5	4
			60.0	0
2	4 V deg^{-1}	Below	56.5	16
		At	56.5	14
			57.0	12
			57.5	10
			58.0	8
			58.5	6
			59.0	2
			59.5	4
			60.0	0
3	2 V deg^{-1}	Below	52.5	16
		At	52.5	15
			53.0	14
			etc	
4	1 V deg^{-1}	Below	45.0	16
		At	45.0	15
			46.0	14
			etc	

What we are doing here is to adjust the amount of correction according to the size of the deviation or 'error'. This is described as **error-actuated control**, and it is what tends to happen in most natural homeostatic systems (see pages 69–70 for an example).

ANTAGONISTIC HORMONES

Let us return to our aluminium block for a moment. Another way we could reduce the temperature fluctuations would be to place some ice in contact with the block when the temperature rises above the set point. The ice opposes the action of the heater and helps to bring the temperature down again. In the same kind of way homeostatic systems often involve two hormones which oppose each other in their actions. For example, in the control of blood sugar, **insulin** lowers the level and **glucagon** raises it; and in the control of calcium, **parathormone** stimulates its removal from the blood whereas **calcitonin** stimulates its release into the blood.

To keep the level of a substance close to the set point there must be an appropriate balance between the two antagonistic hormones. In fact physiological control systems are usually more complex than this. Blood sugar regulation, for example, involves interactions between more than ten different hormones, and the nervous system is involved too. We shall return to this in the next topic.

ADAPTABILITY

The term set point is in some ways misleading since it implies that it is fixed once and for all. In practice, set points change and this is important in enabling organisms to adapt to new situations.

Take hibernation, for example. In summer the body temperature of the European hamster *Cricetus* fluctuates about a mean value of about 34 °C. This is the set point. With the onset of winter the set point drops to approximately 6 °C, the metabolic rate falls by more than 90 per cent and the animal goes into a deep sleep. The following spring the set point returns to its original value, the normal metabolic rate is resumed and the animal becomes active again.

These seasonal changes in the set point involve short periods of **positive feedback**: a drop in body temperature leads to a further drop. This is the reverse of what happens in negative feedback where a deviation from the set point causes a return to the set point. Positive feedback is usually taken to indicate that the system is running away with itself and getting out of control. However, in this case it enables the homeostatic system to adjust to a change in the environment.

Alterations in the set point occur in other situations too. For example, our body temperature varies on a diurnal basis, falling at night and rising during the day; and the level of certain hormones changes during an organism's development.

For consideration

1. Of the four programmes given in Table 13.1, which would you expect to give the smallest temperature fluctuations, and why?

2. Give examples in humans and other animals of the level of a hormone changing during development, and explain why the change is important.

3. Discuss the cause, and possible value to the human, of the elevated body temperature which often accompanies infection by a pathogenic micro-organism.

Further reading

R.N. Hardy, *Homeostasis* (Studies in Biology no. 63, Arnold, 1976)
A useful if rather complicated introduction.

G. Hardin and C. Bajema, *Biology – Its Principles and Implications* (Freeman, 3rd edition 1978)
There is an admirable account of homeostasis in this general biology text.

J.A. Lee, 'Homeostasis – the Modern Concept' (*Biologist*, vol 28, no. 1, 1981)
The author emphasises the dynamic nature of the set point in homeostatic systems.

The control of blood sugar

Blood sugar is regulated by two pancreatic hormones, insulin and glucagon. This statement, however, ignores the fact that many other hormones, not to mention the nervous system, also play a vital part. The aim of this topic is to convey something of the complexities involved, but first we must understand what happens to sugar metabolically once it has got into the bloodstream.

THE METABOLIC FATE OF SUGAR

Most free sugar in the human body is in the form of glucose. The concentration of glucose in the bloodstream may be increased or decreased by various metabolic conversions, of which the main ones are as follows:

1 Glucose may be converted into glycogen in the liver and muscles, the liver conversion being reversible.

2 Glucose may be converted into lipid (fat) in the liver and adipose tissue and then stored in the body's fat depots. The glycerol component of the lipid can be converted back into glucose.

3 Glucose may be metabolised, with the release of energy. This occurs in practically all the body's cells and is of course irreversible. The process is usually aerobic, though in active muscles it may be anaerobic with the formation of lactic acid.

4 Glucose may be synthesised in the liver (and kidneys) from non-carbohydrate sources such as amino acids and glycerol, a process called **gluconeogenesis.**

All these metabolic conversions are influenced by the hormone **insulin**, secreted by the beta cells in the islets of Langerhans, which lowers the level of blood glucose. **Glucagon**, secreted by the alpha cells, has the opposite effect and raises the blood glucose level.

Generally the four metabolic conversions listed above proceed at such relative rates that the concentration of glucose in the blood is held at approximately 85 mg per 100 cm^3 except during a meal when it may rise temporarily to about 120 mg per 100 cm^3.

An elevated blood glucose level (hyperglycaemia) is a symptom of **diabetes mellitus**. This is usually attributable to a malfunctioning pancreas but may sometimes have other causes, as we shall see. Let us now look at some of the hormones, besides insulin and glucagon, which play a part in the control of blood glucose.

EXPERIMENTAL EVIDENCE FOR THE INVOLVEMENT OF THE PITUITARY GLAND

Unlike many endocrine glands the pancreas is not under the direct control of the pituitary gland: there is no 'pancreotrophic' hormone in the sense that there is, for example, a thyrotrophic hormone. Nevertheless the pituitary does play a part in the control of blood glucose, as is seen by what happens if its anterior lobe is removed. This enhances the effect of insulin: blood glucose falls at more or less the normal rate but to a lower level than usual and it rises again more slowly. In fact removal of the anterior pituitary in diabetes reduces the insulin requirement by as much as ten times. The same effect is seen in patients whose pituitary gland is not secreting properly (figure 13.3).

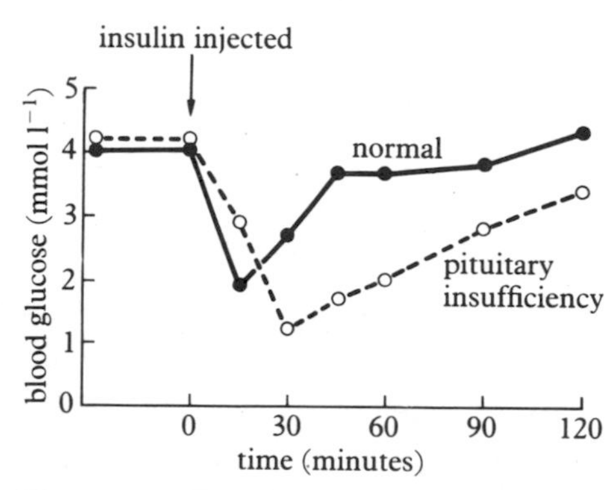

Figure 13.3 The effect on blood glucose of an intravenous injection of insulin in a normal subject and a subject with an under-secreting pituitary. (*Based on Barker and Isles*)

It seems that the anterior lobe of the pituitary secretes a hormone which opposes the action of insulin. Further evidence for this comes from experiments in which anterior pituitary extract is injected into experimental animals: the effects include a marked increase in the concentration of blood sugar – in other words diabetes is induced.

THE GROWTH HORMONE IN GLUCOSE CONTROL

We now know that the active principle in the anterior pituitary extract is the **growth hormone** (**somatotrophin**). Secretion of this hormone into the bloodstream is controlled by two 'neurohormones' which reach the anterior lobe via the axons of nerve cells located in the hypothalamus. One of them (**growth hormone releasing factor**) stimulates the secretion of growth hormone; the other one (**growth hormone release inhibiting factor** or **somatostatin**) inhibits it.

By opposing the action of insulin growth hormone conserves glucose in the body when it is scarce, for example during muscular exercise. This is particularly important to the brain which uses only glucose for respiration; other tissues can fall back on alternative substances such as fatty acids.

OTHER HORMONES INVOLVED IN GLUCOSE CONTROL

There are times, for example during bouts of exercise, when extra glucose needs to be made available for energy release. This is achieved by the hormone **adrenaline**, secreted by the medulla of the adrenal glands. Adrenaline inhibits insulin secretion, raises the basal metabolic rate, stimulates the conversion of liver glycogen to blood glucose and promotes gluconeogenesis.

Another hormone which affects the blood glucose level is **thyroxine**, secreted by the thyroid gland. Thyroxine raises the metabolic rate, increasing the utilisation of glucose. This tends to lower the blood glucose concentration, an effect which is accentuated by the fact that thyroxine, in moderate amounts, accelerates insulin secretion. You might think that this would cause a severe fall in the blood glucose level. However, this does not happen because thyroxine also promotes gluconeogenesis and increases the sensitivity of the tissues to glucagon.

An excessive lowering of blood sugar is also counteracted by another hormone, **cortisol**, which is secreted by the adrenal cortex. This raises the level of blood sugar, mainly by promoting gluconeogenesis and decreasing the basal metabolic rate.

Thyroxine and cortisol are both controlled by hormones from the anterior pituitary. Thyroxine secretion is stimulated by **thyrotrophic hormone,** cortisol secretion by **adreno-cortico-trophic hormone** (ACTH). The overall effect of these hormones is to raise the blood glucose level, compounding the effect produced by pituitary growth hormone.

These are by no means the only hormones which influence the concentration of blood glucose, but they are the main ones. Their actions are summarised in figure 13.4 which serves to illustrate how complex physiological homeostasis can be.

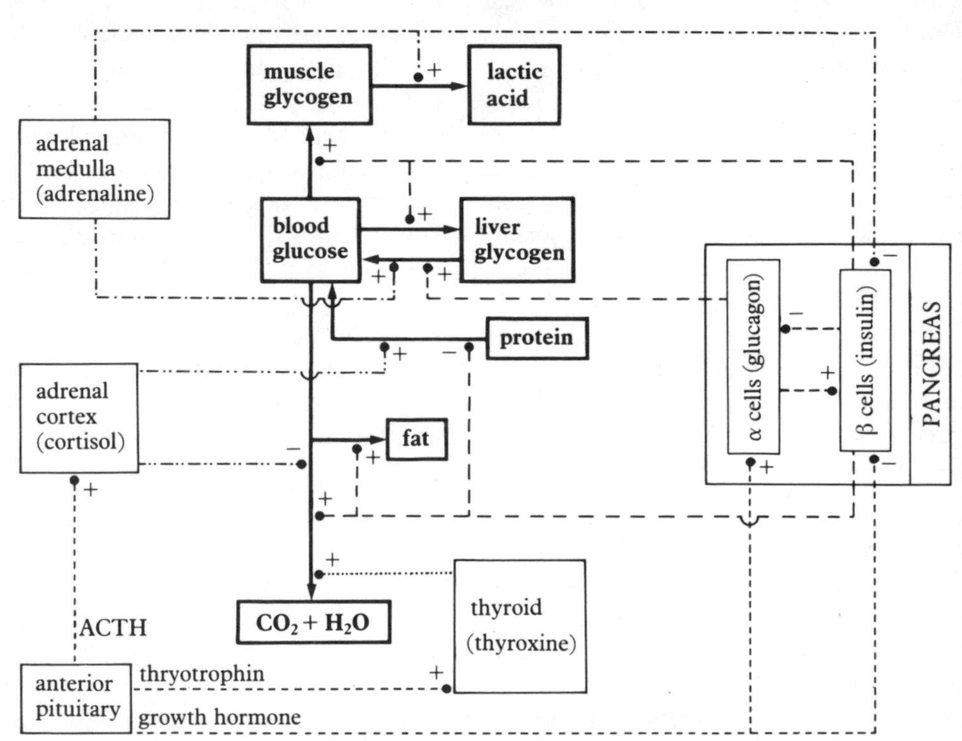

Figure 13.4 Actions and interactions of the major hormones which affect the concentration of blood glucose. The arrows signify metabolic conversions; the broken lines with dots at the end indicate control: + stimulation, − inhibition.

For consideration

1. What contribution does the nervous system make in the control of blood glucose?

2. 'The control of blood glucose is like a set of scales; during a meal insulin prevails and between meals insulin's antagonists prevail.' Discuss.

Further reading

All textbooks of human physiology include accounts of the hormones affecting carbohydrate metabolism. The following may be found particularly useful:

A.C. Guyton, *Physiology of the Human Body* (Saunders, 1984)
A simple and straightforward text.

G.H. Bell, E. Emslie-Smith and C.R. Paterson, *Textbook of Physiology* (Churchill-Livingstone, 1980)
This contains more detail of the individual hormones and their interactions.

14 . Excretion and osmoregulation

Ins and outs of the proximal tubule

Some of the most remarkable features of the kidney reside in the proximal convoluted tubule which leads from Bowman's capsule to the loop of Henle. The function of the proximal tubule is to reabsorb various substances back into the blood. One such substance is sodium which exists mainly in the form of sodium ions (Na^+). In a normal human subject approximately 1000 mmol of sodium are filtered per hour. About 99 per cent of this is reabsorbed, two thirds of it in the proximal tubule. In this topic we shall see how this is accomplished and how it affects the reabsorption of other substances, particularly water.

SODIUM AS THE PRIME MOVER

The reabsorption of sodium is an active process requiring energy. In fact more metabolic energy is expended on sodium reabsorption than on any other activity in the kidney.

The active reabsorption of sodium sets up a concentration gradient which results in water following by osmosis. That osmosis is responsible for water reabsorption is suggested by various lines of evidence. For example, dogs have been given intravenous injections of the sugar mannitol. This is filtered but not reabsorbed. It raises the solute concentration of the proximal tubule which substantially reduces the reabsorption of water, though sodium continues to be reabsorbed.

The reabsorption of sodium ions also creates an electrochemical gradient which favours the passive transfer of various anions. By means of micropipettes, one type of anion has been exchanged for another in the proximal tubule: this has no effect on the reabsorption of sodium.

It therefore seems that sodium is the prime substance to be reabsorbed in the proximal tubule and other substances such as water and anions follow passively. As a result the fluid in the proximal tubule is isotonic with the blood.

EVIDENCE AS TO HOW SODIUM IS REABSORBED

The epithelial lining of the proximal tubule possesses three interesting properties which throw light on how sodium is reabsorbed.

1 It is much more permeable to water and small solute molecules than most epithelia.

2 It is also permeable to somewhat larger molecules, such as the sugar sorbose, which are known to be unable to traverse cell membranes.

3 The electrical resistance across the epithelium as a whole is lower than across individual cell membranes.

These findings have led scientists to suggest that the tubular epithelium must have special features favouring reabsorption; this idea is supported by studies on the fine structure of the epithelium, using the electron microscope.

STRUCTURE OF THE EPITHELIUM

Part of the epithelium of the proximal tubule is shown diagrammatically in figure 14.1. Three features are particularly interesting: the microvilli and basal infoldings, which increase the surface area for absorption and transfer of substances; the numerous mitochondria which provide energy for active transport;

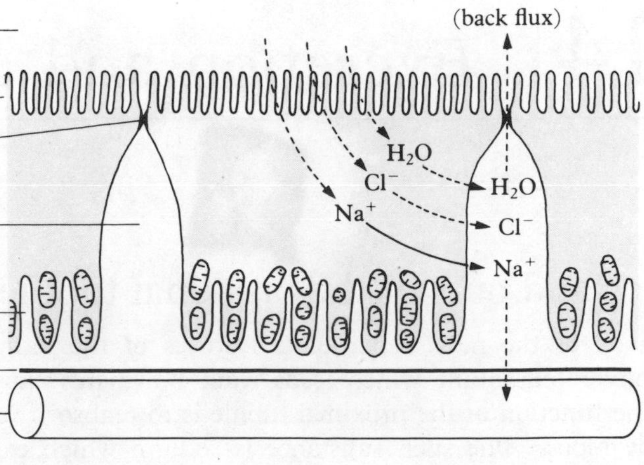

Figure 14.1 Fine structure of the epithelium of the proximal convoluted tubule showing how sodium ions and other chemicals are believed to be transferred from the lumen to an adjacent capillary. Solid arrows, active transport; broken arrows, passive movement.

and – perhaps most intriguing of all – the large spaces between adjacent cells. Only at their apices are the cells in contact. Here the contiguous cell membranes are joined through their protein components, forming **tight junctions** of the kind described on page 20.

WHAT ACTUALLY HAPPENS?

It is believed that sodium ions are actively moved through the epithelium by a sodium–potassium pump of the sort that occurs in other cells and is the basis of the resting potential in nerves (see page 80). As in nerves, the pump is inhibited by ouabain and is adversely affected by lowering the extra-cellular concentration of potassium ions; this suggests that it is a sodium – potassium exchange pump and not just a sodium pump.

The pathway through which sodium moves involves the intercellular spaces just described. The mechanism is summarised in figure 14.1. First the sodium ions diffuse passively from the lumen of the tubule into the epithelial cell down the electrochemical gradient. They are then actively pumped from the epithelial cells into the intercellular space. As a consequence the solute concentration inside the space rises with the result that water flows into it by osmosis. This in turn raises the hydrostatic pressure which drives the contents of the intercellular space into the capillary. If the capillary pressure is too high, the contents of the intercellular space are forced back into the lumen of the tubule via the tight junction.

There is some dispute as to the details of where exactly sodium transport occurs within the cells and the part played by the basal infoldings. Less controversial is that sodium transport facilitates the reabsorption not only of water and anions but also of glucose and amino acids.

For consideration

1. In what respects is the mechanism of sodium reabsorption, as described here, unsatisfactory?

2. Describe experiments which might be performed to test the hypothesis that the reabsorption of sodium ions is coupled to the reabsorption of glucose.

Further reading

G.H. Bell, D. Emslie-Smith and C.R. Paterson, *Textbook of Physiology* (Churchill Livingstone, 10th edition 1980)
This contains a detailed account of the kidney tubules, as do most tertiary level physiology texts.

A.J. Vander, *Renal Physiology* (McGraw Hill, 1980)
An advanced and authoritative account of the kidney.

The contractile vacuole

After the structural complexities of the human kidney, the contractile vacuole – the osmoregulatory organelle of protists – might come as a welcome relief. But paradoxically, although the contractile vacuole is much simpler than the kidney, we known remarkably little about the way it works.

EVIDENCE FOR ITS OSMOREGULATORY ROLE

It is generally agreed that the contractile vacuole expels water which has entered the cell by osmosis. Three pieces of circumstantial evidence support this hypothesis. First, the rate of output is decreased by raising the solute concentration of the external medium and increased by lowering it (figure 14.2). Second, adding cyanide to the medium suppresses the activity of the contractile vacuole and this results in the volume of the cell increasing. And third, marine and parasitic protists, which live in isotonic conditions free from osmotic stress, usually lack a contractile vacuole.

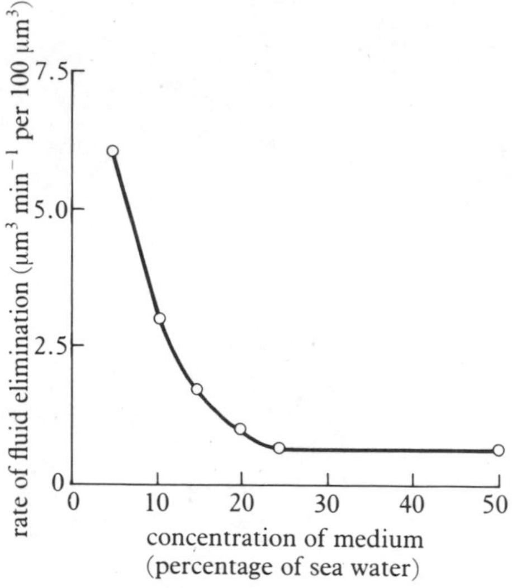

Figure 14.2 Effect of altering the concentration of the external medium on the rate at which fluid is eliminated by the contractile vacuole of *Amoeba lacerata*. The curve is the average of over 100 specimens tested. The output of the contractile vacuole is expressed as cubic micrometres of fluid eliminated per minute by 100 cubic micrometres of protoplasm. (*Data after Dwight L. Hopkins*)

CONTRACTILE VACUOLE FREQUENCY AND CELL SIZE

The contractile vacuole of a small amoeba (*Acanthamoeba* species) was observed under a microscope and its diameter was measured using a micrometer eyepiece. It was found that the contractile vacuole reached an average diameter of 6 μm before discharge and it discharged on average once every 50 seconds. Since the contractile vacuole appeared to be roughly spherical, its pre-discharge volume was taken as 113 μm^3 ($\frac{4}{3}\pi r^3$ where r, the radius, = 3). The volume of fluid discharged was calculated:

Rate of discharge $= 72\ h^{-1}$

Total volume of fluid discharge $= 72 \times 113$

$= 8136\ \mu m^3\ h^{-1}$

The amoeba was then agitated, which caused it to withdraw its pseudopodia. Its diameter was measured, from which its volume was calculated. This turned out to be 3000 μm^3. So the volume of fluid expelled by the amoeba per hour was two and a half times the volume of the body.

This is an exceptionally high figure compared with other organisms, and it may be attributed to the fact that an amoeba is a small organism with a high surface–volume ratio. Water enters all over the cell surface by osmosis. Some water is brought in with food vacuoles and a small amount, probably insignificant compared with the rest, is formed by respiration (metabolic water). The importance of the surface–volume ratio is borne out by the observation that the osmotic influx of water per unit volume is greater in small protists than in larger ones, and the rate of output of the contractile vacuole is correspondingly higher (Table 14.1).

Table 14.1 Output of the contractile vacuoles of two protists of different sizes. The data are the mean values of numerous estimations.

Species	Rate of output ($\mu m^3\ sec^{-1}$)	Cell volume (μm^3)	Time to eliminate equivalent of cell volume (hours)
Amoeba proteus	80	1200×10^3	8.5
Paramecium caudatum	105	305×10^3	0.5

STRUCTURE OF THE CONTRACTILE VACUOLE

The contractile vacuole consists of a membrane-lined sac surrounded by an assemblage of ancillary structures. Comparatively simple contractile vacuoles are found in the amoebas and flagellates. Here the vacuole is surrounded by small vesicles and mitochondria, the latter presumably providing energy for some aspect of the osmoregulatory mechanism. The contractile vacuole itself is a temporary structure: when it discharges, its membrane breaks up into numerous vesicles which then expand and later fuse to re-form the contractile vacuole.

Ciliates such as *Paramecium* have more elaborate, permanent contractile vacuoles with a specific pore in the body surface for discharge. Under the light microscope a ring of radiating canals can be seen round the vacuole (see your main textbook). The electron microscope has shown these canals to have numerous fine tubules opening into them. It is possible that the tubules collect water from the surrounding cytoplasm and channel it via the canals into the contractile vacuole. The mechanism by which this happens is not known.

CONTENTS OF THE CONTRACTILE VACUOLE

Over the years many attempts have been made to obtain vacuolar fluid for analysis. In one such attempt fluid was withdrawn from the contractile vacuole of the giant amoeba *Chaos* by means of a micropipette 2–5 μm wide. The freezing point of the fluid was determined, from which its osmotic pressure was calculated. The sodium and potassium ion concentrations were also found using the technique of flame photometry. Similar determinations were made for the cytoplasm and external medium. The results, displayed in Table 14.2, show that the contractile vacuole is able to separate out a fluid which has a lower osmotic pressure than the surrounding cytoplasm and also contains less potassium but more sodium.

Table 14.2 Osmotic pressure (in relative units) and concentration of sodium and potassium ions in the medium, cytoplasm and contractile vacuole of the fresh water amoeba *Chaos carolinensis*.

	Medium	Cytoplasm	Vacuole	Ratio: vacuole to cytoplasm
Osmotic pressure	<2	117	51	0.44
Na^+ (mmol l^{-1})	0.2	0.60	19.9	33.2
K^+ (mmol l^{-1})	0.1	31	4.6	0.15

FURTHER PROBLEMS

Of the many questions that might be asked about the contractile vacuole, two stand out prominently: how does water pass from the cytoplasm into the vacuole, and how is it expelled to the exterior?

On the first question there are three possibilities. One is that water is pumped into the vacuole in some way. A second possibility is that salt may be actively transferred into the vesicles and thence into the vacuole, water following passively by osmosis. A third idea is that the vesicles contain a fluid which is initially isotonic with the cytoplasm but is subsequently made hypotonic by the active removal of ions across the lining membranes.

On the question of how water is expelled, there are two possibilities: the contractile vacuole might collapse under pressure from the surrounding cytoplasm, or it might actively contract.

For consideration

1. Of the three hypotheses put forward to explain how the contractile vacuole fills up with water, which one is most consistent with the data in Tabel 14.2, and why? Is there anything odd about the data?

2. What observations and/or experiments would you need to make in order to decide whether the contractile vacuole passively collapses or actively contracts?

3. Are you satisfied with the data in Table 14.1 and with the conclusions drawn therefrom?

Further reading

K. Schmidt-Neilsen, *Animal Physiology: Adaptation and Environment* (Cambridge University Press, 3rd edition 1983)
Theories accounting for the filling of the contractile vacuole are briefly reviewed.

J. Laybourn-Parry, *A Functional Biology of Free-living Protozoa* (Croom Helm, 1984)
This contains a short account of the structure and functioning of contractile vacuoles.

D.J. Patterson, 'Contractile Vacuoles and Associated Structures: their Organisation and Function' (*Biological Reviews*, vol. 55, no. 1, 1980)
A comprehensive survey of modern research on contractile vacuoles; useful for finding out how much is *not* known.

Salt regulation in plants

It has been suggested that the problem of food shortage in the third world could be solved by developing crop plants capable of growing in saline conditions. It would then be possible to use sea water for irrigation without having to desalinate it first. The trouble is that most species of flowering plants are harmed by salt (NaCl) concentrations above about 0.5 per cent – and sea water has a salt concentration of 3 per cent. There are, however, a few exceptions, notably those plants that thrive in salt marshes and mangrove swamps. They are known as **halophytes** and if we can understand how they survive in their saline environment we might gain an insight into what special attributes plants irrigated with sea water would need to possess.

WHY IS SALT HARMFUL TO PLANTS?

Salt water may harm plants for two main reasons. Firstly, it might exert an osmotic effect on their cells. The water potential of some saline soils may be as low as −20 000 kPa, whereas the vacuolar sap in the root cells of a typical non-halophytic plant would have a water potential of about −100 kPa. We would expect this to result in a massive loss of water from the roots by osmosis; the cells would become plasmolysed and the plant would suffer from severe dehydration.

The second reason why salt water might harm plants is that the sodium and/or chloride ions might have a direct toxic effect on the cells. Excessive amounts of these ions are known to inhibit plant enzymes *in vitro*, and in the intact living plant they may have the added effect of upsetting the balance between other ions. For example, a high concentrations of sodium ions reduces potassium uptake, thus raising the sodium-potassium ratio.

Various lines of evidence suggest that, of these two potentially harmful effects of salt, toxicity is the more important. Even quite small increases in salinity – too small to have any significant osmotic effects – can poison a plant. Moreover, the same degree of plasmolysis can be induced by solutions of various organic compounds without such devastating effects as are produced by salt solutions. We are therefore led to conclude that the main cause of damage is probably the toxic effect of the salt itself.

HOW DO HALOPHYTES COPE?

Many halophytes actively absorb sodium and chloride ions from the soil, thereby lowering the water potential of their cells so that it is less than that of their surroundings. For example, the glasswort *Salicornia europaea* (figure 14.3), which inhabits muddy tidal marshes, can maintain a water potential of −6500 kPa in its cells when the water potential of the soil water is more like −1600 kPa. The uptake of salt therefore occurs against a steep concentration gradient and must involve active transport. The way ions are transported across plant cell membranes is discussed on page 56.

The accumulation of a high concentration of salt in the cells of a plant like *Salicornia* might be expected to inhibit enzymes, but it does not. What protects them? One possibility is that the enzymes are resistant to the toxic effects of the salt. However, laboratory tests on enzymes extracted from halophytes show that they are just as inhibited by salt as the enzymes of other plants. The explanation has come from experiments, carried out by N. Campbell and W.W. Thomson at the University of California, in which the exact positions of chloride ions inside the cells of the halophyte *Frankenia* (sea heath) have been traced. The chloride ions are precipitated as silver chloride and located by a special technique involving a combination of electron microscopy and X-ray analysis. It turns out that most of the chloride is held in the cell walls and special vacuoles, well away from the living cytoplasm where the enzymes are located.

Figure 14.3 *Salicornia europaea*, a glasswort which grows in salt marshes around the coast. It is an obligate halophyte and can only survive in a saline environment. Many other halophytes are facultative, capable of growing in either a saline or a non-saline environment. *Salicornia* is reputed to be quite good to eat – and you do not need to add salt to it!

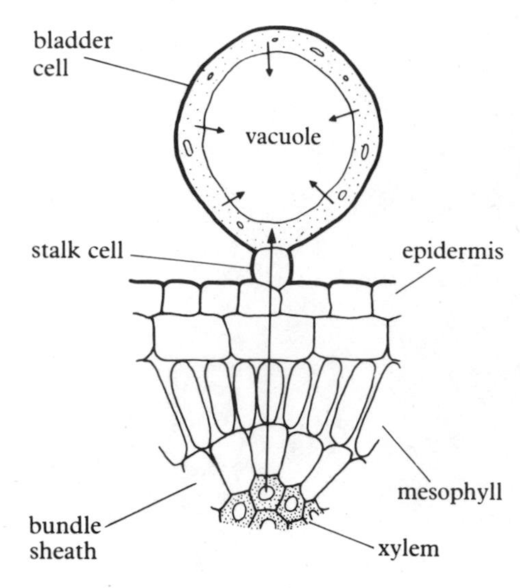

Figure 14.4 Salt bladder, seen in section, on the surface of a leaf of the saltbush *Atriplex*. The bladder is a single cell in whose vacuole salt, received from the vascular tissues, is deposited (arrows). (*After K. Esau*, Anatomy of Seed Plants, *Wiley, 2nd edition 1977*)

But there is still a problem. If the vacuoles contain a high concentration of salt, one would expect water to flow into them by osmosis from the surrounding cytoplasm, thus leading to dehydration. However, this appears to be prevented by the presence in the cytoplasm of a high concentration of organic solutes – amino acids, sugars and the like – whose combined effect is to lower the water potential of the cytoplasm to the same level as that of the vacuolar fluid. Osmotic balance inside the cells is thus preserved.

Salt cannot accumulate in the cells of *Frankenia* indefinitely. After temporary storage in the vacuoles the salt is moved via the cell walls (apoplast) to special **salt glands** at the edges of the leaves. In the cells of these glands the salt appears to be packaged in vesicles derived from the endoplasmic reticulum and then excreted to the exterior.

XEROPHYTES AND SALT

Halophytes are not the only plants to have a salt problem; many desert xerophytes are in the same situation but for a different reason. In this case excessive evaporation from the plant results in a concentrating of the salts within its cells. Many of these plants are succulents and their stored water may help to dilute the salts. Some species, for example the desert saltbush *Atriplex*, have on their leaves bladder-like appendages into which salt is secreted (figure 14.4). Eventually the bladder bursts and the salt spreads out over the surface of the leaf. Other desert plants have complex salt glands from which salt is excreted to the exterior in much the same way as in *Frankenia*.

For consideration

1. 'Plants do not regulate their salt content with the finesse that animals do'. Discuss this outrageous example of zoological prejudice.

2. What other useful adaptations might halophytes possess besides the ones mentioned here?

3. How might one set about developing a crop plant with halophytic properties?

Further reading

W.M.M. Baron, *Organisation in Plants* (Arnold, 3rd edition, 1979)
This general plant physiology text includes information on ion control.

H.E. Street and H. Opik, *The Physiology of Flowering Plants: their Growth and Development* (Arnold, 3rd edition, 1984)
Chapter 7 includes a useful section on salt stress.

J.R. Etherington, *Plant Physiological Ecology* (Studies in Biology no. 98, Arnold, 1978)
The author puts salt control and other physiological processes into an ecological context.

15 . Temperature Regulation

The hypothalamic thermostat

The part of the brain known as the hypothalamus contains, amongst other things, a centre for regulating body temperature. This functions like a thermostat, switching on the body's warming mechanisms when the body temperature falls below the norm (set point), and switching on the cooling mechanisms when the body temperature rises above the norm.

Ingenious research by T.H. Benzinger and his colleagues in the 1950s established that the hypothalamic centre serves not only as the control centre in temperature regulation but also as the receptor, responding directly to changes in the temperature of the blood flowing past it. More recently other experiments have been performed on the thermoregulatory role of the hypothalamus and we shall look at some of them in this topic.

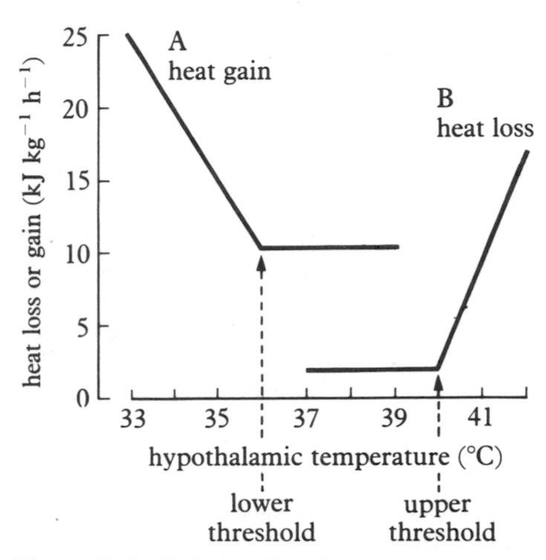

Figure 15.1 Relationship between the temperature of the hypothalamus of a dog and (**A**) metabolic heat gain, and (**B**) heat loss. The environmental (air) temperature was kept at 25°C throughout the experiment. (*Based on data by Hammel*)

CHANGING THE TEMPERATURE OF THE HYPOTHALAMUS

H.T. Hammel, working at Yale in the late 1960s, adopted a direct approach – he changed the local temperature of the hypothalamus and observed the effect on the animal's thermoregulatory responses. Tiny stainless steel tubes, serving as thermodes, were implanted round the hypothalamus of dogs and other mammals; hot or cold water was circulated through the thermodes, thus warming or cooling the temperature regulating centre. The animal was placed in a special chamber through which dry air was drawn at a known rate. After passing through the chamber, the air was analysed for its oxygen, carbon dioxide and water vapour, thus giving a continuous record of the animal's metabolic rate and hence heat-production, and of its evaporative water loss and hence heat loss.

Hammel found that when the hypothalamus was warmed, heat loss rose and heat production fell; conversely, when the hypothalamus was cooled, heat loss fell and heat production rose. This confirmed Benzinger's earlier conclusions regarding the receptive role of the hypothalamus.

Hammel then took his experiment further: he changed the temperature of the hypothalamus gradually, step by step, and recorded the thermoregulatory responses at each temperature. The results, plotted as a graph, are shown in figure 15.1. Notice that heat loss, mainly by panting, commences at a certain threshold temperature (40 °C in this particular case), whereas metabolic heat gain commences at a different, lower, threshold (36 °C here). The dog's optimum body temperature, the set point in the homeostatic control mechanism, lies between the two thresholds – at 37 °C.

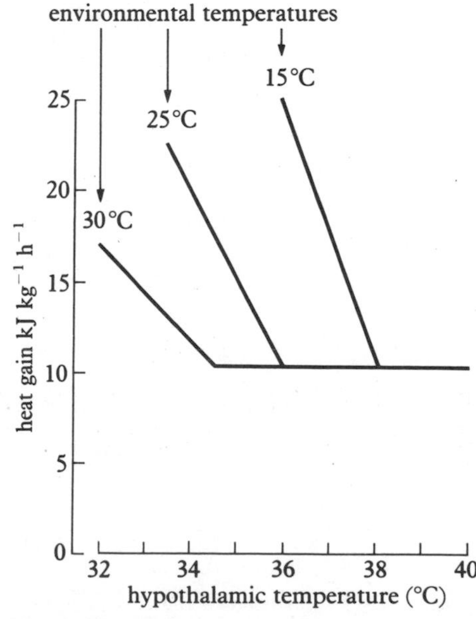

Figure 15.2 Relationship between the hypothalamic temperature and metabolic heat gain of a dog at three different environmental (air) temperatures. (*Based on data by Hammel and Hellstrom*)

SHIFTING THE THRESHOLDS

The experiment described above was carried out at an environmental air temperature of 25 °C. Now if the experiment is repeated at different environmental temperatures an interesting thing happens: the thresholds shift. In figure 15.2, for example, a fall in the environmental temperature has the effect of raising the lower threshold temperature. In consequence the heat-generating response is switched on by a smaller fall in the hypothalamic temperature. Moreover, the response itself is stronger at lower environmental temperatures, as can be seen by comparing the steepness of the curves in figure 15.2. This has been demonstrated in dogs and kangaroo rats, but there is considerable variation in the details in other species.

What is the point of raising the lower threshold and strengthening the response at lower environmental temperatures? The answer is that it enables the corrective mechanism in the homeostatic control of body temperature to swing into action more quickly than would otherwise be the case; in effect the hypothalamus anticipates the change before it actually happens, an important principle of homeostasis which is discussed on page 58.

RECORDING ELECTRICAL ACTIVITY FROM THE HYPOTHALAMUS

Information about the environmental temperature can come to the hypothalamus from only one source, the thermoreceptors in the skin. This has been investigated by recording electrical activity from the hypothalamus in response to changes in the environmental temperature. It was found, as expected, that some neurones show a change in impulse frequency when the environmental temperature changes, though the pattern of the response is by no means simple. Certain neurones respond only when the hypothalamic temperature is changed, and others respond when both the environmental *and* hypothalamic temperatures change. Moreover, certain neurones respond when the hypothalamic temperature is raised, others when it is lowered. The complex pattern of responses has given physiologists plenty of scope for inventing elaborate schemes to explain how the hypothalamus fulfils its thermostatic role.

For consideration

1. From the information given in this topic, can *you* suggest how the hypothalamus fulfils its thermostatic role?

2. Compare the techniques used by Benzinger and Hammel to test the hypothesis that the hypothalamus functions as a thermoreceptor. Whose was the more direct approach, and whose do you find most convincing?

Further reading

C.A. Keele, E. Neil and N. Joels, *Samson Wright's Applied Physiology* (Oxford University Press, 13th edition, 1982)
Recent experimental work on the mammalian hypothalamus is well covered in this general physiology text.

T.H. Benzinger, 'The Human Thermostat' (*Scientific American*, vol. 204, no. 1, 1961)
Benzinger describes his famous experiment in this article.

H.C. Heller, L.I. Crawshaw and H.T. Hammel, 'The Thermostat of Vertebrate Animals' (*Scientific American*, vol. 239, no. 2, 1978)
Recommended if you want to know more about Hammel's experiments.

Behavioural control of body temperature

Australian incubator birds are extraordinary creatures. The female lays her eggs in a mound of rotting vegetation which serves as an incubator. The male keeps the mound at exactly the right temperature by removing some of the vegetation if it gets too warm, or adding more if it gets too cold. The bird monitors the temperature with its beak which functions as an accurate thermometer. As a result the temperature of the mound is kept within 1 °C of the desired temperature.

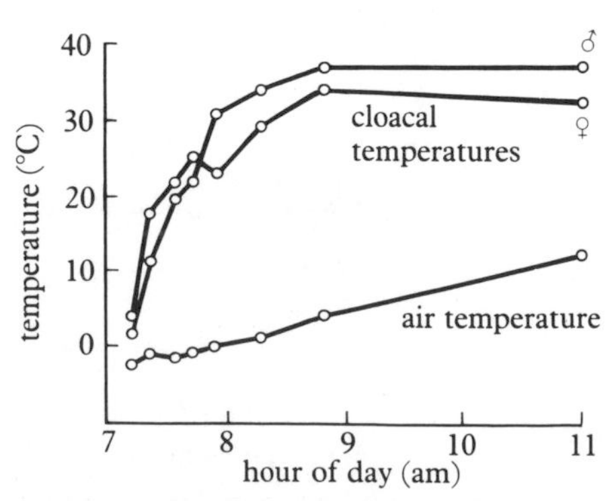

Figure 15.3 Comparison of the cloacal and air temperatures of a male and female lizard (*Liolaemus multiformis*) kept in the sun at an altitude of 4000 m in Southern Peru. The male's mass was 108 g, the female's 48 g. (*Data from O. P. Pearson*)

This is a striking example of behavioural thermoregulation, the maintenance of a constant body temperature by behavioural rather than physiological means. In the case of the incubator bird it is the embryo whose temperature is being kept constant, but many animals – particularly ectotherms – regulate their own body temperature by behavioural mechanisms. There are many aspects to this topic but we will confine ourselves to just one, namely the use which animals make of the sun.

SUN BASKING

Many species of animals gain heat by basking in the sun and absorbing solar radiation. The effectiveness of this has been investigated by measuring the body temperature of lizards kept in the sun at low ambient air temperatures. In some species, such as the one illustrated in figure 15.3, the body temperature rose to more than 25 °C above that of the surrounding air.

Once a suitable body temperature has been acquired, the animal maintains it at a more or less constant value – about 35 °C in the case of the lizards in figure 15.3. This **preferred body temperature** is maintained even if the air temperature exceeds it, as well it might in certain parts of the tropics in the middle of the day.

The preferred body temperature varies from species to species, depending on, amongst other factors, the environment to which the animal is adapted. It is about 36 °C in desert lizards but a mere 14 °C in the Tuatara of New Zealand. Typically a lizard which is active during the day keeps its body temperature to within 3 °C of the preferred temperature. To see how this is achieved we shall consider the marine iguana lizard of the Galapagos Islands, *Amblyrhynchus cristatus*. The same principles apply to many other reptiles.

BEHAVIOURAL THERMOREGULATION IN THE MARINE IGUANA

This animal basks on exposed laval rocks in the full rays of the tropical sun. By midday the air temperature may exceed 50 °C but the iguana's body temperature remains below 40 °C. It achieves this by shifting its position relative to the rays of the sun. In the early morning, or when it has just emerged from the sea, the animal adopts a prostrate position with its long axis perpendicular to the sun's rays, thereby exposing the maximum area to the heating effect of the sun (figure 15.4A). When the preferred body temperature is reached, the lizard reorientates itself so that its long axis is more in line with the sun's rays, thereby reducing the exposed surface and to some extent allowing the head and shoulders to shade the rest of the body. The animal achieves this reorientation by raising the front part of its body, an action which has the added advantage of permitting cooler air to circulate underneath (figure 15.4B).

THE CONTROL OF BEHAVIOURAL THERMOREGULATION

The marine iguana adopts its elevated posture only when the body temperature reaches 39 – 40 °C. What initiates such a change in behaviour? Experiments on other animals suggest that the **hypothalamus** is involved. The animals were given a choice between two environments, one warmer and the other cooler than the preferred body temperature. The hypothalamus was then heated or cooled by means of implanted thermodes (see page 69). When the hypothalamus was heated the animal moved into the cooler environment, and its body temperature fell. When the hypothalamus was cooled, the animal moved into the warmer environment and the body temperature rose. Thus behavioural thermoregulation appears to be controlled by the hypothalamus, appropriate responses being initiated in much the same way as physiological responses are.

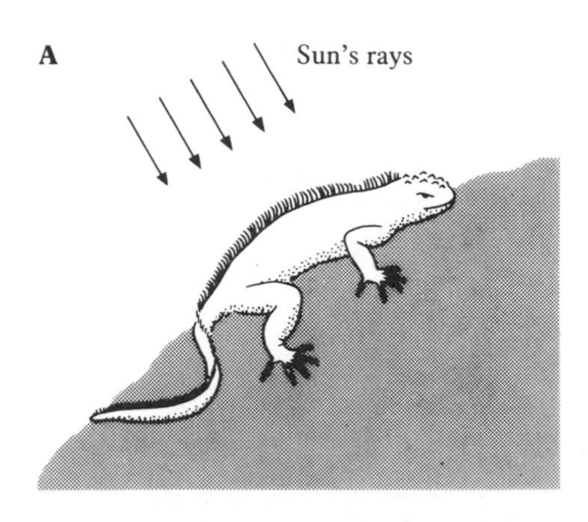

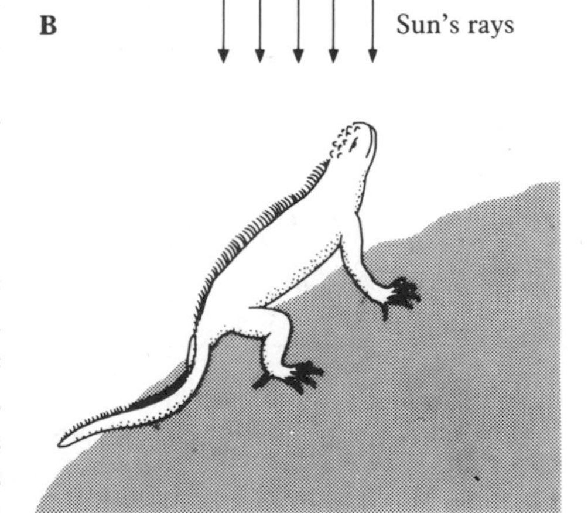

Figure 15.4 The Galapagos marine iguana lizard, *Amblyrhynchus cristatus*: **A** prostrate posture; **B** elevated posture. The elevated posture reduces the amount of heat gained from the sun and substratum.

For consideration

1. A cold marine iguana lizard placed in water which is 20 °C above the body temperature gains heat twice as quickly as a warm iguana loses heat when placed in water which is 20 °C below the body temperature.
(a) How would you verify this phenomenon experimentally?
(b) Suggest a physiological explanation.
(c) Of what use is it to the animal?
2. The experiment whose results are shown in figure 15.3 should have a control. What control would be suitable and why is it necessary? Why do you think the curves for the male and female lizards differ?

Further reading

J.L. Cloudsley-Thompson, *The Temperature and Water Relations of Reptiles* (Merrow, 1971)
This work contains a wealth of information and data about behavioural thermoregulation.

S.A. Richards, *Temperature Regulation* (Wykeham, 1973)
Chapter 5 gives examples of behavioural thermoregulation in different animal groups.

K. Schmidt-Nielsen, *Animal Physiology: Adaptation and Environment* (Cambridge University Press, 3rd edition 1983)
There is a useful summary of modern knowledge in Chapter 8.

16 . Control of respiratory gases

Pin-pointing the stimuli in the control of breathing

Breathing must be geared to the metabolic needs of the body. This necessitates receptors for detecting the partial pressures of respiratory gases in the blood. Information from the receptors is fed to the **respiratory centre** in the hindbrain (see page 37) and the ventilation rate is adjusted accordingly.

Two questions arise:

1 Is the respiratory centre influenced by a change in the partial pressure of carbon dioxide, or oxygen, or both?
2 Where are the receptors, in the brain itself or in the peripheral blood vessels outside the brain?

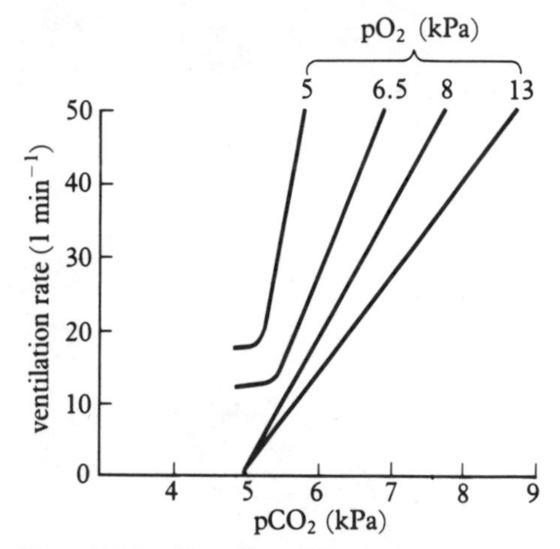

Figure 16.1 The effect of increasing the partial pressure of carbon dioxide on the ventilation rate of a human at four different partial pressures of oxygen. Note that decreasing the partial pressure of oxygen enhances the stimulatory effect of carbon dioxide. (*Data after Cunningham and Lloyd, 1963*)

WHICH IS THE EFFECTIVE STIMULUS?

One way of finding out whether oxygen or carbon dioxide influences the centre is to change the partial pressure of each gas in the alveolar air in turn, and find their relative effects on the rate of breathing, that is the ventilation rate. This has been done and it turns out that raising the partial pressure of carbon dioxide produces a much greater increase in the rate of breathing than lowering the partial pressure of oxygen. From this we may conclude that the former is the more effective stimulus.

It must be stressed, however, that in this experiment no attempt was made to hold the partial pressure of each gas in the bloodstream constant while the other one was being investigated. To do so would be to create an artificial situation which does not prevail in natural conditions. In practice the two respiratory gases influence each other and their effects cannot be separated.

The results of this experiment might suggest that a fall in the partial pressure of oxygen is of little or no importance as a stimulus to breathing. However, experiments have shown that the stimulating effect of an elevated partial pressure of carbon dioxide is enhanced by a simultaneous fall in the partial pressure of oxygen (figure 16.1). A maximum response is therefore produced by both stimuli together.

WHERE ARE THE RECEPTORS?

In the walls of the aorta and carotid arteries there are swellings, the **carotid** and **aortic bodies**, from which nerves pass to the respiratory centre in the brain (see your main textbook). Might these bodies detect changes in the concentration of gases in the blood and signal the relevant information to the respiratory centre? This hypothesis was tested by cutting the carotid and aortic nerves of an experimental animal and noting the effect of this on the animal's ability to respond to changes in the concentration of oxygen and carbon dioxide. It was found that after cutting the nerves the animal could still respond fully to an increase in the partial pressure of carbon dioxide, but not to a decrease in the partial pressure of oxygen. From this it was concluded that the brain itself can sense a change in the partial pressure of carbon dioxide, but not oxygen.

What then is the role of the peripheral receptors? This question has been investigated by isolating the carotid body from its normal blood supply, while leaving its nerve intact. The carotid body was perfused with a series of artificial solutions containing different partial pressures of carbon dioxide and oxygen, and any changes in the ventilation rate were recorded. It turned out that an increase in the partial pressure of carbon dioxide and a decrease in the partial pressure of oxygen both raised the ventilation rate. From the data obtained it was concluded that, while the brain itself responds to an excess of carbon dioxide but not to a lack of oxygen, the peripheral receptors are sensitive to both – particularly oxygen lack.

Further experiments on the peripheral receptors have been carried out by perfusing the carotid body with solutions containing different partial pressures of oxygen and carbon dioxide and recording action potentials from the carotid nerve. It was found that a decrease in the partial pressure of oxygen and an increase in the partial pressure of carbon dioxide can both raise the frequency of action potentials. However, the highest frequencies were recorded when both treatments were given simultaneously, a result which fits in with the earlier discovery that the two stimuli together produce the greatest increase in the ventilation rate.

CARBON DIOXIDE OR HYDROGEN IONS?

We have seen that an increase in the partial pressure of carbon dioxide stimulates the carotid body. However, the effectiveness of carbon dioxide as a stimulus is greatly reduced if the carotid body is treated with a substance which inhibits the enzyme **carbonic anhydrase**. You will remember that this enzyme, which is present in red blood cells, is involved in the carriage of carbon dioxide by the blood: it catalyses the combination of carbon dioxide with water to form carbonic acid which immediately dissociates into hydrogen and hydrogencarbonate ions. We can interpret the results of this experiment by proposing that the carotid body receptors are stimulated not by carbon dioxide as such, but by hydrogen ions resulting from its carriage in the blood. This notion is supported by the observation that all the effects produced by stimulating the carotid body with carbon dioxide are also produced by increasing the hydrogen ion concentration, i.e. decreasing the pH.

For consideration

1. Suggest reasons why a rise in the concentration of carbon dioxide rather than a fall in the concentration of oxygen should be the main stimulus leading to an increase in the rate of breathing.

2. The brain and the carotid bodies are both sensitive to changes in the concentration of respiratory gases. Why have two receptors rather than one?

Further reading

J. Widdicombe and A. Davies, *Respiratory Physiology* (Arnold, 1983)
A readable book with a chapter on the control of breathing.

R.W. Torrance (editor), *Arterial Chemoreceptors* (Blackwell, 1968)
Some of the original experiments on the carotid bodies are described in this collection of papers.

To be a diver

It is recorded in the *Guinness Book of Records* that in 1969 a male Sperm Whale was caught off the South African coast after it had surfaced from a dive lasting one hour and 52 minutes. It was estimated from the kind of food found in its stomach that this whale had been feeding at a depth of over 3000 metres. Although this was an exceptional dive, submergence for long periods occurs in a number of animals, including whales, dolphins, seals and penguins, and many species can stay under water for more than fifteen minutes.

A diving animal faces numerous problems but we shall address ourselves to just two of them: increase in pressure and lack of oxygen.

INCREASE IN PRESSURE

Consider a human diver who descends to a depth of, say, thirty metres. The increased pressure on the body compresses the air in the lungs. This in turn raises the partial pressures of gases in the alveoli, increasing the rate at which they enter the blood and tissues. This particularly applies to nitrogen because it is the most common gas in atmospheric air and, unlike oxygen, is not used up by the tissues.

As the diver returns to the surface, the pressure of water on the body decreases and the partial pressure of nitrogen is reduced. If the diver surfaces too quickly, the nitrogen comes out of solution and forms bubbles in the blood and tissues in the same way that bubbles develop in a bottle of fizzy lemonade when the cap is removed. Such bubbles form whenever the pressure is reduced on a liquid which has been saturated with gas at a high pressure – i.e. a liquid which is *super*saturated.

In the diver the formation of gas bubbles results in **decompression sickness** or 'the bends' which is characterised by an acute pain in the joints and muscles. The bubbles block the smaller blood vessels and tear the tissues, which can be fatal if it occurs in the heart, lungs or brain. A diver who experiences symptoms of decompression sickness as he ascends must dive again immediately and recompress.

Whether or not a diver gets the bends depends on the depth and duration of the dive and the speed at which he surfaces. It takes some time for the body to become supersaturated with nitrogen, so the risk is greatest after a long dive. The only way to prevent the bends is to come to the surface slowly, stage by stage. Decompression tables are available, informing divers how long they should spend at each depth as they ascend towards the surface.

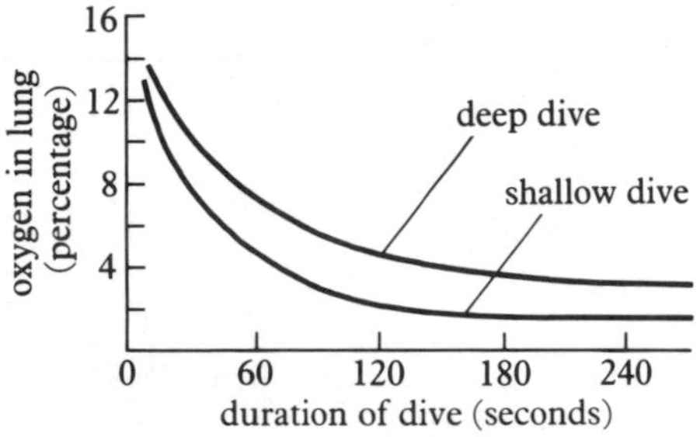

Figure 16.2 Percentage of oxygen in the lungs of the dolphin *Tursiops* during deep and shallow dives. The data were obtained by training a dolphin to dive to a sonar beacon located at different depths and to breathe out into a container before surfacing. The animal's expired air was then analysed. Note that, for a given duration of dive, the percentage of oxygen in the lungs was greater after a deep dive than a shallow dive. The interpretation is that less oxygen was utilised during the deep dive because more of it had been forced out into the dead space. (*After Ridgway, Scronce and Kanwisher*, Science, Vol. 166, *1952*)

HOW DO DIVING ANIMALS AVOID THE BENDS?

Marine mammals not only dive to great depths but they also surface quickly. Why do they not suffer from the bends? The answer probably varies between species but in seals it is related to the interesting observation that these animals *exhale* before they dive. As the animal descends, increasing pressure of water on its body forces the remaining air out of the lungs into the dead space – that is the bronchi, trachea etc. It is difficult to prove that this happens but the ingenious experiment described in figure 16.2 suggests that it does. With no air in the lungs, nitrogen cannot get into the blood and tissues, so decompression sickness is avoided when the animal surfaces later.

Obviously for the lungs to be compressed like this the chest wall must be capable of caving in without damaging the ribs. Diving mammals do in fact have remarkably pliant chests and there are fewer attachments between the ribs and the vertebrae and sternum than in other mammals.

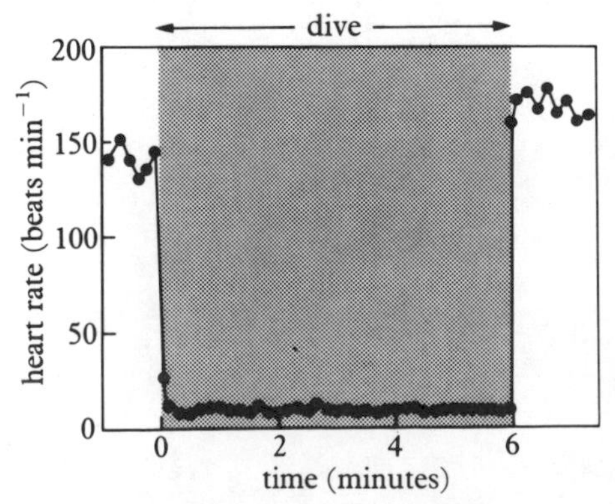

Figure 16.3 Bradycardia in the Harbour Seal. When the animal is under water the cardiac frequency falls from its normal value of about 140 beats per minute to less than 10 beats per minute.

OXYGEN LACK

An animal which performs a prolonged dive would be expected to run short of oxygen, particularly if it has emptied its lungs on the way down. Research on diving birds and mammals has shown that they possess a wide range of physiological and biochemical adaptations which enable them to cope with this situation. For example, the oxygen capacity of their blood is significantly higher than that of non-divers, their muscles contain an abundance of the oxygen-storing pigment **myoglobin**, their metabolic rate falls during the dive so that their oxygen consumption is reduced, and when their oxygen supply is used up they resort to anaerobic respiration. However, we shall concentrate on certain adaptive changes which occur in the circulatory system.

In the 1930s the Norwegian scientist P.F. Scholander showed that when a seal dives (or simply puts its nose under water), a rapid reflex takes place leading to a dramatic reduction in the cardiac frequency (figure 16.3). Slowing of the heart is known as **bradycardia**. At the same time the arteries and arterioles serving the viscera, kidneys and limb muscles constrict, reducing the blood supply to these organs. However, the arteries and arterioles serving the heart and brain remain dilated, so that blood continues to flow to these vital organs, albeit at a slower speed.

An important consequence of shutting down the bloodflow through the muscles is that lactic acid, produced by anaerobic respiration, is held within them and prevented from spreading throughout the body. When eventually the animal surfaces, the arteries and arterioles dilate and the lactic acid is released into the general circulation (figure 16.4). With the resumption of breathing, the oxygen debt is paid off, just as in an athlete after a hundred metre sprint.

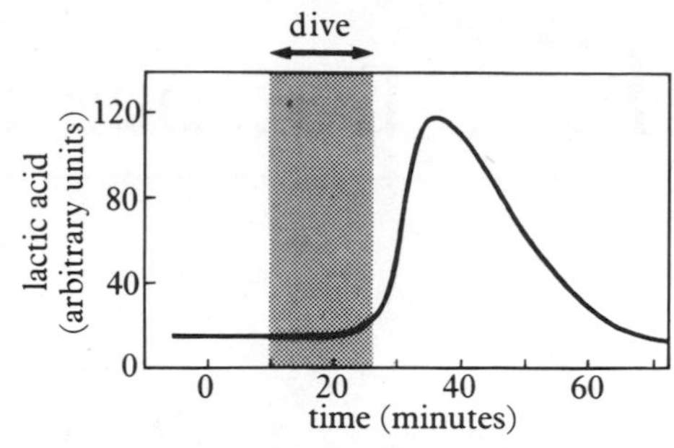

Figure 16.4 Concentration of lactic acid in the blood of a seal before, during and after a dive. (*Data from Scholander, 1962*)

Does bradycardia occur in humans? Scholander investigated this question by measuring the cardiac frequency of a human subject who, at a given moment, immersed his face in a basin of water. The results showed clearly that the heart slows during the period of immersion. By simultaneously recording the circumference of the calf, Scholander was able to show that the blood flow to the lower leg is much reduced at the same time.

Bradycardia, with simultaneous constriction of certain blood vessels, conserves oxygen and ensures that it is sent to the parts of the body that need it most. This important response is by no means confined to diving animals. It is also shown by fishes such as the Grunion and Flying Fish when they are *out* of water and by the mudskipper *Periophthalmus* when it burrows into mud.

For consideration

1. If the graph in Figure 16.4 included a curve for muscle lactic acid as well as for blood lactic acid, what would you expect it to look like? Explain your answer.

2. Compare the problems facing a diving seal, a free-diving human and a SCUBA diver.

Further reading

P.F. Scholander, 'The Master Switch of Life' (*Scientific American*, vol. 209, no. 6, 1963)
This article deals particularly with bradycardia.

H.V. Hempleman and A.P.M. Lockwood, *The Physiology of Diving in Man and other Animals* (Studies in Biology no. 99, Arnold, 1978)
A thorough review of all aspects of diving.

17 . Defence against disease

What generates antibody diversity?

The ability of the immune system to overcome infections depends on the amazing diversity of the antibodies which circulate in the bloodstream. There are some two million million **B lymphocytes** in the human circulation, and between them they make millions of different types of antibody. Each B lymphocyte produces only one type of antibody. When such a cell circulates in the blood and lymph, it may have five thousand identical antibody molecules on the surface of its cell membrane, and if it becomes a plasma cell, it may sit in a lymph node and produce two thousand copies of the same antibody molecule every second.

Different antibodies are almost identical in shape and size, but differ from one another in their amino acid sequences and are coded for by very few genes. How can such a tiny antibody-making kit code for the millions of types of antibodies which can potentially be produced within the same organism? This feature of antibody-producing cells seems incompatible with the traditional concept that one gene codes for one polypeptide chain.

Studies of the structure of antibodies and the genes coding for them have provided the answer. During the development of a B lymphocyte, segments of the DNA rearrange themselves to produce a unique antibody gene set which from then on codes for all the antibody made by the cell. In this topic we shall describe in simplified form how this is done.

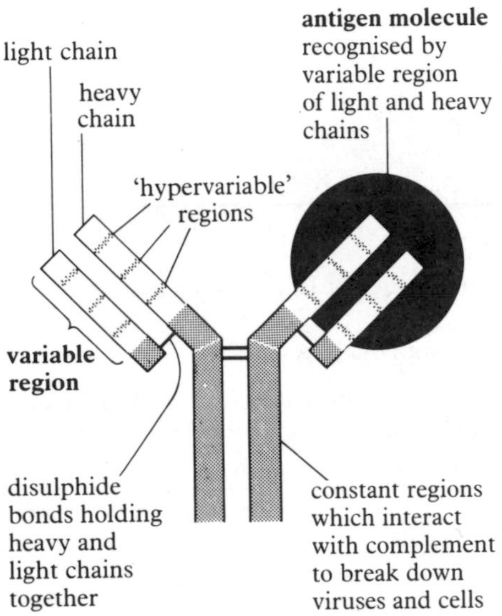

light chain
heavy chain
antigen molecule recognised by variable region of light and heavy chains
'hypervariable' regions
variable region
disulphide bonds holding heavy and light chains together
constant regions which interact with complement to break down viruses and cells

Figure 17.1 The structure of an antibody molecule.

THE STRUCTURE OF AN ANTIBODY

An antibody is Y-shaped and consists of four polypeptide chains – two heavy chains and two light chains (figure 17.1). The light and heavy chains are joined to each other by disulphide bonds. Each chain has a constant region at the base, the same in all circulating antibodies, and a variable region at the top. In a given antibody molecule the two light chains are identical to one another and the two heavy chains are also identical to one another, but the variable regions of the light chains and the heavy chains are not identical. The two arms of the Y react with the antigen, which may be a protein molecule on the surface of a bacterium or a virus, whilst the base of the Y activates **complement**, an enzyme cascade which actually destroys the pathogen.

THE STRUCTURE OF AN ANTIBODY GENE

In humans there are three families of genes which code for an antibody, each on a different pair of homologous chromosomes. All the heavy chain genes are on the same chromosome, but there are two alternative sets of genes which code for the light chains, the **kappa set** on chromosome two and the **lambda set** on chromosome twenty-two. Here we shall concentrate on the light chains only.

The arrangement of the genes for the light chains on chromosomes two and twenty-two in humans is shown diagrammatically in figure 17.2. On each chromosome there are about 150 different genes for the variable region of the light chain of the antibody (**V-genes**). Each of these genes is separated from its neighbours by intervening sequences of DNA. Some distance away, along the same chromosome, there are copies of genes for the constant region (**C-genes**). There is also a range of joining genes (**J-genes**).

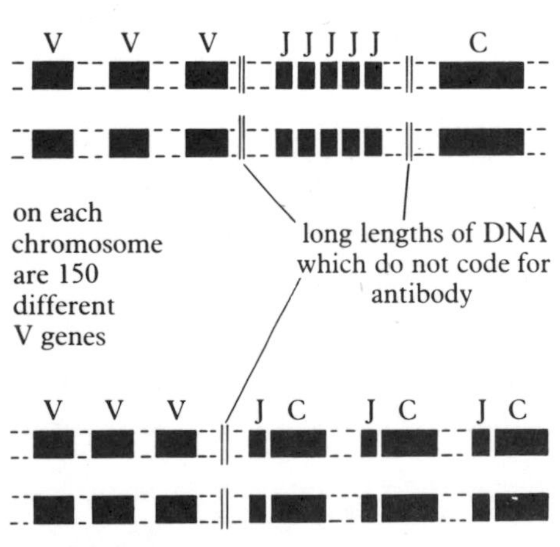

Figure 17.2 Diagrammatic summary of the arrangement on the human chromosomes of the genes for the light chains in an antibody. (*After Robertson*, 1982)

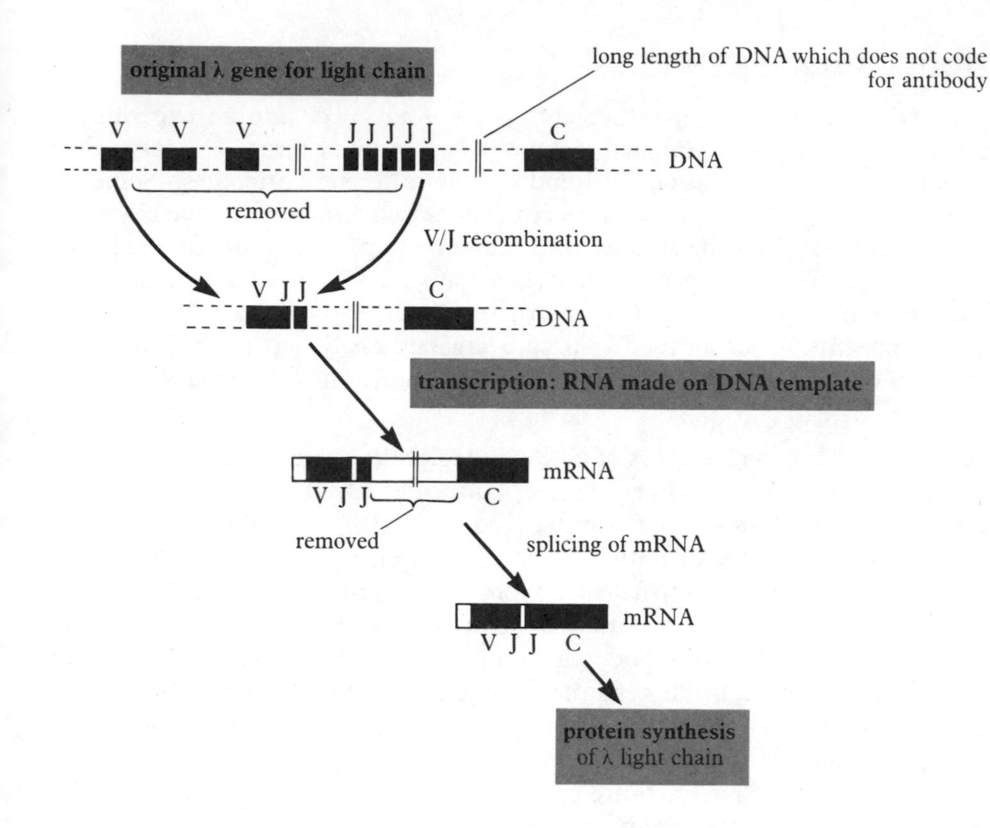

Figure 17.3 The sequence by which the light-chain antibody genes are modified and transcribed to produce a component of an antibody molecule. (*After Leder, 1982*)

GENERATING DIVERSITY

During the maturation of a B lymphocyte, one of the V genes is moved to a position next to one of the J genes on the same chromosome (figure 17.3). This is a 'jumping gene'. The DNA which lay between the two genes is removed, although the exact mechanism involved is obscure. The adjacent V, J and C genes now constitute the gene for the light chains of the antibody to be produced by the cell. This gene is transcribed into messenger RNA from which the unwanted stretch of helix between the J and C genes is removed. The protein chains are then produced at the ribosomes from messenger RNA molecules.

There are 150 types of V gene and any one of them can combine with five types of J gene. In addition the breaking and rejoining of the DNA may happen in different ways, and so can generate about ten different types of amino acid combinations between the V and J regions of the light protein chain. Thus this mechanism produces $150 \times 5 \times 10$ different types of light chain = 7500 possibilities. The same process, happening on a different chromosome amongst the genes coding for the heavy chains, can generate 2.4 million different heavy chains. Assuming that the light and heavy chain genes 'jump' independently, this makes a total of about eighteen thousand million different types of antibody that could be produced by the B lymphocytes in the body. There may also be some somatic mutation within antibody genes giving even greater diversity.

For consideration

When a B lymphocyte is circulating in the blood and lymph its antibody molecules appear on its surface. If it becomes a plasma cell, antibody molecules are secreted into the lymph. The membrane-bound molecule on the circulating cell is slightly longer than the secreted one. Why is there this difference in molecular structure?

Further reading

P. Leder, 'The Genetics of Antibody Diversity' (*Scientific American*, vol. 246, no. 72, 1982)
The author explains the experimental evidence for the hypothesis outlined in this topic, with some excellent diagrams.

M. Robertson, 'How Can Antibodies Be So Diverse?' (*New Scientist* vol. 94, no. 1309, 1982)
A clear introductory summary.

Monoclonal antibodies

The conventional way to produce antibodies is to inject into an experimental animal a heat-killed or attenuated strain of a bacterium or virus. The response of the animal is often to produce hundreds of different antibodies. Some are against different antigens on the surface of the injected microbe, some react with different parts of the same antigen, and sometimes different antibodies are made which attack the same part of the same antigen! When the antibodies are concentrated, they are so similar in structure and relative molecular mass that the components of an antibody mixture are not easily separated. This topic describes both why we need pure antibodies and also how to produce them.

THE VALUE OF PURE ANTIBODY

Large quantities of single types of antibody molecules are potentially very useful in medicine and research. For example, antibodies specific for certain types of pathogenic bacteria and viruses can be used to identify the unknown organisms causing infections. Specific antibodies can be injected to counteract pathogenic micro-organisms or to kill particular types of cancerous tumours. Attached to fluorescent markers, they are used in light microscopy to identify cells with specific antigenic proteins. They can pinpoint specific proteins in the bands resulting from the electrophoresis of a mixture of proteins. Bonded to electron-dense atoms such as gold, they are used in electron microscopy to locate the positions in the cell of certain proteins. When attached to resin in a column of beads down which cell extracts are passed, they can separate from the mixture of compounds specific polypeptide hormones or antigens. The genes for the antigens may then be cloned in bacteria by 'genetic engineering' (see page 139) and used to produce vaccines against disease. The list is endless.

Techniques for purifying antibodies have been known for some time but pure antibodies have only been available cheaply, and on a large scale, since about 1975 when a new means of producing antibodies was developed. The process involves cloning and the antibodies formed are called **monoclonal antibodies**.

THE MANUFACTURE OF MONOCLONAL ANTIBODIES

Antibodies are made by B lymphocytes. These white blood cells are fairly easy to isolate and grow individually but they die in culture after a few days. Cells from some tumours, on the other hand, can exist and proliferate for years in culture and are potentially immortal.

Cancerous tumours of B lymphocytes, known as **myelomas**, continuously secrete large quantities of single types of antibody. The fusion of a B lymphocyte cell which produces the desired antibody with a myeloma cell produces a clone of cells which may churn out the desired antibody for ever.

In practice a mouse is injected with the antigen against which a monoclonal antibody is to be made. After a few days the B lymphocytes which produce antibodies against the antigen have multiplied as plasma cells, and a sample of B lymphocytes, hopefully including those of interest, is taken from the spleen. They are mixed with myeloma cells in solutions to which polyethylene glycol (see page 137) or Sendai virus have been added to fuse the cells together. Individual cells are removed from the culture, placed in separate solutions and left to reproduce. Clones which seem viable and which secrete antibody against the desired antigen are then grown on to produce the monoclonal antibody.

For consideration

1. If a B lymphocyte fuses with a myeloma cell, what determines whether the cell produced will churn out the antibody coded for by B lymphocyte genes, or the antibody coded for by myeloma genes?

2. What other uses does cell fusion have, other than producing monoclonal antibodies?

Further reading

B. Alberts *et al*, *The Molecular Biology of the Cell* (Garland, 1983)
This work contains an elementary introduction to monoclonals.

C. Millstein, 'Monoclonal Antibodies' (*Scientific American* vol. 243, no. 4, 1980)
The discovery and potential uses of monoclonal antibodies.

18 . Nervous and hormonal communication

The membrane potential of nerves

Sometimes the theoretical basis of a phenomenon is formulated long before it is confirmed by practical experiment. Such was the case with the **membrane potential**, the potential difference which exists across the membrane of cells. When Hodgkin and Huxley demonstrated this potential in the giant axon of the squid in the late 1930s, they were confirming a theoretical model based on ideas which had been proposed in the nineteenth century by Walter Nernst. In this topic we shall examine Nernst's model and see how it paved the way towards the discovery of the electrical nature of nerve transmission.

THE NERNST EQUATION

Nernst was a physical chemist who was interested in, amongst other things, the electrical properties of ionic solutions. He considered the situation where two solutions of different ionic concentration are separated in a system which permits a selective movement of ions between the two solutions. From the potential energy of the ions in each solution, Nernst derived an equation for calculating the potential difference (pd) at equilibrium between them. The **Nernst equation**, applied to a cell, is as follows:

$$\text{pd} = \frac{RT}{zF} \log_{10} \frac{C_0}{C_i}$$

C_0 is the concentration of a particular ion outside the cell membrane, C_i is the concentration of the same ion inside the membrane, R and F are constants (the gas constant and Faraday respectively), T is the absolute temperature, and z is the charge on the ion, i.e. its valency.

The axoplasm in the giant axon of a squid can be squeezed out of the cut end and analysed for its ionic content. Similar chemical analysis can be carried out on the blood, which we may assume represents the immediate external medium of the axon in the living animal. The results of such analyses are summarised in Table 18.1. Notice in particular that the concentration of potassium ions is much greater in the axoplasm than in the blood. If we apply the potassium figures to the Nernst equation, we can calculate the membrane potential:

Table 18.1 Concentration of potassium, sodium and chloride ions in the blood and axoplasm of the squid *Loligo*.

	Concentration (mmol kg^{-1})	
Ion	**blood**	**axoplasm**
K^+	20	400
Na^+	550	50
Cl^-	560	120

$$\text{pd} = 58 \log_{10} \frac{20}{400} = 58 \log_{10} \frac{1}{20} = -75 \text{ mV (millivolts)}$$

Thus, on purely theoretical grounds we can predict – as the early neurophysiologists did – that there is a potential difference across the membrane surrounding an axon, the inside being negative with respect to the outside. In other words the membrane is **polarised**.

EXPERIMENTAL WORK ON THE MEMBRANE POTENTIAL

The first evidence to support this theoretical model came from experiments in which two electrodes were placed in contact with a nerve (not a giant axon but a whole nerve), and the potential between them measured with a galvanometer (figure 18.1). If both electrodes are applied to the surface of the nerve, there is no potential difference between them. If, however, the nerve is severed and one of the two electrodes is placed in contact with the cut end, a potential difference is registered. The cut end, where the inside of the nerve is exposed, is negative.

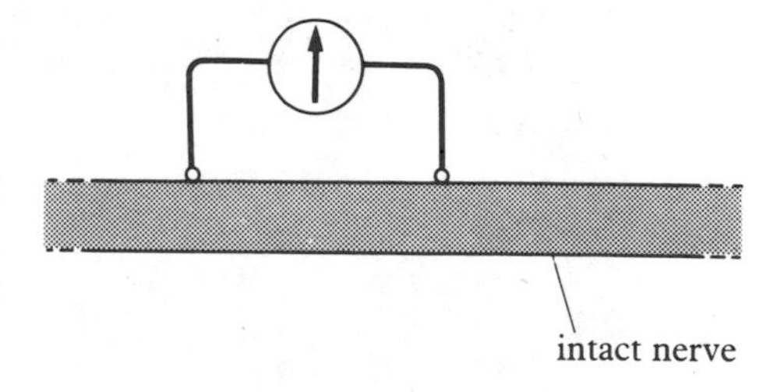

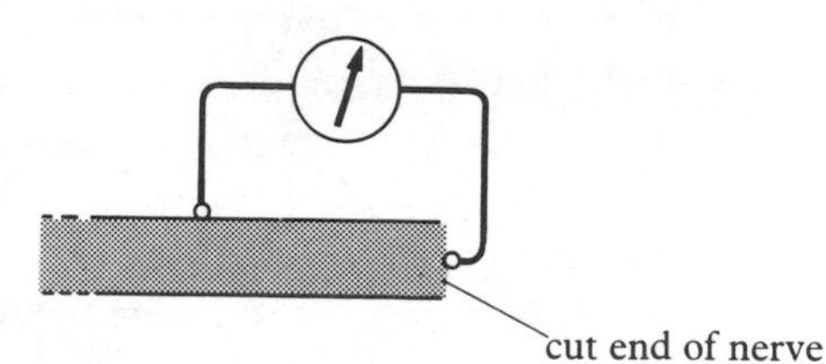

Figure 18.1 One of the earliest experiments demonstrating the existence of a membrane potential in nerves. **A** Intact nerve with electrodes on the surface: no potential registered. **B** Severed nerve with one electrode on the surface, the other in contact with the cut end: a small potential (called the injury potential) is registered.

This is known as the **injury potential** and, although it is nothing like as large as was predicted by the Nernst equation, it supports the hypothesis that a potential exists between the inside and the outside of a nerve.

In 1939 A.L. Hodgkin and A.F. Huxley carried out their classical experiments on the giant axon of the squid. The large diameter of this axon enabled them to insert a microelectrode into its interior, while a second electrode was placed outside. Hodgkin and Huxley demonstrated unequivocally that there is a potential difference across the membrane surrounding the axon, thus confirming the theoretical prediction based on the Nernst equation. However, the membrane potential recorded by Hodgkin and Huxley was slightly smaller than that predicted by the equation: −70 mV as compared with −75 mV.

Why the difference? The reason is that the Nernst equation is based on the assumption that only one type of ion is involved (potassium ions in our discussion) and that it will diffuse down its concentration gradient until equilibrium is reached. But in nerves other ions besides potassium are involved. For example, the axon membrane is slightly permeable to the inward diffusion of sodium ions. In practice there are relatively few sodium ions in the axon because they are continually pumped out in exchange for potassium ions by the sodium-potassium pump. Nevertheless, the few sodium ions which do get in prevent the equilibrium point for potassium ever being reached, and make the membrane potential slightly less negative than it would otherwise be.

A further equation, which takes into account the sodium as well as the potassium ions, was proposed by Goldman in the 1940s. The **Goldman equation** includes the permeability coefficients of the two ions, in addition to their concentrations. When the membrane potential is calculated using this equation, it agrees with the experimental figure obtained by Hodgkin and Huxley.

The membrane potential is a characteristic feature of most cells. In excitable tissue – that is nerve, muscle and sensory tissue – it constitutes the **resting potential**. Momentary reversal of the resting potential, brought about by depolarisation of the axon membrane, creates an **action potential** which is the basis of electrical transmission. That is discussed in the next topic.

For consideration

1. What is meant by the injury potential? Why is it considerably less than the membrane potential as predicted by the Nernst equation?

2. In this account there is no mention of negative ions. What contribution do they make to the resting potential of an axon?

Further reading

R.D. Keynes, 'The Nerve Impulse and the Squid' (*Scientific American*, vol. 199, no. 6, 1958)
This article gives the historical background to the discovery of the resting potential.

R.H. Adrian, *The Nerve Impulse* (Carolina Biology Reader no. 67, 2nd edition, 1980)
A useful overall summary of nerve physiology.

D.J. Aidley, *The Physiology of Excitable Cells* (Cambridge University Press, 2nd edition, 1978)
An exceptionally well written book, advanced but readable.

Figure 18.2 (opposite) Simplified model which has been proposed for the sodium-potassium pump. The grey object is the enzyme ATPase, part of which can undergo a flip-flop movement as shown. When Na^+ is released outside the membrane, the enzyme's conformation changes and it picks up K^+. When K^+ is released inside the membrane, the enzyme resumes its former conformation and picks up Na^+. There are separate binding sites for Na^+ and K^+ and for every two K^+ moved in three Na^+ are moved out. Energy from the hydrolysis of ATP is required to initiate the sequence of events. (*Modified after Alberts et al*, The Molecular Biology of the Cell, *Garland, 1983*)

The sodium–potassium pump and nerve transmission

The membrane potential, discussed in the previous topic, depends on an uneven distribution of potassium and sodium ions being maintained between the two sides of the cell membrane. This is achieved by the **sodium–potassium pump** which expels sodium ions from the axon in exchange for potassium ions. How does the pump work, and what bearing does it have on the transmission of nerve impulses?

The movement of sodium and potassium ions occurs against a steep electrochemical gradient, from which we may infer that it requires energy, presumably from ATP. That this is so was shown by a group of physiologists in the late 1950s. They injected radioactive sodium ions into a squid giant axon and measured their subsequent outflow. A short time after the beginning of the experiment the axon was treated with cyanide, which prevents ATP synthesis by inhibiting cytochrome oxidase. The extrusion of sodium ions was greatly reduced by the cyanide treatment, indicating that metabolic energy is required for the pump. Moreover, when ATP was injected into the poisoned axon, the expulsion of sodium ions started up again, suggesting that ATP is the source of energy.

The researchers also found that the outflow of sodium ions slowed down if there were no potassium ions in the external solution. This suggests that sodium extrusion and potassium uptake are coupled together, a conclusion which has been confirmed by other experiments. In one rather ingenious experiment, the membrane potential was recorded from a nerve cell of a snail and then sodium ions were injected into the cell. On adding the sodium ions the membrane potential became larger, because of the extra sodium ions being expelled; in other words the membrane became *hyper*polarized. Now when this experiment was repeated in a medium lacking potassium ions, no such hyperpolarization occurred. This suggests that potassium ions must be present for sodium ions to be expelled. In fact we now know that three sodium ions are expelled for every two potassium ions that enter.

How then does the pump work? An important clue was provided some years ago by work on ATPase, the enzyme which hydrolyses ATP. It was found that this enzyme works optimally in the presence of sodium and potassium ions. This led to the suggestion that ATPase may be a component of the transport system, possibly serving as the carrier of the two ions. Further evidence for this idea derives from the observation that a poison called ouabain, which competes with potassium ions for attachment to ATPase, also stops the pump. (Ouabain occurs in certain plants and is used as an arrow poison in parts of Africa).

So we may conclude that ATPase serves both as the enzyme catalysing the hydrolysis of ATP and as an essential component of the sodium-potassium pump. Several models have been proposed to explain the mechanism; the one most consistent with the known facts is illustrated in figure 18.2.

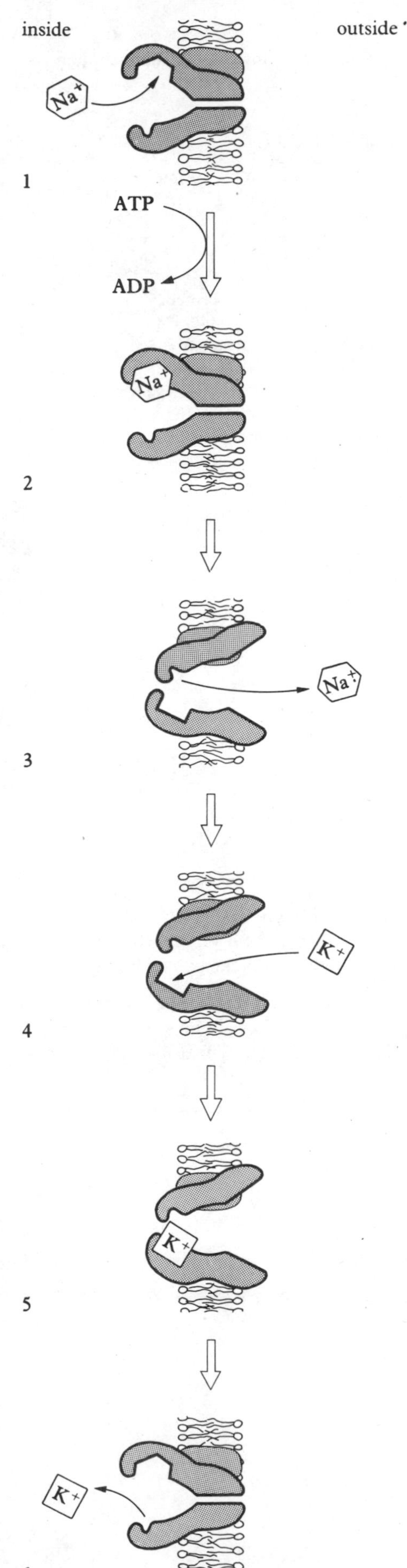

THE ACTION POTENTIAL

In their work on the giant axon of the squid, Hodgkin and Huxley found that when an impulse passes along the axon the membrane becomes **depolarised** and the resting potential is momentarily reversed, creating an **action potential** (see your main textbook). The action potential is brought about by the sudden inflow of sodium ions, followed by a slower outflow of potassium ions which restores the resting potential. The movements of the sodium and potassium ions take place by passive diffusion down the steep concentration gradients created by the sodium–potassium pump.

How do the ions get through the membrane? We can assume that they cannot pass through the phospholipid part of it since the phospholipids are strongly

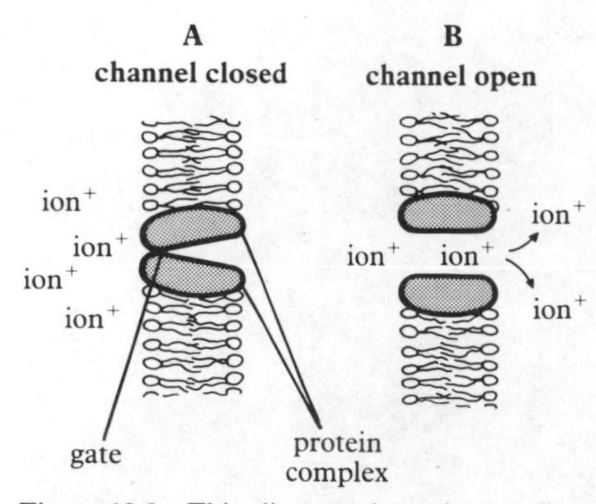

Figure 18.3 This diagram shows how sodium and potassium ions are believed to pass through an axon membrane when the axon is depolarised. Slight movement of the amino acid chains alters the shape of the protein complex in such a way that the gate opens and the ion diffuse through the channel.

hydrophobic. This has led to the suggestion that they may diffuse through special protein-lined channels in the membrane (figure 18.3). Hydrophilic groups on the surface of the protein complexes would permit the ions to pass through unimpeded. It is envisaged that the channels are gated: when the axon is at rest the gates are closed and will not let the ions through. However, when an action potential arrives, the local currents at its leading end bring about a slight change in the potential difference across the axon membrane and this causes the gates to open, letting the ions through. Since the gates are opened by a change in potential, the channels are described as **voltage-dependent**.

Do the sodium and potassium ions share the same channels? It has been found that certain poisons block the inflow of sodium ions but have no effect on the outflow of potassium ions, whilst other poisons block the outflow of potassium ions without affecting the inflow of sodium ions. This suggests that there are separate channels for each type of ion.

The channels remain open for only a brief moment (about one millisecond) and very few ions need to flow through to produce an action potential. The sodium–potassium pump is responsible for returning the ions to their former situations. However, the number of ions involved is so miniscule that this need not happen for further action potentials to be transmitted. Indeed, if the sodium–potassium pump was not working at all, the axon would still be able to transmit several thousand impulses before the system ran down.

The sodium–potassium pump, and the membrane potential resulting from it, are common to cells in general and may have evolved as a mechanism for controlling the volume and osmotic properties of the cell. However, gated channels, and the resulting action potential, are a special feature of excitable cells – that is nerve, muscle and sensory cells.

In the last two topics we have seen that our knowledge of nerves derives in large measure from research carried out on the giant axon of the squid. It is an admirable demonstration of how a combination of the right animal and suitable technology can lead to an important scientific advance.

For consideration

1. Summarise the evidence which suggests that sodium extrusion and potassium uptake by nerve cells are linked.

2. Figure 18.2 is just one of a number of models which have been put forward to explain how the sodium–potassium pump works. Can you suggest any plausible alternatives?

Further reading

R.D. Keynes, 'The Nerve Impulse and the Squid' (*Scientific American*, vol. 199, no. 6, 1958)

B. Katz, 'How Cells Communicate' (*Scientific American*, vol. 205, no. 3, 1961)

These two articles explain the electrical and ionic basis of the nerve impulse.

The visual cortex, an aspect of brain function

The brain is the body's most elusive organ. This is particularly true of the **cerebral cortex**, the nervous tissue that forms the surface layers of the cerebral hemispheres. Consider the problem. The cortex contains a vast number of nerve cells – as many as 20 million in a cubic millimetre of tissue. Thanks to appropriate staining techniques, it has been possible to reconstruct the neurone anatomy of the cortex in some detail. But it is one thing to describe its structure and quite another to say how it works. Here we shall look at advances in our understanding of one particular area, the **visual cortex**.

THE VISUAL CORTEX

This is the part of the cerebral cortex concerned with the sense of sight. Located at the back of the cerebral hemispheres, it is made up of six histologically distinct layers, each of which contains a variable number of nerve cells. The nerve cells are interconnected by synapses mainly in a plane vertical to the surface of the brain; there are relatively few interconnections in the horizontal plane. This suggests that the basic functional units of the visual cortex consist of vertical columns of cells. There would appear to be more opportunity for signals to spread vertically within each column than horizontally between adjacent columns. The visual cortex may therefore be concerned more with receiving signals from the eyes than with integrating the signals into a coherent whole.

FROM THE EYES TO THE CORTEX

The neural pathways along which impulses pass from the eyes to the visual cortex are shown in figure 18.4. The two optic nerves run under the cerebral hemispheres and join to form the **optic chiasma**. Nerve fibres from the inner half of each retina cross in the optic chiasma, with the result that each side of the brain receives information from both eyes: the left side of the brain receives signals from the left half of each retina, and the right side of the brain receives signals from the right half of each retina.

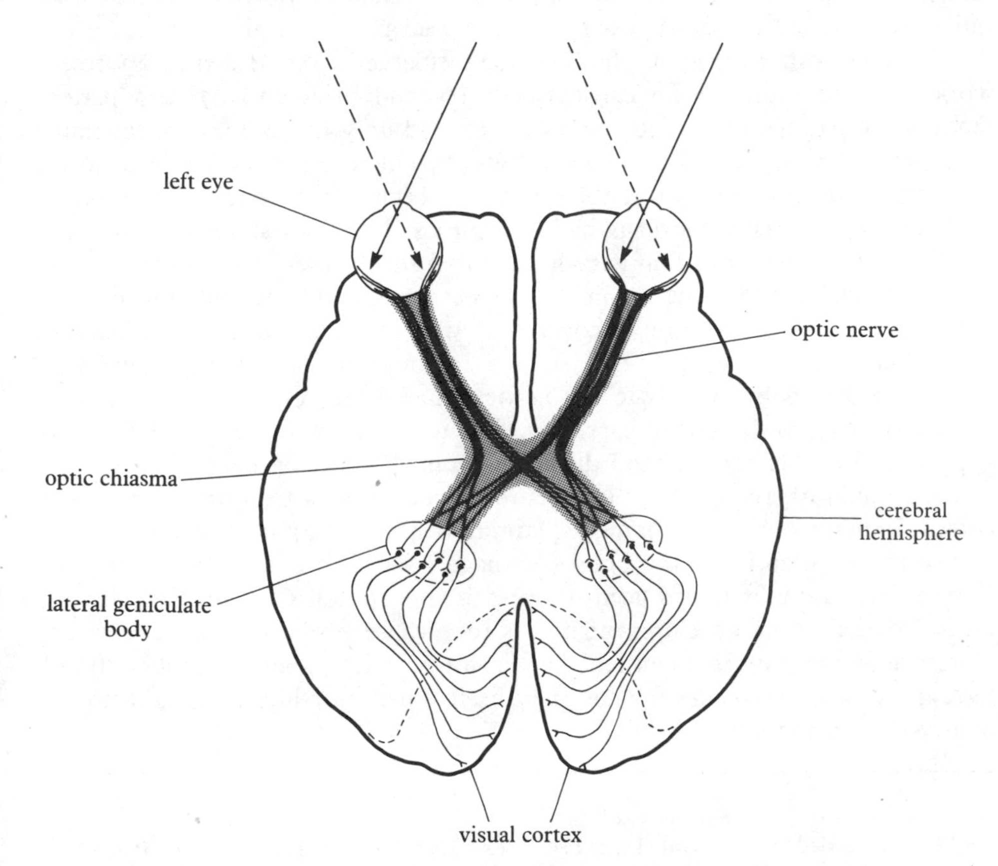

Figure 18.4 Schematic diagram showing the neural pathways which link the eyes with the visual cortex. The brain is here viewed from above as a transparent object so that we can see the optic tracts underneath. The solid arrows represent light rays from objects in the right visual field which are registered by the left side of the brain; the broken arrows represent light rays from objects in the left visual field which are registered by the right side of the brain.

From the optic chiasma the nerve fibres pass, via a pair of **lateral geniculate bodies**, to the visual cortex. This is by no means the end of the line, for the visual cortex is connected to adjacent regions of the cortex which in turn are linked to other parts of the brain. Through these complex interconnections nervous impulses are integrated so as to build up a complete picture of the visual scene.

THE ROLE OF THE CORTEX IN PERCEPTION

It is often the case in biological research that advances have to wait for the appropriate technology to become available. This is certainly true of research on the visual cortex. To understand what is happening we need to record electrical events from individual brain cells, and this has only become possible in comparatively recent years. It requires exceedingly fine microelectrodes, mounted on mechanical manipulators and connected to ultra-sensitive recording equipment.

This approach has been used by David Hubel and Torsten Wiesel of the Harvard Medical School. They have recorded impulses from individual nerve cells in the visual cortex of cats and monkeys in response to a variety of visual stimuli. It was found that the majority of cortical cells do not respond to dots or patches, but only to straight lines. Each individual cell responds best when the line is orientated in a particular way. Thus some cells respond when the line is vertical, others when it is horizontal, and others when it is oblique. A change of orientation by as little as 15° may reduce the response by half. Each cell will respond when either eye is stimulated, but the largest responses are given when both eyes are stimulated together – how would you explain this in terms of the neural connections in the brain?

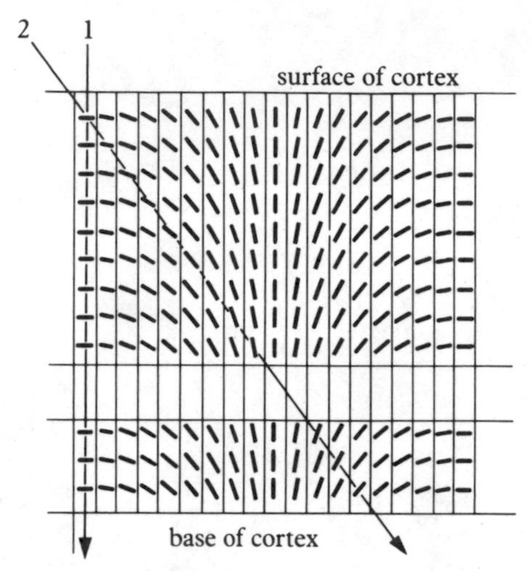

Figure 18.5 Diagrammatic vertical section through part of the visual cortex showing the orientation of visual line-stimuli which give maximum responses in the nerve cells. Notice that identical responses are given by nerve cells in vertical columns. The arrows show the course followed by microelectrodes inserted into the cortex. The electrode represented by arrow 1 encounters nerve cells that respond only to horizontal lines; the electrode represented by arrow 2 encounters, successively, nerve cells that respond to lines at a series of different orientations. The clear layer towards the base of the cortex contains nerve cells which have no preference for a particular orientation. (*Based on Hubel and Wiesel*).

By systematically recording the responses of numerous individual cells, Hubel and Wiesel have shown that nerve cells in the same vertical column share the same preference for a particular orientation, whereas nerve cells in the same horizontal or oblique plane have different preferences (figure 18.5). This may be related to the earlier observation of the neuro-anatomists that the structural units of the visual cortex appear to be organised into vertical columns.

All sorts of other intriguing findings have emerged from Hubel and Wiesel's work. For instance, certain cortical cells respond preferentially to a pattern made up of parallel lines, and some will respond only if the lines are separated by a particular distance. Responses are greatly enhanced if the pattern moves, and some cells only respond if the movement occurs in a certain direction.

Of course our visual world is highly complex, but lines at various angles to one another form a significant part of it. Hubel and Wiesel's work demonstrates that the visual cortex plays an important part in monitoring this information. One of their most interesting discoveries – with important human implications – is that when a young animal such as a kitten is reared in a cage with horizontal bars only, it fails to see vertical lines later. Conversely, an animal reared in a cage with vertical bars only, fails to see horizontal lines later. There appears to be a sensitive period during an animal's development when specific visual stimuli are required for full neuronal connections to form in the visual cortex. In other words environmental stimuli help to determine the development and ultimate potential of the brain (see page 100). Other examples of this are known. For example, a bird needs to hear the full sounds of its own species at a critical period of its development if it is to vocalise properly later, and young mammals brought up in an enriched environment with plenty of sensory stimuli have more dendritic connections in their brains than individuals brought up in a deprived environment.

For consideration

The experiment whose results are shown in figure 18.5 was carried out by recording electrical activity from different regions of the visual cortex. How could you identify the particular nerve cells responsible for each response?
(Hint: radioactive glucose injected into an animal is taken up by active brain cells.)

Further reading

D. Hubel and T. Wiesel, 'Brain Mechanisms of Vision' (*Scientific American*, vol. 241, no. 3, 1979)
In fact, this whole issue of *Scientific American* is devoted to the brain.

How do hormones affect their target cells?

It is sometimes instructive to forget what has, or has not, been discovered and to think about a problem from first principles. Take hormones for instance. How does a hormone evoke a response in its target cell? Any explanation of this must take account of the fact that a given target cell responds to a restricted range of hormones, sometimes only one, and ignores others. We may also predict that the hormone exerts its effect by directly or indirectly triggering the action of appropriate enzymes. In this topic we shall look at some recent research and see to what extent these predictions are fulfilled.

THE ROLE OF CYCLIC AMP

Round about 1960 two American scientists, E.W. Sutherland and T.W. Rall, were investigating the action of adrenaline and glucagon on liver cells. They discovered that these hormones do not enter the cells, but lead to an increase in the intracellular concentration of a compound called **cyclic adenosine monophosphate** (cAMP). The latter then increases the activity of enzymes which break glycogen into glucose. cAMP, a derivative of ATP with a single phosphate group, has been found in a wide range of animal cells and bacteria and is thought to serve as an intermediary in many hormone systems. The hormone itself is described as the **first messenger**, cAMP as the **second messenger**.

The formation of cAMP from ATP is catalysed by the enzyme **adenyl cyclase** which is found in the cell membrane. Research has shown that when a hormone such as adrenaline or glucagon reaches its target cell, it binds reversibly with a receptor site on the surface of the cell membrane. This activates adenyl cyclase which catalyses the conversion of ATP to cAMP in the cytoplasm. The cAMP then activates another enzyme, a protein kinase, which triggers a series of further enzyme activations, resulting ultimately in the breakdown of glycogen to glucose (figure 18.6).

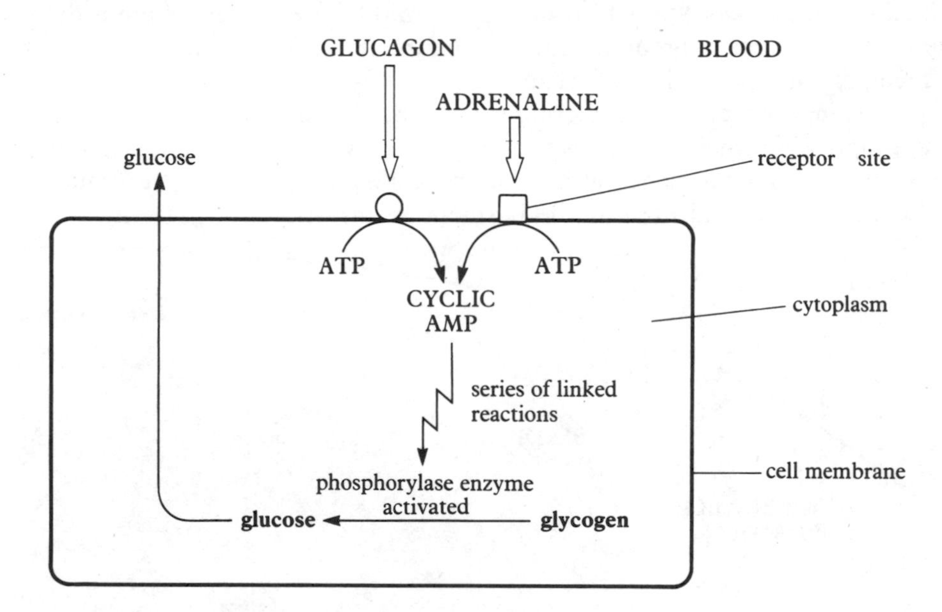

Figure 18.6 Schematic diagram showing how glucagon and adrenaline stimulate glucose production in a liver cell. There are separate receptor sites in the cell membrane for the two hormones. The formation of cyclic AMP from ATP is catalysed by adenyl cyclase, and cyclic AMP's influence over the conversion of glycogen to glucose is achieved through a chain of enzymes which control a series of linked reactions. (*Based on Berridge, 1975*)

One advantage of this system is that it amplifies the response. Each molecule of adenyl cyclase can be used over and over again, as can the enzymes which it activates, so a relatively small quantity of the hormone can lead to the formation of a disproportionately large amount of glucose.

The target cell contains another enzyme, a phosphodiesterase, which *destroys* cAMP. The size of the target cell's response depends on a balance between adenyl cyclase which promotes the synthesis of cAMP and the phosphodiesterase which destroys it.

THE RECEPTOR SITE

To account for the fact that a target cell only recognizes certain hormones, we may assume that each receptor site has a chemical configuration which is specific to the molecule of a particular hormone. We can use this to explain why, for example, glucagon stimulates glycogen breakdown in liver cells but not in muscle: presumably the sarcolemma of muscle fibres lacks the necessary receptor sites. When two or more hormones have identical effects on the same target, as in the case of glucagon and adrenaline on liver cells, it is envisaged that each hormone has its own receptor site all of which are coupled with adenyl cyclase.

The receptor site is thought to be a protein because it can be destroyed by treatment with proteases. Adenyl cyclase is not an integral part of the site and may indeed be situated some distance from it. The fluid-mosaic nature of the cell membrane, allowing sideways movement of proteins, would make it possible for adenyl cyclase to come into close proximity with the receptor protein when the hormone binds with it (see page 11). There is evidence that another protein, activated by guanosine triphosphate (GTP), serves as a 'transducer', linking the receptor protein with adenyl cyclase.

OTHER HORMONE MECHANISMS

When cAMP was discovered it was thought that it might turn out to be a universal second messenger common to all hormone systems. However, lovers of uniformity have had to be disappointed. Some hormones – insulin for example – do not use cAMP as the second messenger. Moreover, steroid hormones, such as oestrogen and cortisol of mammals and ecdysone of insects, do not use a second messenger at all. Instead they pass through the cell membrane and enter the cell. In the cytoplasm the hormone molecule binds with a protein carrier which conveys it into the nucleus. It then associates reversibly with the chromosomal material, switching on the relevant part of the DNA. The latter is transcribed into messenger RNA which moves out into the cytoplasm where it directs the synthesis of the appropriate enzyme (figure 18.7). Producing a response by influencing the expression of genes in this way is presumably slower and longer-lasting than directly triggering enzyme action, and this makes it especially appropriate for long-term responses such as growth and sexual development. Some non-steroid hormones also work by influencing gene expression; they include thyroxine, and the juvenile hormone of insects.

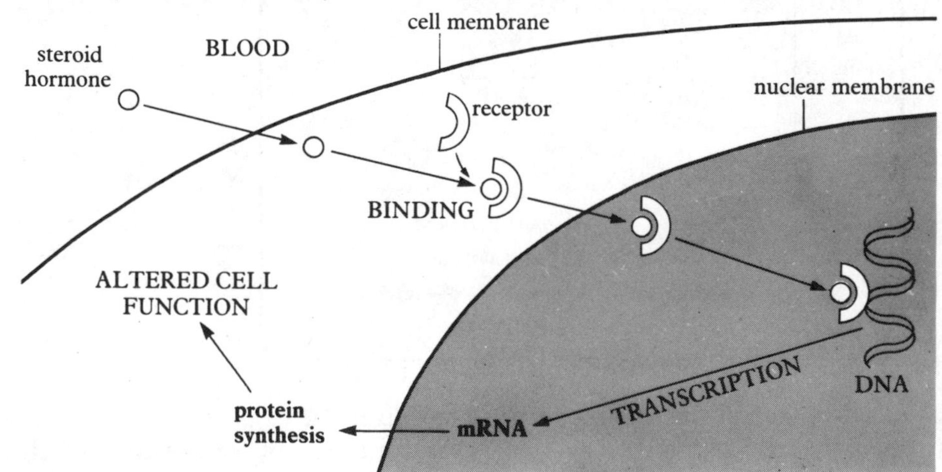

Figure 18.7 Hypothetical, and highly simplified, diagram to show how a steroid hormone may exert its effect on its target cell. (*Based on Schrader and O'Malley, 1978*)

For consideration

1. Suggest ways in which figure 18.6 might be improved. Redraw it, showing your suggested improvements.

2. 'Advances in cell biology have helped to explain phenomena which have been known for many years.' Discuss this with reference to hormones.

Further reading

J.W. Buckle, *Animal Hormones* (Studies in Biology no. 158, Arnold, 1983)
The first chapter includes a brief account of how hormones work.

A.R. Rees and M.J.E. Sternberg, *From Cells to Atoms* (Blackwell, 1984)
This work consists of diagrams with brief explanatory text in a large-page format; there is an account of hormone action.

G.J. Goldsworthy, J. Robinson and W. Mordue, *Endocrinology* (Blackie, 1981)
Not a long book; a detailed chapter deals with the cellular aspects of hormones.

19 . Reception of stimuli

Complexity in the retina

The eye enables us to see things, but perception in its broadest sense involves much more than just our eyes. The nervous pathways leading from the eyes to the brain are interrupted at certain strategic points by synapses which permit all sorts of complex interactions. Some of these interactions occur centrally in the brain (see page 82); others occur in the retina. It is with the latter that we shall be concerned in this topic.

ELECTROPHYSIOLOGICAL WORK ON THE RETINA

The sensory cells in the retina – the rods and cones – are connected to **bipolar neurones**; these in turn are connected to **ganglion cells** from which arise the fibres that make up the optic nerve. Individual ganglion cells have been impaled with microelectrodes and their electrical activity recorded when the retina was stimulated with beams of light. The ganglion cells discharge impulses at a low but steady frequency even when the retina is unstimulated.

However, the resting discharge of an individual ganglion cell can be altered by stimulating a small, clearly defined area of the retina. This is referred to as the ganglion cell's **receptive field**. The receptive fields of the ganglion cells are roughly circular, and are organised into two concentric areas which give opposite responses. For some ganglion cells the central part of the field is excitatory (the 'on' region) and the peripheral part is inhibitory (the 'off' region), whereas for others the central part is inhibitory and the peripheral part excitatory (figure 19.1). The two parts of the field therefore oppose each other, and if both are stimulated together, only a very weak response is given.

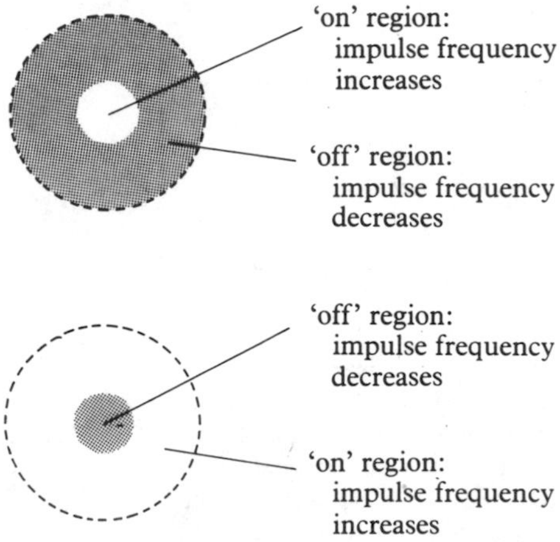

Figure 19.1 Receptive fields of two retinal ganglion cells. They differ in which parts of their receptive fields are excitatory ('on') or inhibitory ('off').

Ganglion cells differ in the size of their receptive fields. As you might expect, the fields are very much smaller in the central (foveal) part of the retina where visual acuity is high, than further out where visual acuity is relatively low. Also adjacent fields show considerable overlap in the more peripheral parts of the retina.

Another feature of the peripheral part of the retina is that the receptive fields are not fixed once and for all but can change according to the general level of illumination. At low levels of illumination the inhibitory effect of the 'off' region becomes weaker and the sensitivity of the 'on' region increases. Most ganglion cells show a sensitivity typical of either cones or rods but not both; however, if the illumination is reduced, the sensitivity of some of the ganglion cells may change from that typical of cones to that typical of rods.

From these physiological observations certain predictions may be made concerning the structure of the retina. For example, one would expect to find extensive connections between adjacent cells; and presumably rods and cones share the same ganglion cells at least in the more peripheral parts of the retina.

STRUCTURE OF THE RETINA

Research on the structure of the retina has confirmed these predictions and has revealed many other intriguing details of how the retina is organised (figure 19.2). Several receptor cells – rods, cones or both – may converge onto a single bipolar neurone; and additional neurones link adjacent retinal cells laterally.

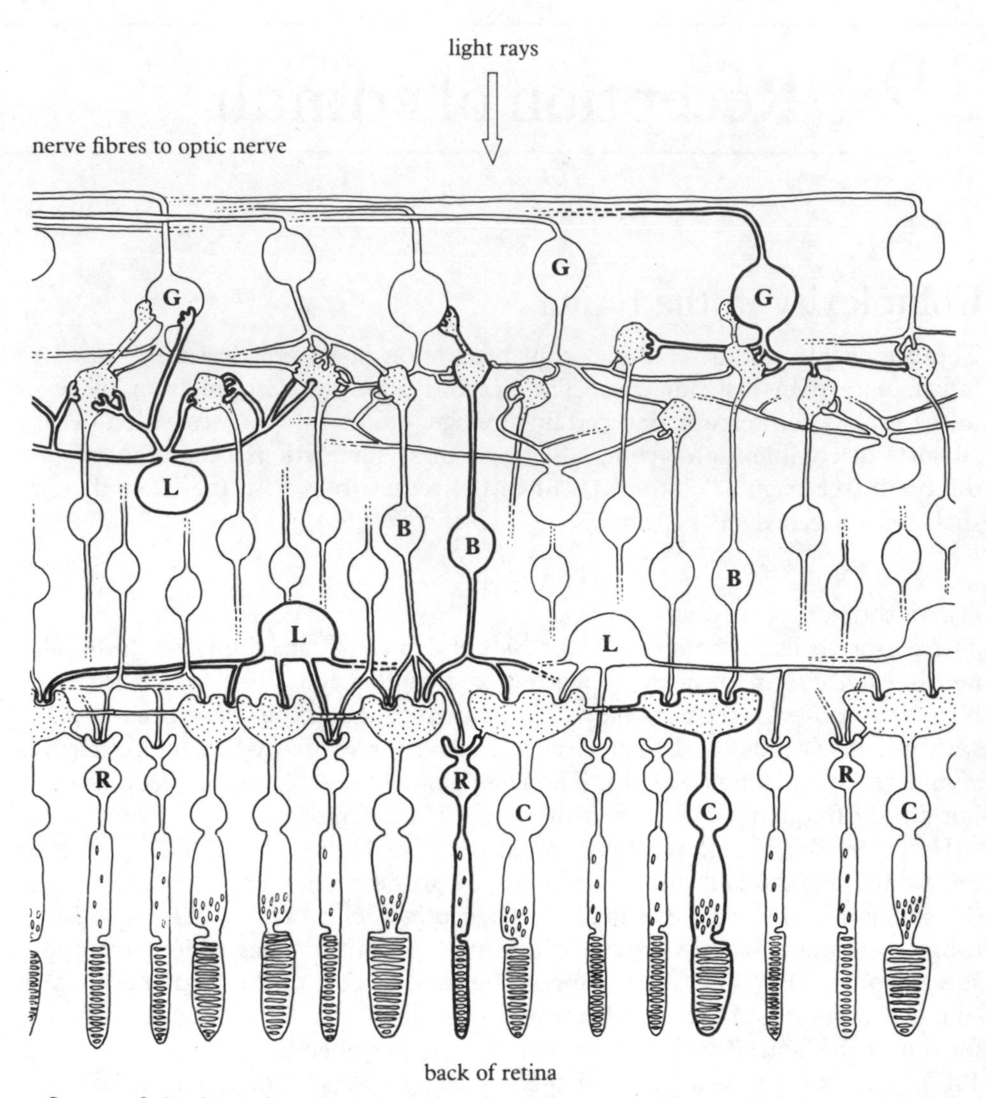

Figure 19.2 Diagram based on electron microscopy showing the organisation of a representative part of the retina. R, rod; C, cone; B, bipolar neurone; L, lateral neurone; G, ganglion cell. (*After Dowling and Boycott*, Proceedings of Royal Society B. *Vol. 166*, 1966)

Some of the lateral neurones make synaptic connections with the bases of the rods and cones. They are thought to be responsible for the interactions between the 'on' and 'off' regions of the receptive fields, and for sharpening the boundaries between adjacent fields as a result of mutual (lateral) inhibition (see your main textbook). Other lateral neurones run between the bipolar neurones and ganglion cells and are possibly related to the ability of some ganglion cells to respond to movement or sudden changes in illumination.

It has also been suggested that interactions between adjacent parts of the retina may in some instances lead to displacement and distortion of neurally recorded images, resulting in optical illusions of various kinds. However, it is probable that illusions also arise centrally in the brain where fusion of signals from the eyes takes place. In any event the retina is highly complex and considerable integration takes place within it before the signals reach the brain.

SUMMING UP

The behaviour and structural organisation of the ganglion cells suggests that they are designed to respond to *contrast*, i.e. contrasting levels of illumination and also contrasting colours. This is achieved by complex interactions, which we have barely touched on here, between rods, cones, bipolar neurones and ganglion cells. Embryologically the retina is part of the brain, and understanding how the retina processes information may give us an insight into how similar processing occurs in the brain. It is interesting that retinal processing takes place by the development of localised potentials in the various neurones and does not necessitate the use of propagated action potentials.

For consideration

1. Classify the different nerve cells in figure 19.2 according to their synaptic connections. Suggest a function for each type.

2. The diagram in figure 19.2 is based on the extra-foveal part of the retina, that is the part peripheral to the fovea. How would you expect the foveal region to differ from it structurally and why?

Further reading

R.A. Weale, *The Vertebrate Eye* (Carolina Biology Reader no. 71, 2nd edition, 1978)
The last few pages are relevant to this topic.

S.W. Kuffler, J.G. Nicholls and A.R. Martin, *From Neuron to Brain* (Sinauer, 2nd edition, 1984)
Chapter 2 contains a definitive account of the retina; easier going, though, is the 1st edition by S.W. Kuffler and J.G. Nicholls.

The organ of Corti

The organ of Corti is the 'business' part of the ear. It is the place where mechanical vibrations resulting from the transmission of sound waves through the ear are transformed into nervous signals. In this topic we shall consider how this particular transformation takes place. But first we must look at the structure of the cells responsible.

THE CELLS OF THE ORGAN OF CORTI

The organ of Corti runs the whole length of the spiral cochlea. In addition to various supporting cells it contains a series of **hair cells** which lie between the basilar and tectorial membranes (see your main textbook). The hair cells are arranged in rows, one row towards the inner side of the cochlea (i.e. towards the centre of the spiral), and between three and five rows towards the outside.

The electron microscope has been used to study the structure of the hair cells (figure 19.3). The top end of each one is flattened and bears over 100 tiny hairs ranked in several rows of graded height. The hairs point upwards towards the tectorial membrane, and the longest ones may be embedded in it. Each hair is a simple structure containing actin filaments which are rooted in the main body of the cell. Adjacent hairs of successive rows appear to be connected because if the longest one is moved with a fine probe the others move with it.

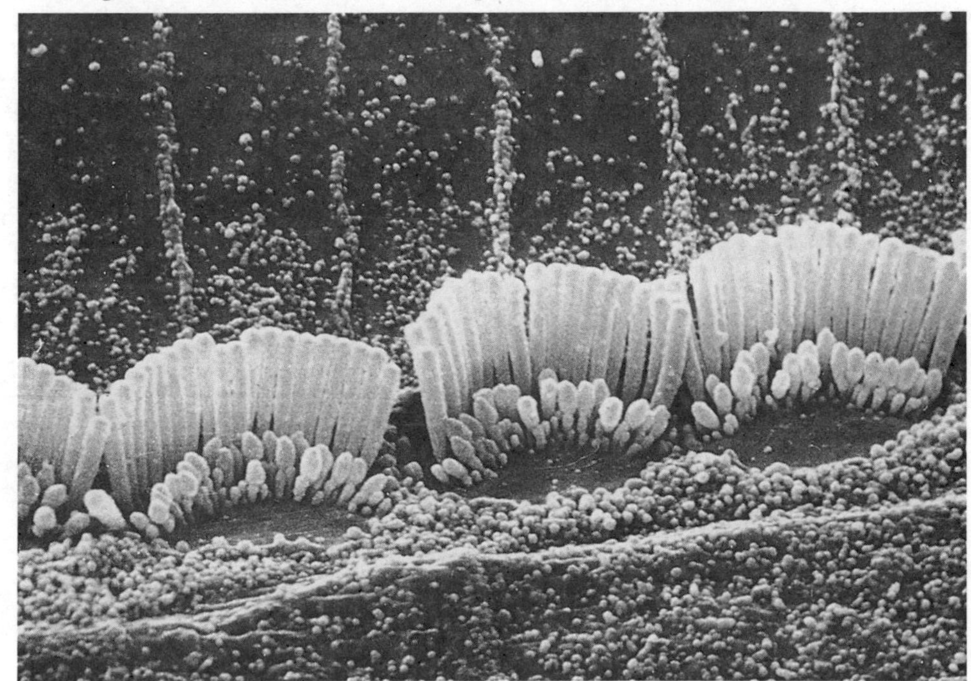

Figure 19.3 **A** Scanning electron micrograph of the surface of part of an organ of Corti showing hairs projecting from the hair cells. (×6200)
B A hair cell from the organ of Corti reconstructed from electron micrographs. Notice the hairs projecting from the top of the cell. Inside the cell numerous mitochondria and an extensive smooth endoplasmic reticulum may be involved in transforming mechanical deformation of the hairs into electrical activity. (*After I. Friedman*, The Human Ear, *Carolina Biology Reader No. 73, 2nd edition 1979*)

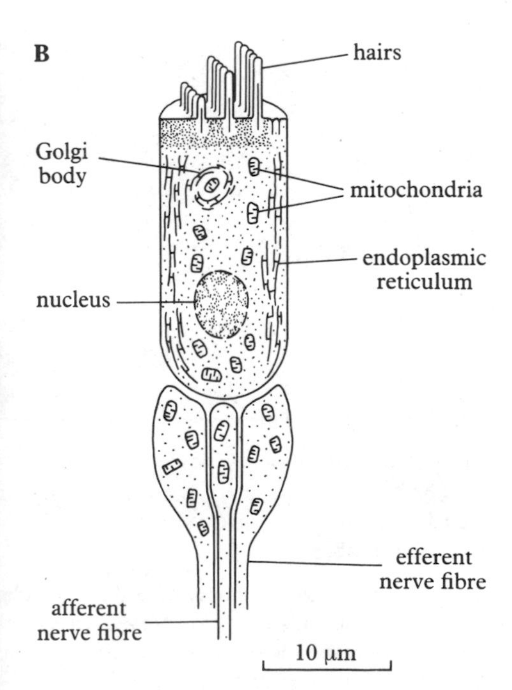

Each hair cell makes a synaptic connection with an afferent nerve fibre. The nerve fibres from all the hair cells up and down the cochlea are gathered together to form the cochlear branch of the auditory nerve through which impulses are transmitted to the brain.

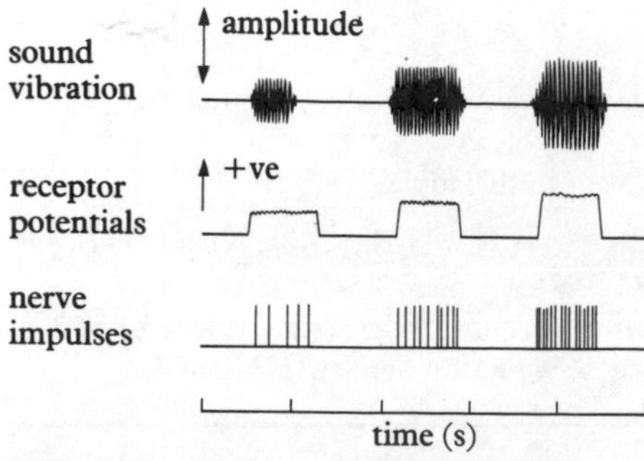

Figure 19.4 Receptor potentials recorded from the cochlear hair cells, and action potentials recorded from the nerve fibres of the cochlea nerve, in response to sounds of different intensity (amplitude). (*After Rosenberg*)

HOW ARE THE HAIR CELLS STIMULATED?

With the aid of microelectrodes, electrical activity has been recorded from the hair cells and the nerve fibres that lead from them. When the basilar membrane is vibrated a receptor potential develops in the hair cells and impulses are discharged in the nerve fibres. The frequency of the impulses depends on the amplitude of the vibrations (figure 19.4).

But what exactly excites the hair cell and causes it to develop a receptor potential? Over the years many suggestions have been put forward. The one most generally favoured at the present time is that when the basilar membrane moves, the hairs are subjected to a sideways (shearing) displacement. This alters the surface properties of the cell in such a way as to increase the flow of ions between the inside of the cell and the surrounding endolymph. As a result the cell membrane becomes depolarised and a receptor potential develops inside the cell. One circumstantial piece of evidence which supports this hypothesis is that the endolymph has an exceptionally large positive potential compared with the interior of the hair cells: the total potential difference is of the order of 140 mV which is considerably greater than the potential difference between the inside and the outside of a typical nerve cell. This would steepen the electrochemical gradient, facilitating the flow of ions when the membrane is depolarised. Thus the responsiveness of the cell is increased.

THE DAMAGING EFFECT OF EXCESSIVE NOISE

It has been known for many years that people subjected to a continuous loud sound of more than approximately 95 dB eventually become deaf to sounds of that particular frequency (pitch). This defect of hearing is called **tonal gap** and it is an occupational hazard amongst people who work in certain types of industry.

When the ears of such people are examined *post mortem* they turn out to lack some of the hair cells in the part of the cochlea which normally responds to that particular frequency. It seems that excessive vibrations of the basilar membrane – that is, vibrations of too great an amplitude – damage the hair cells. So the cells cease to function and subsequently degenerate, together with their nerve fibres. This kind of deafness has also been reported in individuals who continually listen to very loud pop music, particularly through earphones.

A HOST OF UNCERTAINTIES

Many problems remain unsolved. For example, do the inner and outer hair cells perform different functions and, if so, what are they? Why are the rows of hairs projecting from the hair cells of different length? And which particular hairs, if any, are embedded in the tectorial membrane? Research is beginning to provide possible answers to these questions, but there is still much that we do not understand about the ear.

For consideration

1. The hair cells in the organ of Corti are connected to efferent as well as afferent nerve fibres (see figure 19.3B). Suggest a function for the efferent fibres.

2. In this topic we have only considered how individual hair cells respond to a given sound. Put forward a hypothesis to explain how the cochlea responds to sounds of different pitch. How could your hypothesis be tested experimentally?

Further reading

I. Friedman, *The Human Ear* (Carolina Biology Reader no. 73, 2nd edition, 1979)
This book contains a useful account of the organ of Corti, particularly its structure.

M.E. Rosenberg, *Sound and Hearing* (Studies in Biology no. 145, Arnold, 1982)
Chapter 4 contains a summary of how the hair cells may work.

20 . Effectors

Some factors affecting muscle performance

Adrian Horridge, the neurophysiologist, was once doing an experiment on reflex action when his professor walked into his laboratory. Looking at the tangled mass of wires and electronic gadgetry, the professor asked, 'Is there an animal there somewhere?' The story reminds us that, although much biological research is carried out on little bits of animals ('physiological preparations' as they are euphemistically called), the ultimate aim of research is to understand the working of the whole animal.

In studying muscle numerous experiments have been carried out on isolated nerve-muscle preparations. In this topic we shall consider several observable properties of such preparations and see how they help to explain movement in the intact animal.

FINE AND COARSE MOVEMENTS

Suppose we record the responses of a muscle whose motor nerve is stimulated with a series of electrical shocks of gradually increasing intensity. What we get is a series of graded contractions of the kind shown in figure 20.1.

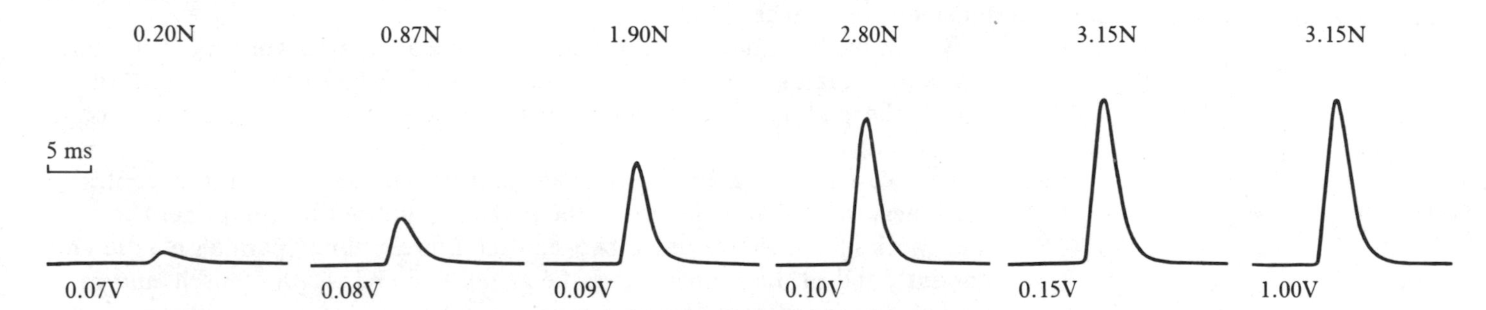

Figure 20.1 Twitches given by one of the flexor muscles from the hind limb of a cat in response to a series of single electrical pulses of increasing strength. The force, in newtons, achieved by the contractions is written above each recording, and the voltage underneath. (*After Buller and Buller*)

At first sight this might appear to contradict the all-or-nothing nature of nerve and muscle. How can we explain it? A motor nerve may contain over 100 axons, each of which supplies a group of perhaps 150 muscle fibres. The muscle fibres innervated by a single axon constitute a **motor unit**. When we stimulate a nerve with shocks of increasing intensity, successive shocks excite more and more axons so that a greater number of motor units respond each time – hence the graded contractions seen in figure 20.1. The all-or-nothing principle applies to individual axons and muscle fibres, not to whole nerve-muscle preparations.

Through variations in the number of motor units activated, a muscle may give small or large responses as the case may be. The larger the motor unit, the greater the strength of the contraction. Muscles such as those of the larynx and eyes, which perform very delicate movements, tend to have small motor units consisting of as few as two or three muscle fibres. On the other hand, muscles such as those that work the limbs, which are concerned more with power than precision, have much larger motor units consisting of several hundred muscle fibres.

LENGTH VERSUS TENSION

The recordings shown in figure 20.1 were obtained by attaching the muscle to a transducer which converts the tension developed by the muscle – that is the force of contraction – into an electrical signal of proportional size. The latter is recorded with an oscilloscope or electronic chart recorder. The muscle is held in a fixed position so that its length does not change.

If a muscle is held at a constant length and develops tension when it is activated, the contraction is described as **isometric**. If, however, the muscle maintains a constant tension and changes in length when activated, the contraction is described as **isotonic**.

In the body both types of contraction occur, depending on circumstances. The sustained contractions necessary for maintaining posture involve mainly isometric contractions, with antagonistic muscles pulling against each other. On the other hand, when the body moves (such as in walking), contractions are at least partly isotonic. Rarely, if ever, would they be entirely isotonic, for the load on the muscles varies from moment to moment, particularly when for instance we lift something up.

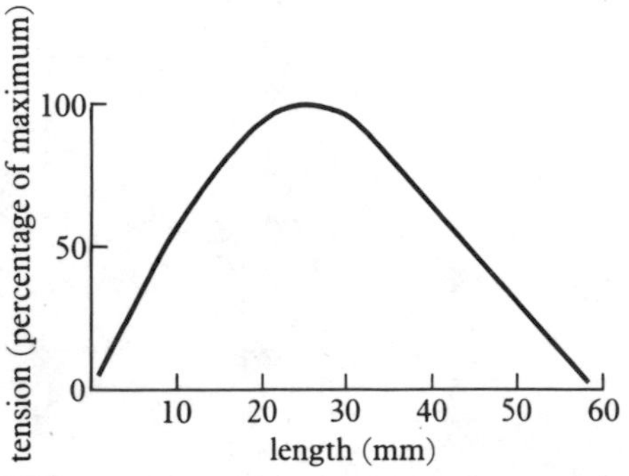

Figure 20.2 Graph showing the tension developed by the sartorius muscle of a frog or toad at different lengths. The curve shows the tensions achieved during maximum sustained contractions produced by tetanizing the muscle with trains of high-frequency stimuli. Notice that sub-maximal tensions are given when the muscle is either too short (i.e. under-stretched) or too long (i.e. over-stretched).

ELASTICITY

It is easy to see that an isolated muscle, such as the frog gastrocnemius, is elastic. Even tendons, which are commonly described as unstretchable, have a certain amount of elasticity, though nothing like as much as ligaments. If our muscles and tendons were not elastic, running would consume more than twice as much metabolic energy as it does.

When a muscle contracts isometrically, the tension developed depends on the length of the muscle, in other words on how stretched it is. This can be shown by recording the sustained contractions of a muscle held at a series of different lengths. If you look at figure 20.2 you will see that the tensions achieved get greater as the length of the muscle is increased, but above a certain length the tensions get smaller again.

Within the body our muscles tend to be held at lengths which permit them to develop maximum, or near maximum, tension when they contract. In this regard the pulling effect of antagonistic muscles against each other is of the utmost importance.

The elasticity of a muscle resides in three places: within the contractile machinery of the muscle fibres themselves (**contractile component**), in the meshwork of connective tissue between the muscle fibres (**parallel elastic component**), and in the connective tissue at the two ends of the muscle and in the tendons (**series elastic component**).

A considerable amount of energy can be stored in these elastic components, as is seen if you attempt to perform a high jump. It is better to bend the knees just before jumping than to start from a stationary position with the knees already bent. Flexing the leg at the knee stretches the extensor muscles and their tendons which consequently become loaded with elastic energy. This phenomenon is not of course confined to jumping; the stretching of muscles immediately before contraction occurs in most kinds of movement.

For consideration

1. How could you prove that when you stimulate a nerve with shocks of increasing intensity, successive stimuli excite more and more axons?

2. Study figure 20.2. How would you explain the relationship between muscle tension and length in terms of the sliding filament theory of muscle contraction?

3. When performing a high jump it is customary to take a run at it. In what way does this help?

Further reading

A.J. Buller and N.P. Buller, *The Contractile Behaviour of Mammalian Skeletal Muscle* (Carolina Biology reader no. 36, 2nd edition, 1980)
The second half deals with the mechanical properties of muscle.

R. Margaria, *Biomechanics and Energetics of Muscular Exercise* (Oxford University Press, 1976)
A useful book in that it relates the properties of muscle to overall performance.

The molecular basis of muscle contraction

It is often said that every new discovery throws up a host of new problems to be solved. Such was the case with muscle contraction. When, in the late 1950s, H.E. Huxley and his colleagues discovered that a muscle shortens by bundles of thick and thin filaments sliding between each other, the question immediately arose: how do the filaments move? The only clue at that time was the observation that in electron micrographs there appeared to be numerous bridges extending from the thick to the thin filaments. Could it be that these were functioning as ratchets, pulling the thin filaments towards the thick filaments when the muscle contracted? In recent years the ratchet hypothesis has received considerable support from studies on the molecular structure of the filaments.

THE THICK FILAMENTS

The thick filaments are made of the protein **myosin**. In high resolution electron micrographs a single myosin molecule can be seen to consist of a compact 'head' and an elongated 'tail'. Biochemical techniques have shown that the head is a double structure composed of a globular protein, whilst the tail is a fibrous protein made up of two α-helical polypeptide chains coiled round each other (figure 20.3).

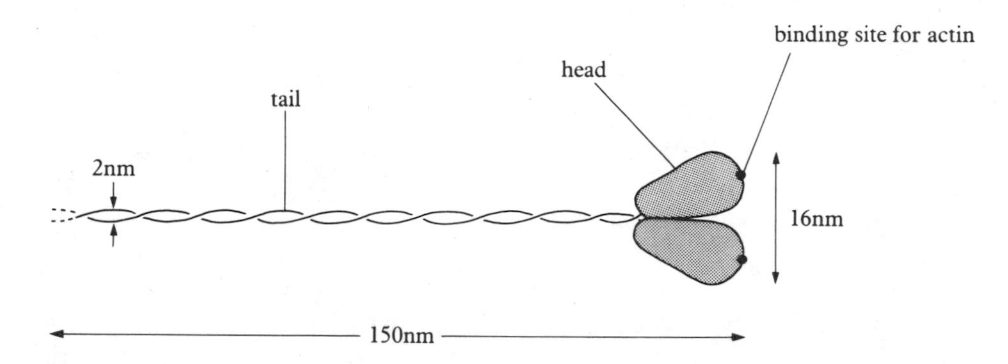

Figure 20.3 The structure of a single myosin molecule.

How are the myosin molecules are arranged in the filaments? It is interesting that in weak ionic solutions they aggregate together. This could well be the way they are organised in the intact muscle. The globular heads are the bridges seen in electron micrographs of myofibrils, and they point outwards towards neighbouring thin filaments in regions where the thick and thin filaments overlap.

The globular head has two important features which enable it to associate with the thin filaments: it can hydrolyse ATP which provides the energy for contraction, and it can bind reversibly with **actin** which is the main protein in the thin filaments. When ATP is hydrolysed, the myosin head becomes attached to the actin with the formation of **actomyosin**. This is immediately followed by a tilting movement of the head which pulls the thin filament towards the thick filament. In the presence of ATP, the myosin head becomes detached from the actin. The ATP is then hydrolysed and the cycle is repeated.

What causes the head to tilt? The current theory is that the products of ATP hydrolysis (ADP and inorganic phosphate) are retained inside the head until the actomyosin link has been formed. The ADP and inorganic phosphate are then released and it is this that causes the tilting. The sequence of events is illustrated in figure 20.4. In effect, the myosin heads 'walk' along the thin filaments.

Figure 20.4 (below) Hypothetical sequence of events which may occur when a myosin head, the bridge between the thick and thin filaments, goes through its cycle. **A** A molecule of ATP enters the myosin head. **B** Hydrolysis of ATP causes the myosin head to attach to the actin filament. **C** Expulsion of ADP and inorganic phosphate (iP) from the head causes it to tilt. **D** Unsplit ATP molecule causes head to detach itself from actin filament and regain the configuration seen in *A*. (*Based on Wilkie*)

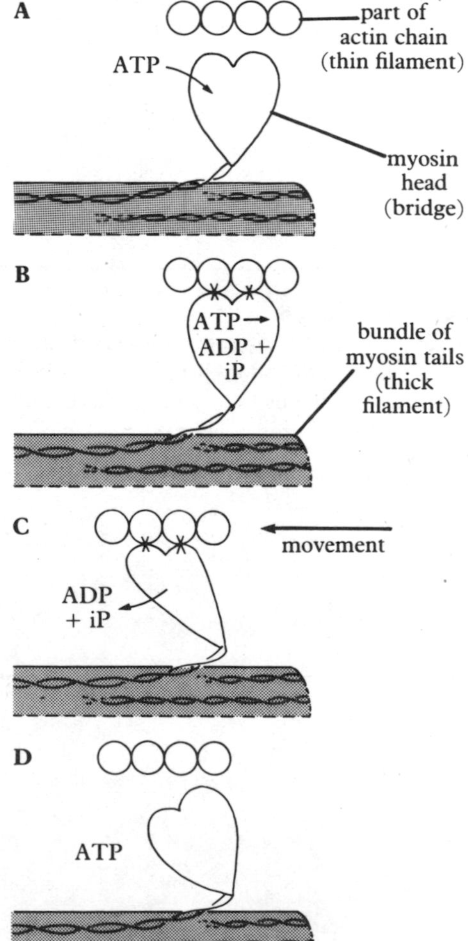

THE THIN FILAMENTS

A thin filament is composed of two chains of globular actin molecules wound round each other like a double-stranded plaited rope. Each actin molecule has a binding site for a myosin head. Running alongside each actin chain is a fibrous protein, **tropomyosin**, which consists of two α-helical polypeptide chains tightly coiled. Yet another protein, a globular complex called **troponin**, is attached at regular intervals to the tropomyosin.

What part does this complicated structure play in the contraction process? An important clue is provided by the observation that calcium ions, which are known to initiate contraction, are able to bind reversibly with troponin. This is believed to happen when the muscle membrane is depolarised, on the arrival of a nerve impulse. The troponin – calcium complex then undergoes a conformational change which moves the tropomyosin strand in such a way that the binding sites on the actin molecules are exposed (figure 20.5).

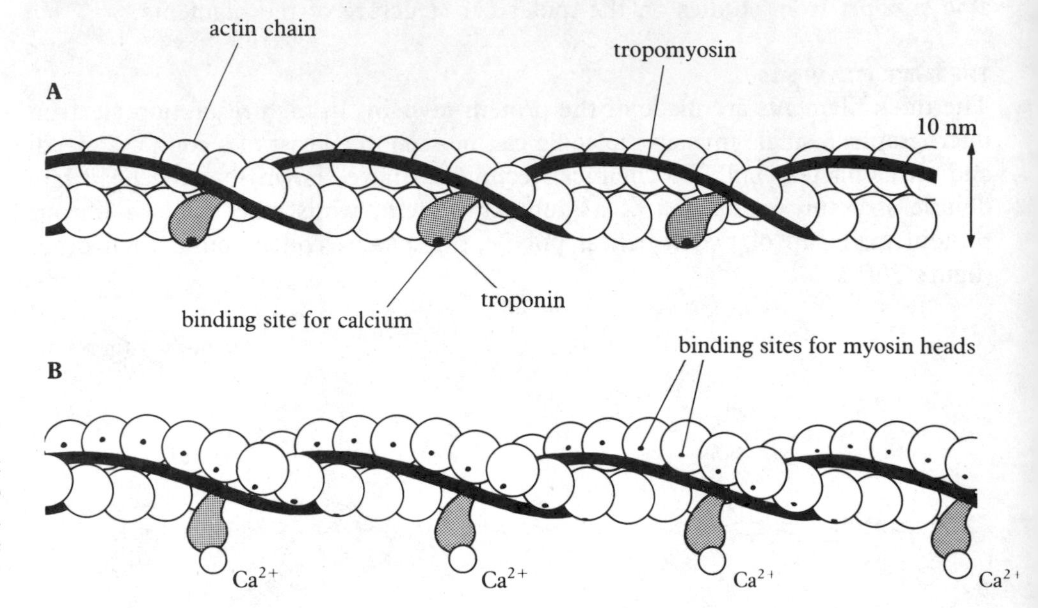

Figure 20.5 Molecular structure of a short length of a thin filament showing the relationship between actin chains, tropomyosin and troponin (**A**) in the absence of calcium ions, and (**B**) in the presence of calcium ions. (*After P. G. Kohn*)

There are still many unanswered questions about how muscle contracts. For example, why should the expulsion of ADP and inorganic phosphate cause the myosin heads to tilt? And what provides the motive force that pulls the thick and thin filaments together? But we have come a long way since the sliding filament hypothesis was first put forward.

For consideration

Rigor mortis is the stiffening of the body after death. It sets in some hours after death and eventually wears off. It happens particularly quickly if strenuous exercise took place immediately before death. Explain these events in terms of the molecular biology of muscle.

Further reading

H.E. Huxley, 'The Mechanism of Muscular Contraction' (*Scientific American*, vol. 213, no. 6, 1965)
Some of the earlier work on the role of the bridges is described in this article.

D.R. Wilkie, *Muscle* (Studies in Biology no. 11, Arnold, 2nd edition, 1976)
The whole of chapter 2 and the end of chapter 4 are relevant.

W.F. Harrington, *Muscle Contraction* (Carolina Biology Reader no. 114, 1981)
A useful summary of how muscle contracts.

21 . Locomotion

Aerofoils

One of the joys of biology is that many of the things we study are beautiful. For instance, right in the middle of the town where I live in Scotland, I can watch gulls soaring around the tenements, pigeons performing acrobatics above the streets, and the delicate flight of terns as they fish along the Water of Leith.

Gulls, especially when soaring or gliding, look just like aircraft. This is no coincidence, since they both achieve buoyancy by using an **aerofoil**. An aerofoil is a surface over which air flows in such a way as to provide a useful reaction. The sail of a yacht is another aerofoil. One ingenious yachtsman tried using an upright aircraft wing in place of the mast and sail – and it worked.

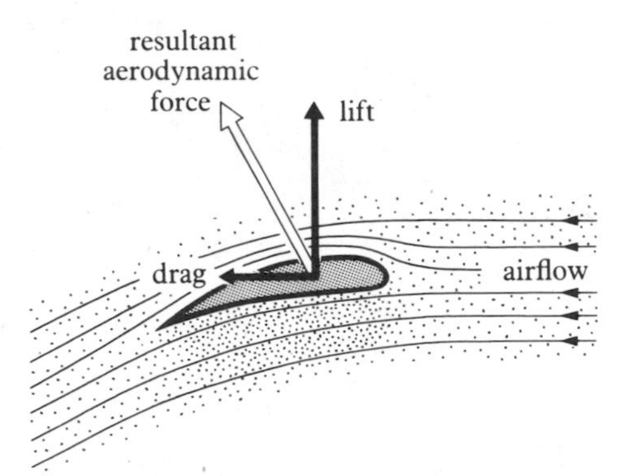

Figure 21.1 Cross-section of a stationary aerofoil in a strong flow of air which is moving from right to left. The streamlines show the direction the air takes. Intensity of dotting shows the air pressure – decreased above the aerofoil and increased below it.

There are two main ways of keeping a heavy object above the ground, other than by using a crane. One is to push air or some other gas downwards, as in a rocket. Because the rocket pushes the gas downwards, the gas exerts an equal force on the rocket upwards. The second way is to reduce the pressure above the object, so that the pressure underneath pushes the object up.

A typical aerofoil such as an aircraft wing uses both methods. Figure 21.1 shows a cross-section of one and how it acts if placed in a horizontal airflow. The aerofoil is at a positive angle of attack, in other words with the thick fore edge above the hind edge. Streamlines show how the air follows the upper and lower surfaces of the aerofoil. When the air leaves the hind edge, it is lower than it was originally. This means that the air is being pushed downwards; so it simultaneously forces the aerofoil up. This simple reaction to the displacement of air makes up about one third of the total upward force or **lift** on the aerofoil.

But this is not all. As the air passes over the upper surface, its velocity increases, and this in turn reduces the pressure. Why this happens is not easy to understand, but the principle is known as **Bernoulli's theorem**. Similarly the air slows down as it passes under the aerofoil, and the pressure correspondingly increases. The pressure difference between upper and lower surfaces exerts an upward force which accounts for the remaining two-thirds of the lift.

One way of demonstrating the velocity effect is illustrated in figure 21.2. Try holding the edge of a piece of paper horizontal and taut, and then blow along the upper side. The paper should be pushed up into the area of reduced pressure.

The airflow not only creates buoyancy; it also gives rise to another force on the aerofoil known as **drag**. Drag is the backward-acting resistance of the air on the aerofoil. The resultant **aerodynamic force** on the aerofoil is therefore both upwards (lift) and backwards (drag), giving a motion similar to that of a kite.

Figure 21.2 Blowing along the top of a piece of paper demonstrates Bernoulli's theorem.

HOW A PIGEON FLIES

An aerodynamic force upwards and backwards is alright for aircraft, since the aerofoil can be pushed forwards by a jet or propeller engine. The aerofoil is only used for buoyancy. Birds however must use their wing aerofoils not only to stay up in the air but to create a force pushing them forwards.

They have solved the problem in the simplest way. If the aerofoil in figure 21.1. is rotated clockwise by about 30°, the aerodynamic force now points upwards and forwards, rather than upwards and backwards. To achieve this in practice, the bird must drive its aerofoils downwards at quite a steep angle. Figure 21.3A shows a pigeon in three positions during a downstroke; the forward descent of the aerofoil is due to the downstroke combined with the

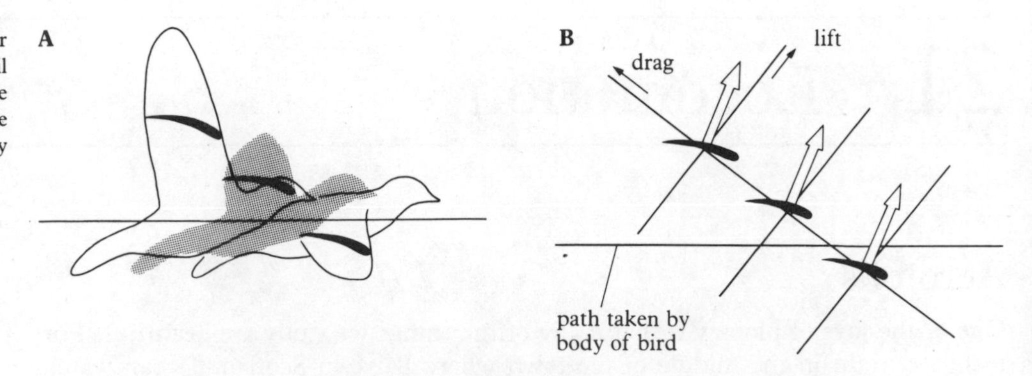

Figure 21.3 **A** A pigeon flying through the air during a downstroke; at each position an aerofoil section is shown half-way along the wing. **B** The same downstroke, showing how lift, drag and the resultant aerodynamic force are determined by the direction in which the aerofoil is moved.

bird's forward movement. The same sequence appears in figure 21.3B and it can be seen how the aerodynamic force causes forward movement as well as buoyancy.

It is worth adding one point to this simplified account. Figure 21.3 implies that the whole wing acts as a single aerofoil, and this is indeed true for the inner part of the wing. During slow flight, however, most of the buoyancy is created by the wing tips, because they sweep through the air much faster than the inner part of the wing. The wing tip itself may not be a particularly good aerofoil; instead each individual feather acts as one. So these feathers must be very strong. If they were weak, the powerful upward forces on the wing tips would merely bend them up, rather than lift the bird's body.

For consideration

1. Think of reasons, both aerodynamic and otherwise, why a bird's wing might be of more use than a bat's or pterodactyl's wing.

2. Penguins 'fly' under water, but not in the air. Supposing their ancestors could fly, suggest what evolutionary changes occurred and why. (NB: some birds can 'fly' both in the air and under water, for instance the auks.)

Further reading

J. Gray, *Animal Locomotion* (Weidenfeld & Nicholson, 1968)
Chapter 9 is on bird flight.

T. Weis-Fogh, 'Unusual mechanisms for generation of lift in flying animals' (*Scientific American*, vol. 233, no. 5, 1975)
Well illustrated; it discusses aerofoils and how small insects hover.

The case of the missing wheel

A wheel is round, it rotates, and it is part of a larger object. Simply from the number of wheels we use in our vehicles and other machines, we might guess that they are extremely useful. One of their advantages in locomotion relates to resistance: the small rolling resistance of the wheel is better than the larger sliding resistance of the leg joints. Continuous motion of the wheel also makes it more efficient, since the leg's back-and-forth action means that kinetic energy is alternately gained and lost.

Energy consumption in locomotion may be measured as joules used for every gram of body mass and kilometre travelled. A person uses 3.2 kJ kg^{-1} while walking one kilometre but uses only 0.6 kJ kg^{-1} when covering the same distance on a bicycle – an obviously much more efficient way of travelling.

So why do multicellular organisms not use wheels, either on land or in the form of propellers? It is true that pangolins roll down hills, dung beetles push balls of dung, and tumbleweeds are blown along by the wind in desert areas; but these are not wheels.

ARGUMENTS AGAINST WHEELS

The traditional argument has been that a wheel could not function in a multicellular organism. This is because blood vessels and nerves could not cross the rotating joint, between the body and the wheel, without breaking or being twisted. The argument might seem to be supported by the few examples we have of biological wheels. Bacterial flagella (see page 8) involve a true wheel; and some flagellate protozoans (eukaryotes) can indefinitely rotate the anterior part of the cell relative to the rest of it. In these unicellular organisms, there is no need for blood vessels or nerves since molecules can cross the rotating joint by diffusion.

In fact, the argument about blood vessels and nerves is not convincing. A biological wheel might not need them at all, especially if it is made of non-living material like chitin or keratin, with the 'motor' on the other side of the joint. More recently a new argument has been proposed. Wheels are all very well on roads, or the smooth hard gravel of some deserts. But have you tried cycling through soft sand? Or up a flight of steps? If so, did you find it easy to do?

The point is that wheels are efficient on hard surfaces, but pretty useless where the surface 'gives' – such as mud, sand, and damp earth. Again, wheels cannot easily surmount obstacles higher than about a third of the wheel's diameter. So an organism would not gain advantage from travelling on wheels unless it spent most of its time on hard smooth surfaces. How many natural habitats can you think of which offer these conditions?

By contrast legs are wonderfully versatile. They are not only adequate for hard smooth surfaces, but are far better than wheels on soft substrates and for stepping over tree trunks. With legs you can also jump across streams and climb trees. The conclusion therefore seems to be that multicellular organisms lack wheels for locomotion simply because they aren't much use in many situations.

Rather than develop walking machines to cope with bumpy or soft terrain, engineers have been more successful with caterpillar tracks. Because it exerts a low pressure on the ground, a tracked vehicle can cross soft peat or mud impassable to wheeled vehicles; and a large army tank is adept at getting over ditches or high obstacles. Some organisms also use tracks. Certain diatoms glide along surfaces, apparently by exuding a stream of mucus along an external groove in the cell wall and withdrawing the mucus at the other end of the groove. Amoeboid movement is an even better example.

For consideration

1. The argument in this topic has mainly concerned terrestrial organisms. Does it equally apply to air-borne and aquatic organisms?

2. What forms of animal locomotion have not been imitated in human transport? Suggest why they haven't.

Further reading

J. Diamond, 'The biology of the wheel' (*Nature*, vol. 302, no. 5909, 1983)
A fairly technical article.

S.J. Gould, *Hen's Teeth and Horse's Toes* (Norton, 1983)
Chapter 12, 'Kingdoms without wheels', is a typically engaging essay from Stephen Jay Gould.

S.S. Wilson, 'Bicycle technology' (*Scientific American*, vol. 228, no. 3, 1973)
If you are not yet an admirer of the bicycle as an invention, this will change your mind.

22 . Behaviour

Iron in the head: bird migration and orientation

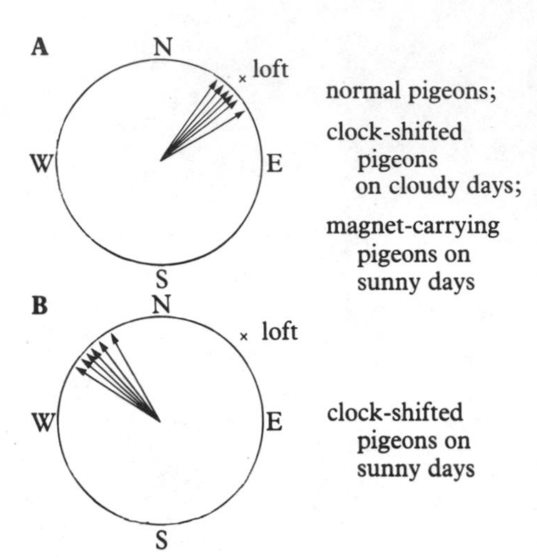

Figure 22.1 Each compass diagram shows the direction taken by homing pigeons when released hundreds of kilometres south-west of their home loft. The clocks of the 'clock-shifted' pigeons are six hours fast.

The mystery of bird migration is not so much why it happens – birds increase their chances of surviving the winter by migrating to a distant region which offers warmth and good feeding. Nor is the mystery in how these small birds fly such enormous distances. What is amazing is their ability to navigate their way to a relatively small target area thousands of kilometres away.

Imagine you are a young cuckoo, about to fly to Africa for the first time. All the adult cuckoos have already departed, and in any case you have probably never seen your parents. Would *you* know where to go and how to get there?

At the least you would need to have a map, and know your position and destination on it. Birds must have some substitute for this, and we call it **map sense**. Positions on a map make it possible to calculate the direction to go in, that is the **orientation**. But to head in the right direction requires a compass, or something that will do the same job. It has been known for a long time that birds can find north from the stars at night, and also that they use the sun as a compass when it is visible during the day. Unlike the stars, the sun cannot be used on its own; the time must be known as well. So birds need at least three things for navigation – map sense, a 'compass', and a 'clock'.

Some interesting experimental work has been done on homing pigeons. They were chosen because they have excellent powers of navigation and, unlike many migratory birds, are active in daytime and will perform at any time of the year. It is also possible to alter pigeons' internal clocks so that they run several hours fast or slow; this is achieved by keeping them in an artificial day–night cycle the required number of hours out of step with the real cycle. Suppose we have a homing pigeon whose clock in six hours fast; then if it is released in sunny weather, it sets off 90° left of the homeward direction (figure 22.1B). This fits the theory that it uses a sun-and-clock compass for orientation.

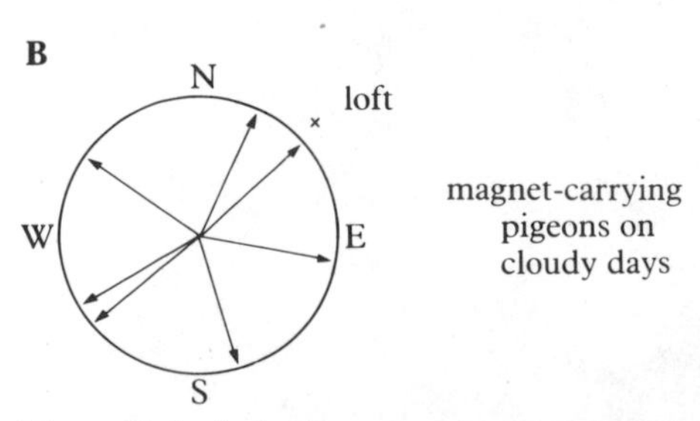

Figure 22.2 **A** A pigeon carrying a magnet. **B** The direction taken by magnet-carrying pigeons when released on a cloudy day.

MAGNETISM

It used to be thought that birds use only the sun, stars and an internal clock. However, when migrating birds were observed by means of radar, it became clear that they could navigate on cloudy days, with no view of the sun or stars. The orientation of human navigators, of course, depends on the response of a compass to the earth's magnetic field. It was naturally asked whether birds could also detect magnetism.

Let us return to the pigeon whose clock is 6 hours fast. In cloudy weather, unlike sunny weather, the pigeon sets off for home in the correct direction (figure 22.1A). Apparently a different kind of compass is being used – one not dependent on a clock. One possibility is that pigeons are sensitive to the earth's magnetic field. To see whether magnetism was involved, William Keeton of Cornell University released other pigeons each carrying a small bar magnet strapped to its back (figure 22.2A). As a control, other birds carried instead a piece of non-magnetic brass. On sunny days all the pigeons orientated correctly (figure 22.1A), while only the control ones did so when the sun was hidden by cloud. The result supported the idea that pigeons use a magnetic compass when they cannot see the sun; the magnets were upsetting the compass (figure 22.2B).

Other organisms, too, respond to magnetism; including bacteria, fish, amphibians, honey bees, and possibly humans. This response may serve a variety of functions. A north-seeking bacterium, for example, swims both northward and downward if it follows the earth's magnetic field in the northern hemisphere. Accordingly it has been suggested that the behaviour is merely a way of finding the mud at the bottom of the pond. On the other hand, in elasmobranch fish the responses to magnetism could be concerned with their ability to detect electric fields.

There are two quite different ways of detecting magnetism. One involves moving an electrical conductor through the magnetic field; the result is a current in the conductor. Some elasmobranch fish use this method. Pigeons on the other hand do not, and it is likely that they contain small bar magnets somewhere in the body, acting like compass needles. In fact particles of the magnetic substance **magnetite** (Fe_3O_4) have been found in pigeons' brains, as well as in bees and bacteria. More research will be needed to show whether the magnetite is actually used by the pigeons to detect magnetic fields.

For consideration

1. The experiment in which pigeons carried magnets tells us that their magnetic navigation is more likely to be based on magnets than on moving a conductor through a field. Why?

2. Suggest how you might investigate whether humans have a magnetic sense.

Further reading

R.P. Blakemore and R.B. Frankel, 'Magnetic Navigation in Bacteria' (*Scientific American*, vol. 245, no. 6, 1981)
The story of how a student made an exciting discovery.

J.L. Gould, *Ethology: the Mechanisms and Evolution of Behaviour* (Norton, 1982)
Chapters 13 and 14 of this undergraduate textbook are on navigation.

S.J. Gould, *The Panda's Thumb* (Penguin, 1983)
Essay 30 is on bacteria.

W.T. Keeton, 'The Mystery of Pigeon Homing' (*Scientific American*, vol. 231, no. 6, 1974)
A fascinating account of Keeton's experiments.

Where does behaviour come from?

Soon after a chick hatches from a hen's egg, it begins to peck. Over the course of only four days its aim becomes more accurate. You might wonder where this improved skill comes from. Perhaps the chick is learning by watching where its pecks go. But this is not the case. The pecking is innate, and its accuracy improves automatically as the chick matures. It will be better several days later even if the chick is prevented from practising.

So discovering how an organism acquires its behaviour is not always easy. Table 22.1 is an attempt to classify behaviour in terms of its origins. **Innate behaviour** is genetically inherited; the organism's environment and experience have little effect on its development. However all other kinds of behaviour are influenced by both genes and experience.

Table 22.1 One way of classifying behaviour. The five kinds grade into each other. Why is one place left blank?

	Innate	Programmed Learning	Plastic Learning
Individually acquired	Chick pecking	Bees' recognition of flowers	Learning changes in your local area
Socially or culturally acquired	—	White-crowned Sparrows' song	Tits and milk bottles

Somewhat like innate behaviour is **programmed learning**. For instance bees visiting a flower for the first time learn enough about it to recognise it again. But the way they learn its features is stereotyped: the flower's colour is learnt only in the few seconds before landing, the scent only while the bee is on the flower, and the surrounding landmarks only while leaving. Landmarks cannot be learned during arrival, nor colour on departure. This kind of learning is inflexible, and its rigid timing is genetically controlled.

On the other hand, **plastic learning** is highly flexible. What the organism learns, and how or when, are difficult to predict. During the lifetime of a mouse, its habitat changes as plants grow, holes appear in the ground, and spiders spin webs across paths. Yet the mouse notices these changes whenever they occur, and learns to jump over the holes, avoid the webs, and eat any tasty bit of new greenery. You presumably do much the same thing every day as your environment changes. An animal needs the correct genes to do this kind of learning, but the outcome depends largely on its experience.

These examples so far involve the organism learning on its own. But other members of the same species may sometimes be needed, in the case of **social** or **cultural learning**. White-crowned Sparrows learn their song only between the second and fourth weeks of life, by hearing adults. But they do not themselves start singing until 21 weeks old. If these American birds fail to hear the adults singing between the second and fourth weeks, they can never learn to sing. A rather different kind of cultural learning concerns tits in Britain; once a few had learnt to open milk bottles and drink the cream, millions of others started copying the habit.

EXPLAINING THE DIVERSITY OF BEHAVIOUR

Even if the ethologist knows where behaviour comes from, it does not explain why there are different kinds. One approach is to ask what sort gives the organism the best chance of surviving and leaving offspring. Consider the pecking: there is no advantage in the chick wasting time and energy in learning, when the behaviour can be coded in the genes. So it might as well be innate. Generally speaking, innate behaviour is most appropriate where the environment is predictable and the organism has insufficient time or opportunity to learn. This is especially true of many invertebrates, which have a short life cycle and never see their parents.

In contrast, many mammals live a long life in an unpredictable environment. They are also immature for an extended period. For them plastic learning is important, and they have the opportunity to acquire it. In some species, including humans, the young ones are not restricted to learning from their own experience; they can also learn from the experience of others – cultural learning.

The important idea to emerge is that each kind of behaviour has its advantages. It is not that one kind is 'superior' over another. On the contrary, each organism appears to behave in the way best suited to its circumstances. In fact most species use several kinds of behaviour types, mixed in various proportions.

For consideration

1. Try listing and classifying your own kinds of behaviour. How many belong to each of the five kinds in Table 22.1?

2. What kinds of organisms would you expect to benefit from plastic cultural learning, and in what circumstances?

Further reading

A.P. Brookfield, *Animal Behaviour* (Nelson, 1980)
All the main types of innate and learned behaviour are covered.

J.L. Gould, *Ethology: the Mechanisms and Evolution of Behaviour* (Norton, 1982)
Learning is covered in chapters 16, 17, and 18 of this undergraduate textbook.

23 . Cell Division

Cell cycle: a summary of modern knowledge

The **cell cycle** is the sequence of events, in dividing cells, from one cell division to the next.

Very little of the cell cycle can be discerned under the microscope, except for mitosis. Instead biochemical analysis is an important tool. To find out what chemicals are present at a particular stage of the cycle, you need thousands of cells all at the same stage together, in other words cycling **synchronously**. There are ways of making cells do this, but a valuable natural example of synchronous cycling is the myxomycete plasmodium (see page 16). A single plasmodium may contain 10^8 nuclei and 20 mg of protein – plenty for analysis.

FOUR STAGES

Modern research into the cell cycle began in the early 1950s, when the first methods for producing synchronising cells were introduced. Another important development concerned DNA; synthesis of new DNA was found to take place in a short period of the cycle well before mitosis. Accordingly the cycle was divided into four stages: M for mitosis, S for DNA synthesis, G_1 for the gap between M and S, and G_2 for the gap between S and M (figure 23.1).

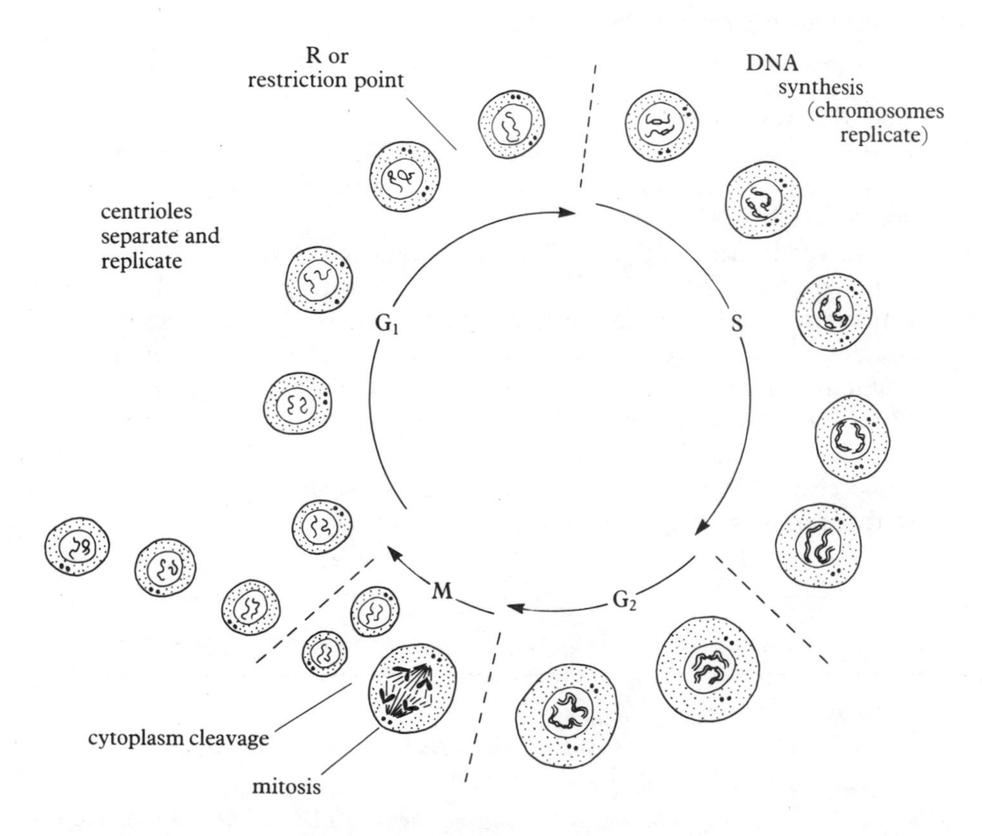

Figure 23.1 A typical animal cell cycle. For simplicity only two chromosomes are shown. The large dots just outside the nucleus are centrioles. Chromosomes are only visible during M. In multicellular organisms, the cells would not normally be spherical, and would remain attached after cell division. In plants, the centrioles would not be present. If a cell stops cycling, it normally does so at the restriction point.

The length of the cycle in human cells may be anything from eight hours to 100 days or more. A fairly typical timetable, such as that of a root tip or human cancer cell, would be about 6 hours in G_1, 7 hours in S, 4 hours in G_2, and 1–2 hours in M. G_1 is much more variable in length than any of the other stages.

There are many departures from this typical pattern. Myxomycete plasmodia miss out G_1 and cytoplasm cleavage. Other cells go through the cycle amazingly rapidly. For instance early embryos of the fruit fly *Drosophila* take only ten minutes per cycle. The reason may be that most of a normal cycle is needed to synthesize proteins and other materials, while the early embryo has enough material for several thousand cells already in the zygote. Cells undergoing meiosis also have a modified cycle.

IS THERE A STRICT SEQUENCE OF EVENTS?

On average, all the components in a typical cycling cell must double before the next cell division. The doubling of the various components might be independent events, or happening simultaneously, or starting and stopping at random. Another possibility is that the events can only take place in a definite sequence.

Some cellular components appear to increase continuously through the cycle: this applies to certain individual proteins, for example, and to the total protein content of the cell. RNA also builds up fairly steadily through G_1, S and G_2, although not during mitosis. However some proteins are synthesised only in particular stages; thus histones, which form part of the chromosomes, increase in quantity mainly in S. DNA synthesis, mitosis, and cytoplasm cleavage also have their own positions in the cycle, and the centrioles move apart during G_1.

So certain events in the cell cycle usually occur in a set order, one stage often finishing before the next begins. For instance centriole duplication proceeds once the nucleus has divided, which in turn can only start once the DNA has replicated, and DNA replication needs the prior duplication of the centrioles. The strange thing is that the most obvious event in the cycle – cytoplasm cleavage – is not essential; it can be missed out altogether.

WHEN DOES THE CYCLE START?

When a cell stops cycling, it nearly always stops in G_1. Moreover when a cell has passed a particular point in late G_1, called the **restriction point** or R, it normally continues through S, G_2 and M to arrive in G_1 again. It is almost as if in many cells S, G_2 and M form a tightly co-ordinated sequence which cannot be stopped once it has started. The reason G_1 is so variable in length is partly that cells spend different lengths of time resting at R; in addition, small cells may take more time in G_1 than larger ones so as to grow to a normal size before S.

One aim in cell cycle research is to reconstruct the sequence of events within the cell during the cycle. Another aim is to find out what external stimuli stop and start the cycling in a cell. This is particularly important in the fight against cancer, because it is the ability to escape the normal controls which *prevent* cycling that turns ordinary cells into cancerous ones.

For consideration

1. Why would you expect some events in the cell cycle to occur in a definite sequence, and others to be gradual?

2. Why do you think the restriction point is in G_1 and not in G_2?

Further reading

B. Alberts *et al*, *Molecular Biology of the Cell* (Garland, 1983)
Chapter 11 is on the cell cycle.

B. John and K.R. Lewis, *Somatic Cell Division* (Carolina Biology Reader no. 26, 1981)
A good general account.

D. Mazia, 'The Cell Cycle' (*Scientific American*, vol. 230, no. 1, 1974)
Well illustrated.

Fair shares: the mitotic spindle

Mitosis is a highly efficient way of sharing out chromosomes so that each daughter cell receives a complete set. But finding out how it works is another matter. Despite recent advances there are many unanswered questions.

One approach has been to use **immunofluorescence** (see page 5). When substances which bind to actin or to tubulin are applied to a dividing cell, in both cases the spindle gives out light under ultra-violet radiation. So it appears that the spindle contains microfilaments as well as microtubules.

Another method uses **colchicine**. This is a carcinogenic alkaloid extracted from the Autumn Crocus. It binds to tubulin molecules and so slows down their polymerisation and speeds up depolymerisation. Added to dividing cells, colchicine causes the spindle to disappear, suggesting that microtubules form a major part of it. This is why colchicine can induce autopolyploidy in a cell.

Colchicine also suggests what microtubules do *not* do. Despite its presence, the chromosomes may still condense and the nuclear membrane disintegrate in prophase, the centromere of each chromosome may split, and the reformation of the nuclear membrane and elongation of chromosomes may take place. These nuclear events operate quite independently of the spindle.

TWO SETS OF MICROTUBULES

Before mitosis begins, the microtubules are spread throughout the cell, forming part of its normal internal 'skeleton'. They mostly radiate from a region near its centre. Just before the cell divides, the microtubules depolymerise to form a 'pool' of tubulin monomers.

The role of **microtubule organising centres** was mentioned on page 6. A well known example is the basal body of a cilium, which may act as a template for the growth of the axoneme (see page 7). In mitosis, two kinds of organising centre are involved.

It might be thought obvious that the centrioles are one of these centres, since they have the same structure as the cilium basal body, and are found at the poles of the mitotic spindle of animal cells. However, they do not exist in plant cells, where the microtubules arise from material that appears in electron micrographs merely as a diffuse smudgy mess (figure 23.2). Even in animal cells, the microtubules do not actually touch the centrioles. It is thought that around the centriole there is a **pericentriolar material** which acts as the true organising centre.

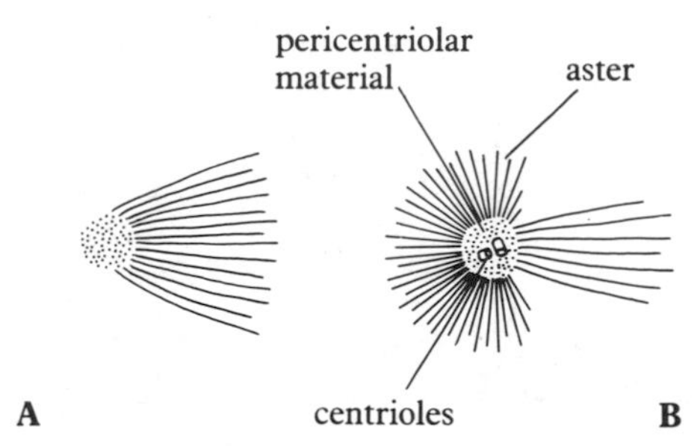

Figure 23.2 Typical half-spindles of (**A**) a plant cell and (**B**) an animal cell.

Microtubules called **polar fibres** grow from each organising centre to produce a half-spindle. In plant cells the pole is diffuse, while animals' cells have a more concentrated one. In the latter there may be an **aster** of microtubules radiating out in all directions, although it is sometimes poorly developed.

Two organising centres also appear on each chromosome, one on each side of the centromere. These are called **kinetochores**, and in early metaphase give rise to a bundle of microtubules sticking out at right angles to the chromosome (figure 23.3). The number of microtubules attached to each kinetochore varies from just one, in a yeast cell, to more than a hundred.

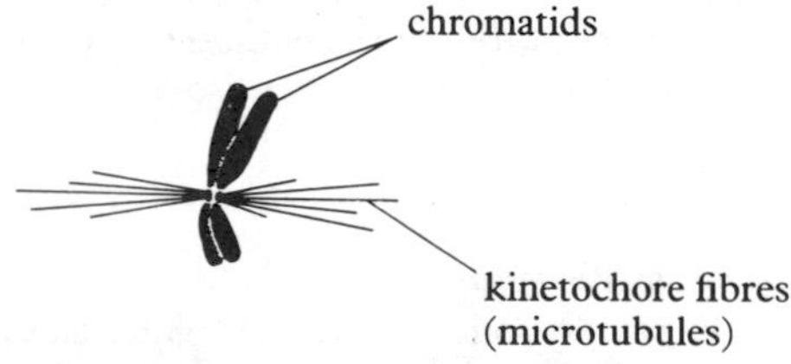

Figure 23.3 Kinetochore fibres emanating from the kinetochores on the centromere of a chromosome.

METAPHASE AND ANAPHASE

Once the nuclear membrane has broken up, the chromosomes and spindle fibres come into contact. At this time the chromosomes jerk around in an agitated manner until they are securely arranged on the metaphase plate. This characteristic movement is probably caused by the interference of polar and kinetochore fibres – only when they are more or less parallel with each other do they become stable.

Although the chromosomes are now secured, this does not mean nothing is happening. Each chromosome is being pulled strongly towards both poles at once, and it is the balance of the two forces which holds it in place.

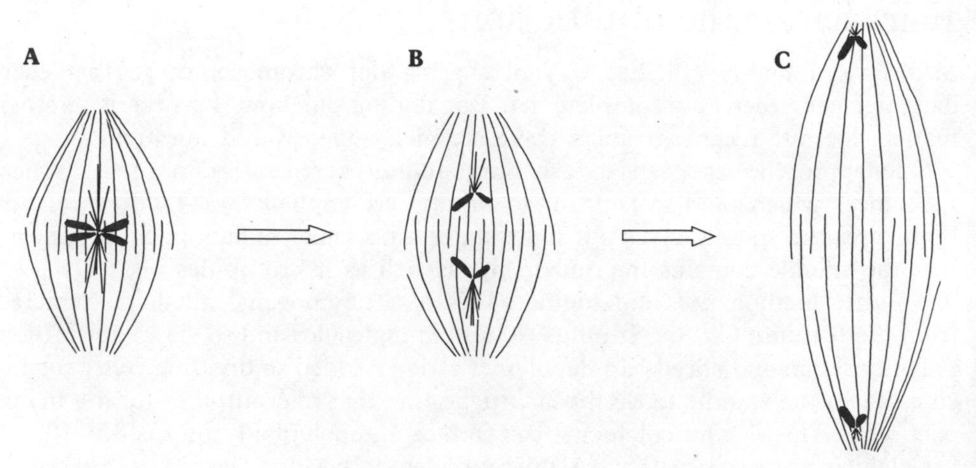

Figure 23.4 Separation of chromatids in mitosis. **A** Overlapping polar fibres interact with the kinetochore fibres to make the chromosome line up on the metaphase plate. Only one chromosome is shown. **B** The centromere then divides, and the chromatids separate. **C** The poles are pushed apart by the polar fibres.

Eventually the centromere of each chromosome divides. One explanation is that it is the only piece of DNA that did not replicate in the S stage of the cell cycle, and now its replication takes place. Immediately, each chromatid is pulled towards the pole its kinetochore faces at about 1 μm min^{-1} (figure 23.4). The mechanism is not known for sure, but as the kinetochore fibres approach the pole they shorten. Treatment with inhibitors of dynein (see page 7) does not affect the movement greatly, but colchicine speeds it up. This suggests that the force pulling the chromatids to the poles is caused partly by depolymerisation of the kinetochore fibres near the pole.

During anaphase the poles move further apart. Where the two half-spindles overlap (figure 23.4 B, C), the microtubules grow longer; at the same time the half-spindles slide past each other. Treatment with a dynein-inhibitor does have an effect here: it stops the movement. So the force may be generated by a dynein-like molecule causing the microtubules to slide along each other, pushing the poles apart.

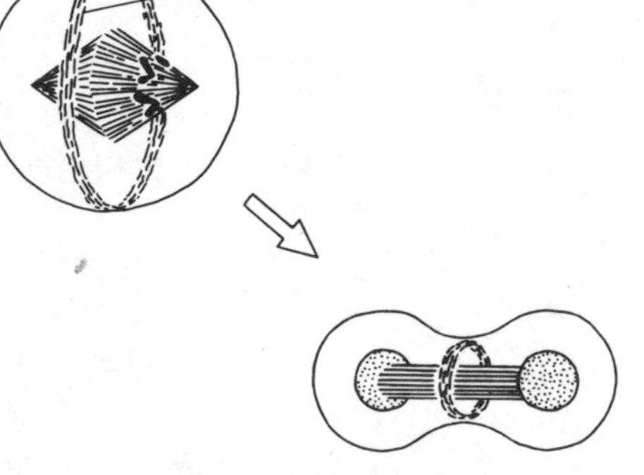

Figure 23.5 Cytoplasm cleavage. In animal cells a contractile ring of microfilaments forms in early anaphase, and tightens to separate the cells in telophase.

CYTOPLASM CLEAVAGE

The movements already described probably do not involve microfilaments and myosin because inhibitors of actin and myosin have no effect. Inhibitors do however stop cytoplasm cleavage. Immunofluorescence shows that a band of microfilaments encircles the cell, in the plane of the metaphase plate, just under and attached to the plasma membrane (figure 23.5). Animal cells divide by a tightening of this **contractile ring**, and the movement is probably generated by myosin acting on actin. In plant cells, cell division follows a different course, and this is described on page 116.

For consideration

1. Suggest reasons why the cell's internal skeleton of microtubules breaks down just before mitosis.
2. (a) Suggest a possible function for the centrioles in animal cells.
   (b) During the cell cycle the centrioles move apart until they are at opposite ends of the nucleus. How might this happen?

Further reading

B. Alberts *et al*, *Molecular Biology of the Cell* (Garland, 1983)
There is a useful section on mitosis.

B. John and K.R. Lewis, *Somatic Cell Division* (Carolina Biology Reader no. 26, 1982)
A good general account of mitosis.

24 . Reproduction

What is an individual?

We usually think of organisms as members of a particular kingdom – as plants, animals or fungi, for instance. There is another way in which we can classify organisms, however, and that is as **unitary** or **modular**. This distinction has important implications for our understanding of reproduction, natural selection and ecology, and requires us to revise our traditional concept of what constitutes an individual.

INDIVIDUALS AND MODULES

In this discussion we shall regard an individual as the product of a zygote. It has therefore resulted from sexual reproduction. In the case of a human or an elephant it is obvious what an individual looks like. Each time these **unitary organisms** reproduce sexually, the resulting individuals are distinct units, different in genotype from both their parents.

On the other hand many organisms, such as plants and colonial coelenterates, are built of a series of repeated units known as **modules**. A tree can be regarded as a population of modules, each consisting of a stem, bearing a leaf, with a bud in the axil. Many plants, such as Bracken and couch grass, are equivalent to trees on their sides, because they spread sideways by underground stems known as rhizomes, producing modules at intervals as they grow. If these modules become detached from one another, as when the rhizome between two modules dies, different parts of the same individual become separated. Since sexual reproduction has not occurred, no zygotes have formed, and no new genotypically distinct individuals have been produced. It is therefore possible, indeed frequent in some habitats, for large areas of the landscape to be dominated by a single genotypic 'individual', whose modules have separated. Think of duckweed spreading over the surface of a pond, for example, or reeds invading the edge of a lake.

THE BIOLOGICAL PROPERTIES OF MODULAR AND UNITARY ORGANISMS

Unitary organisms, such as humans, keep growing only until they reach a certain size. Individuals move about, sensing their surroundings. Usually, separate male and female individuals can find each other so as to reproduce. The fertility of an individual tends to reach a peak and then declines. When resources in the environment are in short supply, individual unitary organisms may die.

However, modular organisms, like many higher plants, could be said to appear potentially immortal. They can usually continue to grow throughout life by producing more and more modules. Modular organisms tend to be static and to lack sophisticated sensory mechanisms. In general, such organisms possess organs which produce both male and female gametes. The fertility of a modular organism may increase throughout its life without peaking, and the organism does not obviously 'grow old'. When resources are in short supply, module production may slow down, or the mortality of modules within the organism may increase, but the organism as a whole lives on. Besides plants, modular organisms include corals, bryozoans, and sea squirts.

DANDELIONS AND APHIDS

We can illustrate the properties of modular organisms by concentrating upon two rather extreme examples, dandelions and aphids.

Firstly, consider the dandelion, a rosette plant with a yellow inflorescence that you might find growing on a bare patch of lawn (A rosette plant has a group of leaves which radiate out from a very short stem.). Most dandelions develop without fertilization, so they produce seeds which contain embryos identical to the 'mother' plant. The modules which produce seeds and the modules resulting from germination of the seeds are therefore parts of the same plant. Most dandelion rosettes on a lawn are separate parts of a 'fragmented individual'. It is like a very large dandelion tree, with no investment in trunk, major branches or perennial roots, which might live for thousands of years.

Secondly, consider the aphid. Between spring and autumn, most aphid species reproduce by diploid parthenogenesis. In the circumstances, 'reproduce' may be the wrong word. Since new modules result from mitosis, as in dandelions, what appear to be separate aphids are really parts of the same genotypic 'individual'. This 'individual' spreads its parts out thinly over suitable food plants. This has the advantages that the modules can all feed without killing the host plants, potential predators are unlikely to find all the scattered modules and the modules are small enough to get from plant to plant without incurring massive haulage costs. Looked at this way, an aphid could be regarded as the world's biggest insect, a single 'super-organism'.

How can dandelions and aphids evolve? Occasionally there might be a mutation in the genotype which might alter the phenotype, and very occasionally, by chance, proper meiosis and fertilization may occur. Aphids can produce new zygotes each autumn and spring, when males and females reappear in the population.

This situation presents a conundrum to plant ecologists who are investigating populations of modular species. Whereas unitary organisms are easy to count, the different rosettes of say, a buttercup, scattered across a pasture might all be modules derived from the same colonising seed. It is almost impossible to tell how many different genotypes there are in such a population. Ecologists interested in how the population size changes with time would be justified in counting leaves or buds rather than rosettes.

For consideration

1. Are you happy with the definition of an individual as the product of a zygote?

2. Which of the following organisms are unitary and which are modular? Yeast, Mucor, Portuguese Man-of-War, Ivy, Hydra, earthworms, flatworms. Give reasons for your answers.

Further reading

M. Begon, J.L. Harper, C.R. Townsend, *Ecology – Individuals, Populations and Communities* (Blackwell Scientific, 1986)
The section covering pp 124–30 deals with unitary and modular organisms.

Mechanisms which promote cross-fertilization in flowering plants

Various structural and genetic features of flowers tend to increase the proportion of zygotes formed by cross-fertilization, that is between gametes from different genetic individuals. Correspondingly, these features also decrease the proportion of zygotes formed by self-fertilization, that is between gametes from the same flower or from different flowers on the same individual. Some of the features merely reduce the chances of self-fertilization, and others prevent it. In this topic we shall look at some of them in detail.

MECHANISMS WHICH REDUCE SELF-FERTILIZATION

In many hermaphrodite flowers, for instance plantains, grasses and members of the daisy family (Compositae), the anthers ripen and release their pollen a day or two before or after the stigmas in the same flower are receptive to effective pollination. This is known as **dichogamy**. There are two sorts of dichogamy, **protandry** and **protogyny**. In the daisy family and the *Arum* lilies the stigmas are exposed before the stamens dehisce (protogyny), but in plantains and grasses the stamens dehisce first (protandry). Of course in a dense inflorescence of flowers, like a daisy capitulum, some flowers are in the male stage whilst others are in the female stage, and pollen transfer between the flowers might result in self-fertilization.

A few species, widely scattered amongst different families, have separate male and female flowers borne on the same plant, often produced on different parts of the plant. Common examples include hazel, cucumber and maize. This, especially if combined with dichogamy, reduces the proportion of pollen from the same plant which lands on each stigma.

MECHANISMS WHICH PREVENT SELF-FERTILIZATION

The ultimate structural mechanism which prevents selfing in flowering plants is when, as in higher animals, there are separate sexes. This is **dioecy**. In some plants, it has a chromosomal basis, just as in animals. Dioecious species include nettles, willows, holly and bladder campions.

Dioecy, however, occurs in less than one in two hundred of all flowering plant species. Most angiosperms are **monoecious** with hermaphrodite flowers. Nevertheless, self-fertilization is prevented in about seventy per cent of all angiosperms, by a genetic mechanism known as **incompatibility**. This depends on the fact that a female gamete in the ovule is separated from the pollen grains which land on the stigma by maternal diploid tissue, the stigma, style and ovary wall. The maternal tissue "decides" which pollen grains can grow down the style and fertilize the female gamete. It can discriminate against grains from the same genetic individual and hence prevent self-fertilization.

The simplest type of incompatibility occurs in some economically important plants such as tobacco, grasses, apples, pears, plums, cherries and clover. Each apple tree, for example, is heterozygous for two alleles, known as **s-alleles**, out of the hundred or so which exist. If a particular tree was genotypically s_2s_{56}, its haploid pollen grains would either have the genotype s_2 or the genotype s_{56}. If either of these grains met the s_2s_{56} stigma, its pollen tube would grow very slowly indeed or stop growing down the style altogether. In the race to the female gamete, these tubes are usually passed by grains of different genotype, which must have come from other apple plants. As a result, apple seeds are always heterozygous for the s-alleles, and therefore probably heterozygous at many other loci as well.

This incompatibility system presents problems for apple growers. If apple pips are germinated from, say, a Cox's orange pippin tree, they will not grow into plants of the same variety. The new plants have half the Cox's genes and half the genes from another apple strain. This means that tasty apple varieties have to be propagated vegetatively, by taking cuttings from existing trees. In fact all the Cox's trees in Britain are really parts of the same genetic individual (see page 105) and originated from a single plant which arose in Mr Cox's garden. In that case, all the Cox's trees in an orchard have the same genotype, and successful pollination of one tree by another is impossible. In the past this problem was overcome by growing an occasional row of a different apple variety amongst the plants of the more important crop strain. Bees would transfer pollen from the flowers of one strain to the flowers of the other. Nowadays, however, self-compatible apple strains have been developed and are used commercially on an increasing scale.

Figure 24.1 Features of heterostyly in the Primrose, *Primula vulgaris*.

Figure 24.2 Pollinations in the fox tail grass *Alopecurus pratensis*: **A** compatible pollinations; **B** incompatible pollination. Both slides were stained with aniline blue and photographed under the fluorescence microscope, in which case any callose deposited will fluoresce. In the compatible pollination the tube (T) has emerged from the grain (PG) and has penetrated the cuticle on the stigma before growing on towards the ovary. In the incompatible pollination the tube (T) has stopped at the cuticle of the stigma and callose has accumulated both in the pollen tube and in the pollen grain. (×500)

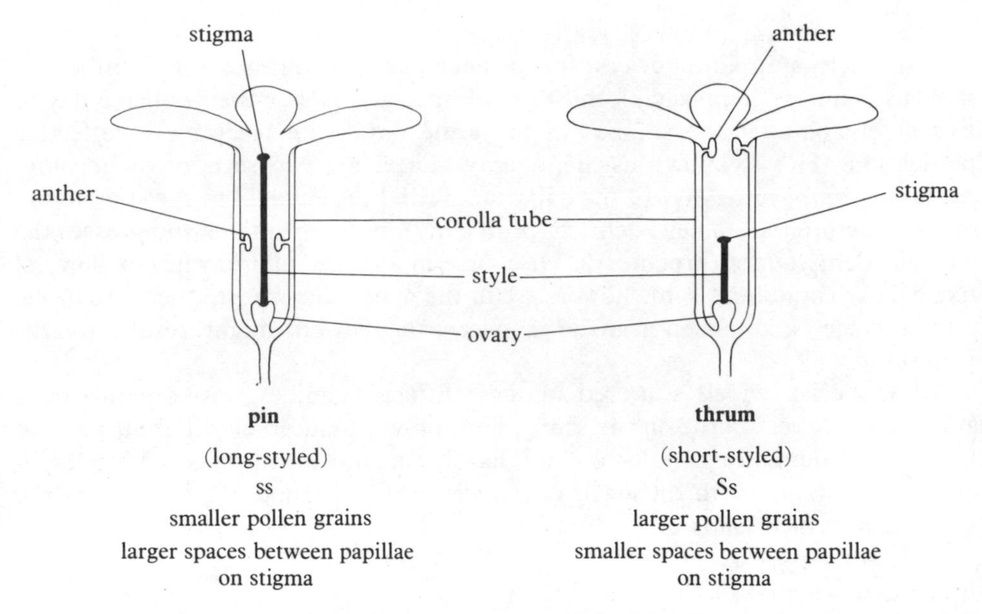

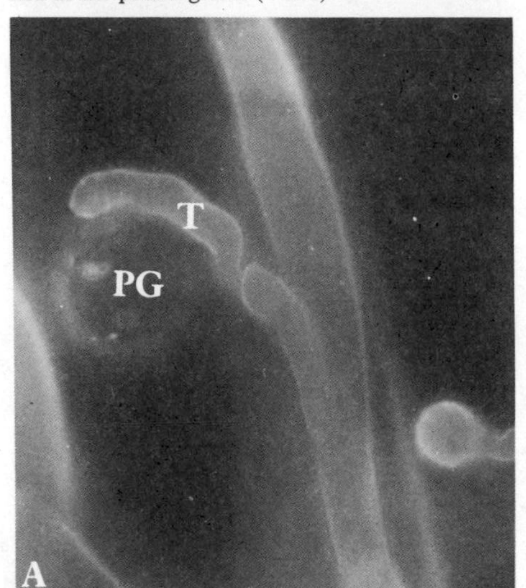

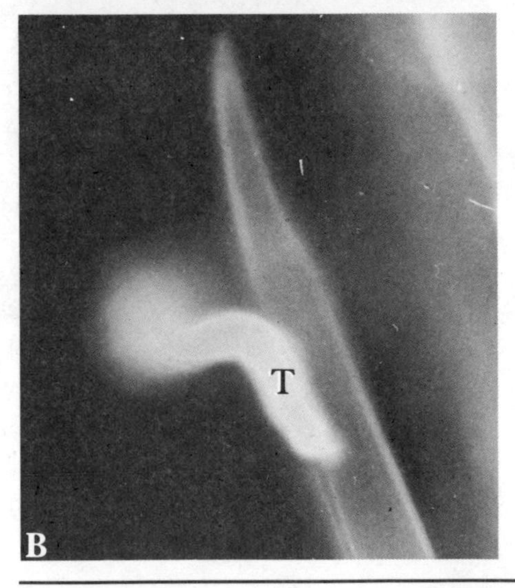

One other mechanism which prevents self-fertilization is **heterostyly**, but only because it is usually combined with some form of incompatibility. In the Primrose, for example, half the plants produce **pin flowers**, with the stigma at the mouth of the corolla tube and the stamens half way down; and the other half are **thrums**, with the stigmas half way down the pollen tube and the stamens above them (figure 24.1). Charles Darwin suggested that this might promote cross-pollination by a long-tongued insect. For example, pollen grains from a pin flower might become attached halfway down the insect's tongue when it probed for nectar, and transferred to the thrum stigma at the same level on a subsequent visit. However, despite careful searching, a long-tongued insect which visits Primroses and has the right credentials has never been found.

Thrums are heterozygous for the gene which controls this particular dimorphism, Ss, and pins are homozygous recessive, ss. Thrum pollen cannot grow down a thrum style and pin pollen grows very slowly indeed down a pin style. Half the thrum pollen has the genotype s and all the pin pollen has the genotype s. Yet the styles and stigmas can distinguish between them. This mechanism depends on compounds which were deposited in the outer surfaces of the pollen grains during pollen maturation. The pin pollen carries a different range of compounds from the thrum pollen. When pollen grains land on a stigma of the same form, a reaction rather like an antibody-antigen reaction occurs and the carbohydrate callose (a polysaccharide) is deposited in the pollen tube, preventing its growth (figure 24.2).

Incompatibility reactions like these have probably made a marked contribution to the success of angiosperms, because cross-fertilization promotes genetic variety between individuals, which in turn favours rapid evolution.

For consideration

1. If all apple fruits are formed as a result of cross-pollination, why do all the apples on the same tree look and taste similar?

2. Suggest the advantages of self-pollination over cross-pollination in some situations.

3. In gymnosperms, such as pine trees, the pollen tubes can reach the egg cells without encountering maternal tissue. In angiosperms the pollen tubes have to grow through the maternal tissue of the style before they reach the egg cells. What advantage does this give the angiosperm?

Further reading

J. Heslop-Harrison, *Cellular Recognition Systems in Plants* (Studies in Biology no. 100, Arnold, 1978)

R.B. Knox, *Pollen and Allergy* (Studies in Biology no. 107, Arnold, 1979)

D. Lewis, *Sexual Incompatibility in Plants* (Studies in Biology no. 110, Arnold, 1979)

25 . The Life Cycle

How many offspring?

Which can eat a dead horse the faster: a lion, or a couple of flies? In the opinion of the eighteenth century Swedish naturalist Linnaeus, it was the flies.

This may have been just a frivolity at the eighteenth century equivalent of a cocktail party. But more recently biologists have given serious thought to this question. What Linnaeus meant, of course, was that the flies can reproduce very quickly. In a few generations the flies could give rise to multitudes of maggots, which might indeed consume the horse faster than the lion could. In this topic we shall relate these sorts of considerations to the life cycles of organisms.

OPPORTUNIST SPECIES

The two flies are a typical example of an **opportunist species**. Locating a suitable place to lay their eggs – in this case a carcass – is a chancy business and depends on an opportunity arising. Many females may die without ever finding a suitable place. On the other hand, when a carcass *is* found and the eggs *are* laid the supply of food and space is unlimited, at least for the time being. The way flies cope with their uncertain environment is to multiply rapidly at any opportunity. This increases the chance that at least a few adults will find another carcass before they all die.

Opportunist species tend to share a number of characteristics. They are usually small and develop rapidly. A large number of eggs are laid, and often they all appear at once ('big bang' reproduction). Much energy is devoted to producing offspring, and comparatively little to maintaining the adult's body for prolonged survival.

Suppose a couple of flies lay 200 eggs, and these offspring become mature and lay eggs after ten days. If this rate is kept up, the population will be 100 times larger every ten days. You might care to work out how large it would be after six months, and compare it with what a couple of lions could achieve.

Opportunist species are sometimes said to be ***r*-selected**; they tend to reproduce quickly and die suddenly – the 'boom and bust' strategy.

Suppose N is population size and t is time. Then the speed of population growth is proportional, for low values of N, to the size the population has already reached. Put mathematically,

$$\frac{dN}{dt} = rN$$

The constant of proportionality, r, describes how fast the population is increasing. In an opportunist species, it is supposed that individuals with genes for high values of r contribute more individuals to future generations, so that eventually the whole species has the capacity to reproduce quickly. This is why these species are called r-selected.

EQUILIBRIUM SPECIES

The flies never come to an equilibrium with their immediate environment, the dead horse. This is because the horse is eventually consumed and the fly population moves elsewhere. The lions however are in a different situation. In some areas their prey may be available at all times, and the weather is rarely so

A

B

Figure 25.1 **A** The House Fly (top) lays large numbers of small eggs but the Tsetse Fly *Glossina* (below) lays a full grown larva every nine days or so. **B** Thus the Tsetse Fly is an unusual organism, a *K*-selected insect. The larva pupates almost immediately and an adult emerges a month later.

bad that the lion population is reduced drastically. In these circumstances, we expect the lion population to increase to a size which is in balance with the environment. In such an **equilibrium species** competition for space or food prevents the population increasing any further.

This population size represents the **carrying capacity** of the environment. In mathematical equations it is given the letter K. If the lions had a multiplication rate anything like that of the flies, it would clearly be wasteful. The environment cannot 'carry' any more lions, so on average only one young lion survives for each one that dies. Instead of rapid multiplication, what is important for each lion is to keep itself alive and to survive the competition from other lions.

Individuals surviving best in these conditions have characteristics quite different from those of the flies. Body size tends to be larger, development is much slower, and the individual lives much longer. Only a few young are produced in a mother's lifetime, and these may be spread out over an extended period. In such animals there is often parental care. So each individual is given the best chance of surviving and leaving one or two replacements in the next generation. Accordingly these equilibrium species are said to be ***K*-selected**, that is, adapted to the competition which exists when the population is at its maximum size and the full carrying capacity of the environment has been reached. (*K*-selection should not be confused with kin selection which is a different phenomenon.)

ARE ALL SPECIES *r*- OR *K*-SELECTED?

Some species are clearly opportunist. Flies and bacteria are adapted to an uncertain food supply. Annual weeds can similarly cope with the uncertain availability of disturbed ground. In contrast with these unpredictably fluctuating conditions, most birds and mammals enjoy a relatively stable or predictable environment, and hence are equilibrium species. Comparable examples among plants are forest trees, especially those of tropical rain forests.

Many species however are neither opportunist nor equilibrium species. So it is best to regard the two types as extremes of a continuum.

In this topic we have considered the quantitative aspects of life cycles. The next one deals with the reasons why some life cycles are so complex.

For consideration

The lion and the fly are different in size and not closely related taxonomically. However, the House Fly and Tsetse Fly (figure 25.1) are similar in size and closely related but different in terms of *r* and *K* selection.

(a) How would you explain the difference?

(b) Can you think of other closely related species which differ in this respect?

Further reading

S.J. Gould, *Ever since Darwin* (Penguin, 1980)

Chapters 10 and 11 describe some extraordinary examples of *r*-selection in insects and bamboo.

Life cycles

No one has yet found an immortal organism. A 4000 year old Bristle Cone Pine in the mountains of California may seem everlasting compared with a bacterium that lives only a few minutes, but even the pine eventually dies.

Because all organisms die others must come into existence, if species are not to become extinct. As generations succeed one another, there is a repeating pattern of events which we call the **life cycle**. As we shall see, it sometimes takes several individuals to complete one cycle.

COMPLICATING THE LIFE CYCLE

By comparison with many life cycles, that of humans is really rather drab. It represents the simplest possible pattern (figure 25.2A): individuals start as

propagules, then grow larger until as adults they produce one or more new propagules. A propagule is a single cell, group of cells, or larger part of an organism which can grow into a new individual. Clearly the human zygote is a propagule – but of course not all organisms start from a zygote.

By contrast there are those holometabolous insects which have a pupa – fleas, flies, bees, and so on. The young stages, larvae, are always different from the adults. In Figure 25.2B, this is shown by dividing the lifespan into two parts. The transformation from larva to adult is called **metamorphosis**. In these insects it takes place in the pupa, although in frogs the metamorphosing individual – the tadpole – remains active. Metamorphosis is one important way of complicating a life cycle.

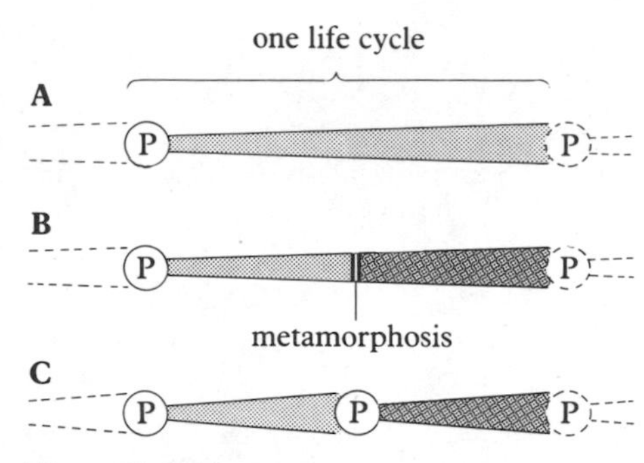

Figure 25.2 Two main ways in which life cycles may be complicated. **A** The simplest pattern, with the individual developing gradually to maturity from a propagule, marked P. **B** Metamorphosis divides the life cycle into two different body forms. **C** Alternation of generations also enables the life cycle to contain two body forms.

ALTERNATION OF GENERATIONS

The next complication to consider involves going one step further than metamorphosis. In a tadpole, most of the larval body becomes the frog, while the gills and tail are absorbed. Starfish also have a swimming larva (figure 25.3), but initially only a small part of it turns into the starfish. Although in fact the rest of the larva is eventually absorbed by the starfish, it would have been much the same in the end if the starfish had been cast off by the larva, rather as a *Hydra* bud separates from the parent *Hydra*. In this case the larva would have produced a propagule, which would have become a new individual – the starfish.

In many organisms this is in fact what happens. A typical coelenterate life cycle involves two individuals – **polyp** and **medusa** (figure 25.4A). When full grown, the polyp casts off propagules in the form of small jellyfish or medusae. When sexually mature these medusae produce more propagules, zygotes this time, which grow into polyps.

This pattern is comparable to **alternation of generations** which one sees in ferns (figure 25.4B) and with certain modifications in mosses and liverworts. There can be yet further metamorphoses and generations in such life cycles, making them more complicated still. In addition, meiosis and fertilisation occur at various points, and asexual reproduction may do so too. In one case, namely cellular slime moulds, there is even a sort of negative reproduction in which individuals of one generation join together to make the next (see page 15). Perhaps fertilisation can also be regarded as negative reproduction. A whole book would be needed to describe the fascinating variety of life cycles.

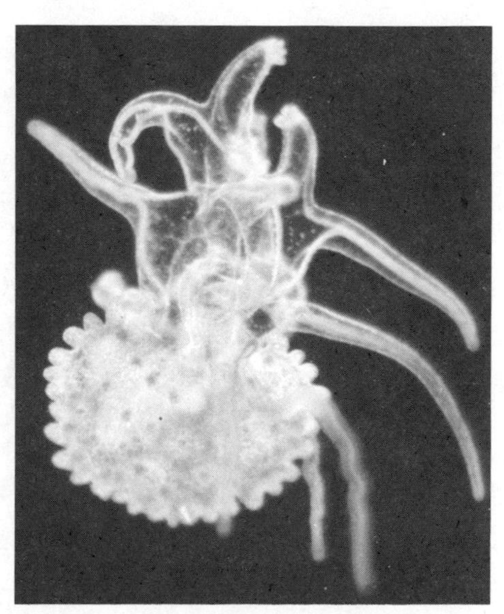

Figure 25.3 The planktonic larva of a starfish. The young starfish grows initially from just one part of the larva, and later absorbs the rest of it.

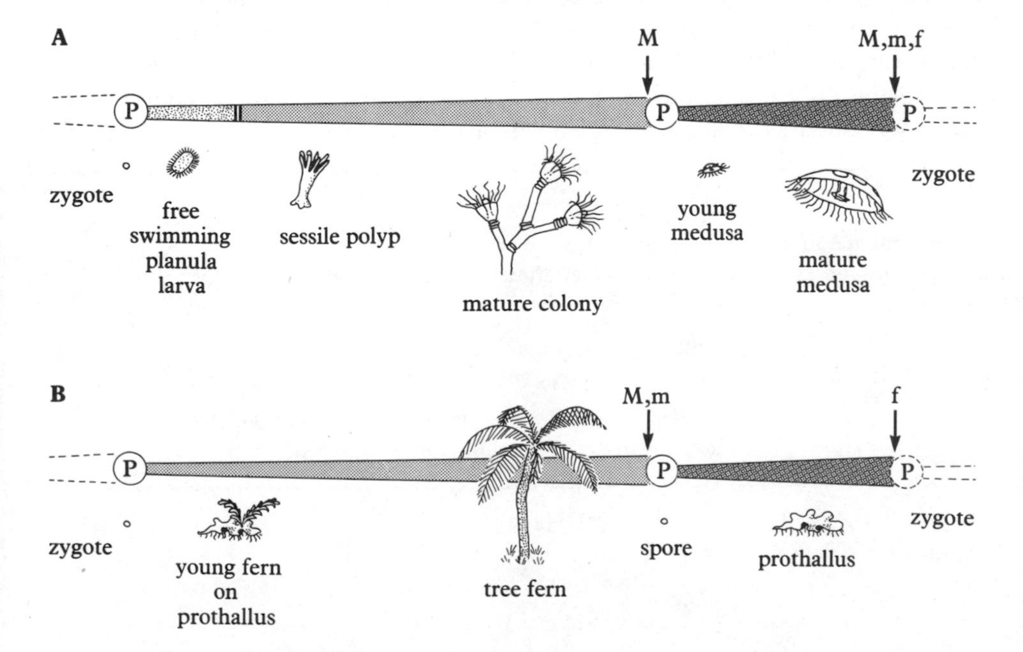

Figure 25.4 Alternation of generations. The various stages are not drawn to the same scale. P, propagule; M, multiplication; m, meiosis; f, fertilisation. **A** In the common coelenterate *Obelia*, the reproductive polyp produces tiny medusae by budding. **B** In the tree fern, as in all ferns, one generation is diploid and the other haploid. This kind of alternation of generations is called **haplo-diplonty**. Multiplication occurs mainly at spore formation, since each prothallus generally produces only a single sporophyte (the main fern plant).

WHY ARE SOME LIFE CYCLES COMPLICATED?

We have concentrated on the number of different kinds of organism found within a cycle. Why is it that humans and grasshoppers have only one, dragonflies and frogs have two, and the liver fluke *Fasciola hepatica* has about five?

Consider the tree fern (figure 25.4B). Being the size it is, it competes for light with other tall plants. Spreading its leaves at a considerable height is essential if it is to receive enough sunlight for photosynthesis in a forest. On the other hand, as in other ferns, its sperm swim by means of flagella, and need water to reach the female's archegonium. So photosynthesis and fusion of gametes are two functions that the plant must perform, and it makes sense to have a separate plant form specially adapted for each function. This is what we find: a tall tree in one generation, and a semi-aquatic prothallus in the other.

Here we have the key to understanding life cycles. What functions must the organism perform during the life cycle, and how can it best perform them? Four important functions are photosynthesis or feeding, dispersal, multiplication, and genetic variation (shuffling genes into new arrangements by means of meiosis and fertilisation). In addition some species must survive a period of unfavourable weather. Is the organism best able to compete with other organisms by having one general-purpose body form which performs all these functions? Or is it more efficient to use different body forms, each specially adapted for just one or two functions?

The tree fern and *Obelia* use the second approach. Most of the photosynthesis or feeding is performed by the fern and polyp stages. The dispersal stages are the fern spore, and *Obelia*'s planula and medusa. Multiplication occurs at spore formation in the fern, and at two stages in *Obelia*'s life cycle. Genetic variation depends on meiosis and fertilisation; these are well separated in time in the fern life cycle, while in *Obelia* they occur in quick succession.

If one body form changed gradually into the next in a complex life cycle, the intermediate stages might be inefficient at performing one (or more) of the important functions. They might even have difficulty staying alive, and avoiding being killed by predators or competing organisms. For instance an immature butterfly with half-formed wings and mouthparts adapted neither for eating leaves nor for sucking nectar might not survive long. So in complex life cycles, the body forms tend to change rapidly from one to another by means of metamorphosis or the production of a new generation.

Complex life cycles serve well for not only the fern and *Obelia* but also a multitude of other organisms. A complex life cycle is like having a combine harvester for collecting food, a car for dispersal, and a word processor for genetic variation. By contrast a simple life cycle is like having a single machine to do all these jobs. The puzzle is not why some life cycles are so complex, but why others such as mammals' and birds' are so simple.

For consideration

1. Why is it useful to think of the life cycle as the fundamental unit in biology? Suggest one or two other aspects of living things, other than the life cycle, that might be called 'the fundamental unit'. How would you argue in favour of each?

2. Think of an organism which has different body forms in the life cycle, but which changes from one to the other *gradually*. How does it manage it?

Further reading

J.T. Bonner, *Size and Cycle* (Princeton University Press, 1965)
Professor Bonner argues that the life cycle, rather than the individual, is the fundamental unit in biology.

R. Buchsbaum, *Animals without Backbones* (Penguin, 1951)
Beautifully written, with clear accounts of invertebrate life cycles.

T.J. King, *Green Plants and Their Allies* (Nelson, 1983)
Many life cycles are included in King's book.

26 . Patterns of growth and development

Larvae

As described in the last chapter, larvae are juvenile forms following the egg stage in complex life cycles. Larval and adult stages are separated by metamorphosis. We normally think that only animals have larvae, but some plants have juvenile stages which are essentially the same as larvae. For instance the **protonema** of a moss might be considered larval, especially where only a single moss plant arises from the protonema. Better examples are some flowering plants with modified leaves. Here the seedling may have the 'normal' leaves characteristic of closely related species; then it suddenly changes, with all new leaves being quite a different shape (figure 26.1).

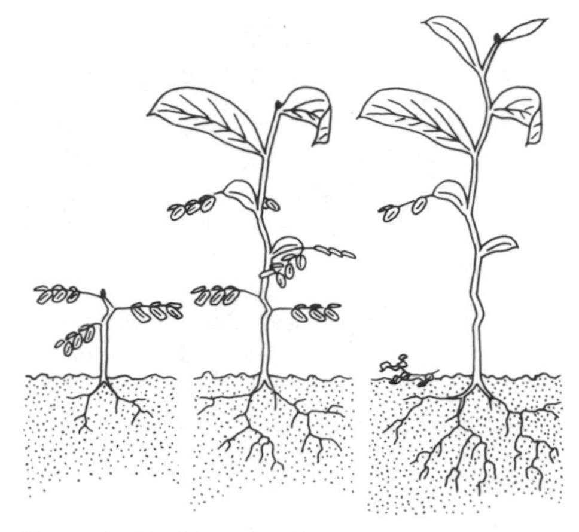

Figure 26.1 Seedling of an acacia. The first few leaves after the cotyledons (not shown here) are pinnate, like many other members of the pea family. Then there is a rapid transition to the adult 'leaf', which is actually the sideways-flattened leaf stalk.

LARVAL FUNCTIONS

The larva and adult of an organism follow different ways of life. But which does each follow?

Four important functions to consider are feeding, dispersal, multiplication, and genetic variation. In animals, genetic variation always coincides with multiplication, so that they can be considered a single event for present purposes. Moreover any stage of a life cycle producing propagules is by definition an adult stage; so multiplication must occur at the end of a larva–adult sequence. This leaves us with feeding and dispersal. Whereabouts in the life cycle is it best to perform these functions?

Perhaps the most familiar larva–adult cycle is that of butterflies and moths (figure 26.2A). The caterpillars are the feeding stage; the winged adults disperse. This is not to say that the butterflies and moths do not feed. In most cases they do, but the nectar they use is probably only sufficient to sustain their activity and produce gametes. Some adult insects, such as mayflies, do not feed at all.

Now let us consider marine life. If a plankton net is towed through the surface waters in the ocean, a considerable proportion of the plankton collected will be crustaceans. Some will be the adult forms of small species, but many others are the larvae of barnacles, crabs, and other groups whose adults live on the bottom. The crab's **zoea larva** is a typical example. As you can see in figure 26.2B (overleaf) it is tiny, and has two long spines. Like the spines on some planktonic diatoms and protists, their main function is probably to act as parachutes rather than providing defence. By increasing the zoea's surface area, they create a large resistance to movement; so, although denser than seawater, the zoea sinks only slowly.

The zoea, then, is adapted to a planktonic life. It drifts with the ocean currents and may cover large distances. When still small it passes through another larval stage and then undergoes metamorphosis into a tiny crab. Most of the growth in the life cycle occurs in the crab stage. Thus, compared with the moth, the crab does things in reverse order – dispersal stage first, feeding stage second.

Why do moth and crab larvae have such different functions? There is probably no simple answer to this. In some ways it does not matter in what order feeding, dispersal, and multiplication occur. The important thing is that each occurs at least once in each cycle. We might put the problem as follows. Dispersal is necessary since old food sources become exhausted and new ones

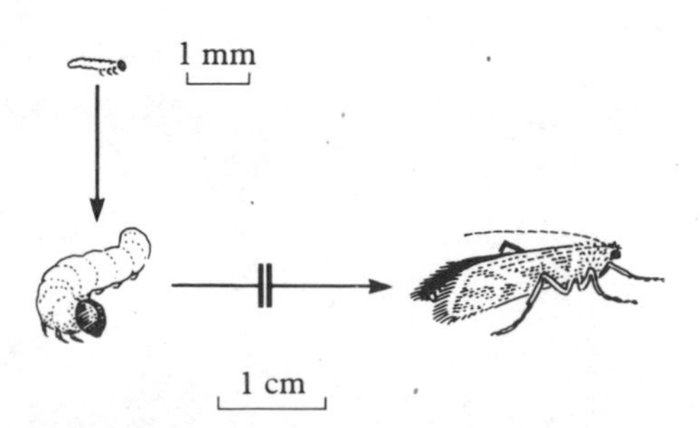

Figure 26.2A Larval and adult stages of the moth. The broken arrow indicates where one or more metamorphoses occur. Most feeding and growth occurs in the caterpillar, while the moth is the dispersal stage.

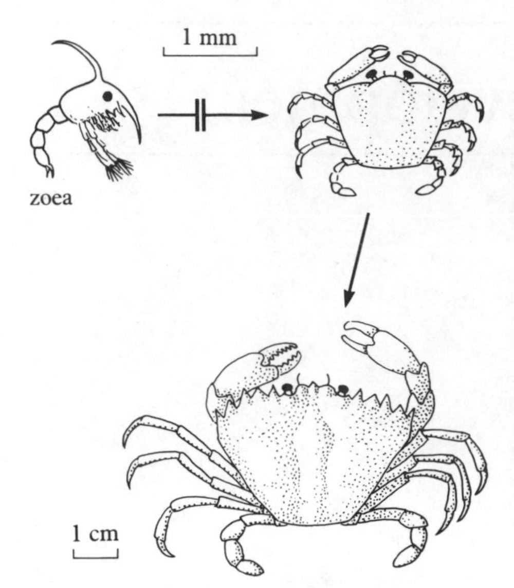

Figure 26.2B Larval and adult stages of the crab. The broken arrow indicates where one or more metamorphoses occur (the crab has several larval stages). Most feeding and growth occurs in the crab while the zoea larva is the dispersal stage.

must be found. Why, then, is it the adult moth which disperses, while in the crab's case it is the larva?

Crabs have a ready-to-hand transport system, in the form of ocean currents rich in planktonic food. The larva is the appropriate stage for travelling in the plankton because its small size gives it a high surface–mass ratio. Hence it does not sink too fast. However moths are in a different situation. The caterpillars feed on only one or a few particular plant species, often located in small patches widely separated from each other. By relying merely on drifting in the wind, the chances of finding a new food patch might be negligible. It would be like crossing the Sahara in a hot air balloon, hoping that you will be blown directly over an oasis. The moth can locate new food sources far more reliably by flying and searching. For this the dispersal stage must be powerful enough to cover considerable distances and control its movements despite winds. So it is the large adult which is best suited for dispersal.

But as we said there may be no simple answer to the question. This is just one plausible explanation.

POSTSCRIPT

Here is a thought to end with. Have you ever wanted to call your younger brother or sister a rude name? Some names of course are banned in polite society, but how about this – 'Pass the marmalade, you loathesome little larva'. If reprimanded you could point out with an air of authority that in our own development several rapid changes take place simultaneously – something like a metamorphosis. The male's voice breaks, and in both sexes body shape and behaviour change, hair comes forth in sundry places, and so on. The point is that the difference between a larva and any other juvenile form is only a matter of degree.

For consideration

1. What other explanation might there be for the adult moth being the dispersal form?

2. Do you think it is valid to regard the young seedling of an *Acacia* plant, described at the beginning of this topic, as a larva?

3. Think of several complex life cycles. Try to decide in each case what functions the adult and larval stages play. Amphibians, parasites, and marine invertebrates will provide examples.

Further reading

Sir Alistair Hardy, *The Open Sea, its Natural History: Part I, The World of Plankton* (Collins, 1956)
Beautifully illustrated; chapters 6, 9 and 10 deal with planktonic larvae.

Scaling

As humans we are obsessed with braininess. Perhaps this is because we regard ourselves as brainier than any other animal. But is it true, and in any case what does it mean? To find out we must resort to some mathematics.

Let us start with frogs. Once a tadpole has metamorphosed, the tiny frog looks remarkably like an adult. Suppose it grows to adult size with no further change in shape: then as the body length doubles so also will the femur length and all other linear measurements. Plotted on a graph, the femur (y) and body length (x) measurements will form a straight line through the origin. In other words they are proportional. Put in mathematical language, $y \propto x^1$. Normally we do not bother to write the exponent '1' in the equation, but you will see the point of doing so in a moment.

By contrast, surface area (y) varies approximately as the square of the length (x): $y \propto x^2$. And the frog's volume or mass (y) varies approximately as the length (x) cubed: $y \propto x^3$. In these two cases, plotting y against x will give curves (figure 26.3A). Now curves are more difficult to analyse than straight lines, so it is fortunate that these particular curves can be straightened. Converting each side of the equation to logarithms:

$$y \propto x^2 \text{ becomes } \log y = \text{constant} + 2 \log x$$
$$\text{and } y \propto x^3 \text{ becomes } \log y = \text{constant} + 3 \log x$$

Plotting these equations using logarithms gives straight lines (figure 26.3B). The line's gradient, which is 2 or 3 in these examples, tells you the size of the exponent.

SURPRISING EXPONENTS

Where the exponent is anything other than 1, these graphs are called scaling relations or **allometry**. You are familiar with the surface area, volume, and mass examples, which involve exponents of 2 and 3. What is surprising, though, is that many scaling relations have exponents which are not whole numbers.

For example, comparing various endothermic species, metabolic rate (y) and body mass (x) conform to $y \propto x^{0.75}$. The same is true in comparing ectothermic species with each other, and also if one analyses unicellular organisms. No convincing explanation has been found for this unexpected finding. Another scaling relation is seen in such things as the durations of life cycles, breathing cycles, wing beat, and heart beat: these tend to vary with body mass (x) as $y \propto x^{0.25}$.

Many if not most organisms change shape rather obviously as they grow. You could not for instance mistake the silhouette of a young child for that of an adult, even though you had no clue as to the person's size. Change of shape occurs when two measurements are not proportional. In some cases they fit a scaling relation. One study of tree species in the United States showed that the average diameter of the trunk at the base (y) varied with the species' height (x) as $y \propto x^{1.5}$. In other words, in larger trees the trunk diameter is greater than you would expect if the measurements were proportional. This is called **positive allometry** because the exponent is larger than 1.

BRAININESS

Brain size provides a case of **negative allometry**. Unlike the tree trunk, brain mass in larger organisms is smaller than you would expect if brain size (y) and body mass (x) were proportional. This is true, not only comparing one species with another, but also during the development of an individual mammal. In individual development the exponent is about 0.3: $y \propto x^{0.3}$. One reason why the child's silhouette cannot be mistaken for the adult is that the child's head looks so large.

Since the number of neurones is fixed at an early stage, in the embryo, the subsequent growth of the brain is largely due to the individual cells increasing their size. They do not maintain a constant proportion of the body mass; in fact they keep pace much more nearly with the cube root of the body mass.

Comparing different species of mammal, adult brain mass (y) is still allometrically related to body mass (x), but now the exponent is larger: $y \propto x^{0.66}$. Incidentally it is curious how the exponents in these scaling relations are often close to simple ratios – one quarter, three quarters, three halves, and now two thirds. The reason for the two thirds exponent is another unsolved problem; perhaps it has something to do with body surface area (y), which varies with body mass (x) as $y \propto x^{0.67}$.

We are now in a position to consider human braininess. Just taking absolute brain mass is clearly no good: elephants and whales have larger brains than we do. Surely, you may say, that is because they are larger animals. So how about using *relative* brain mass? – that is, brain mass per kilogram of body. Now we find that shrews and mice have a larger relative brain mass than we do, because their bodies are so small. This follows from the scaling relation mentioned above.

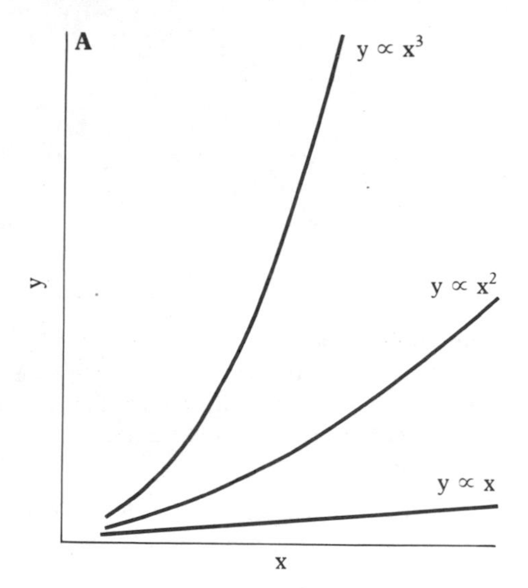

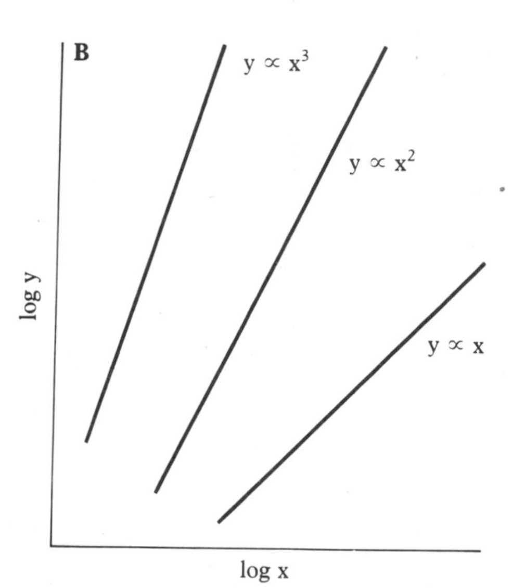

Figure 26.3 **A** The shape of a proportional relation ($y \propto x$) and two non-proportional ones. **B** The same measurements on logarithmic axes give straight lines. $y \propto x$ gives a line with gradient $= 1$; $y \propto x^2$, gradient $= 2$; $y \propto x^3$, gradient $= 3$.

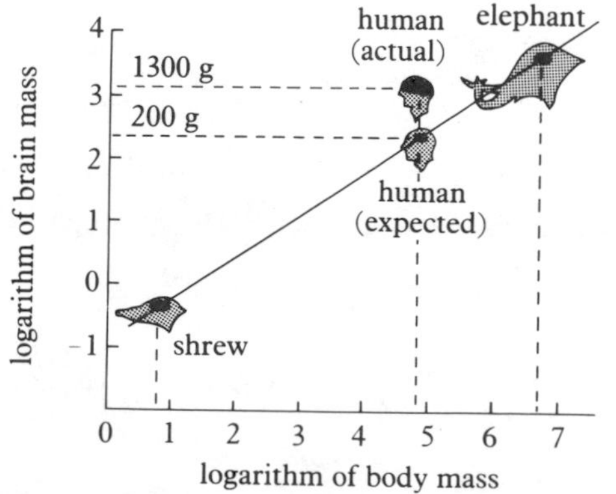

Figure 26.4 The graph shows the scaling relation of brain (y) and body (x) mass, both in grams, in mammals: $y \propto x^{0.66}$. From the graph human brain mass is expected to be about 200 g, while in fact it averages 1300 g.

Well, you will be relieved to know that there is a way to get the conclusion we want. Figure 26.4 shows the scaling relation for mammals' brains. The straight line tells us the normal brain mass for a mammal of any given size. For instance, a mammal of mass 70 kg is expected to have a brain of 200 g. Humans are mammals weighing about 70 kg, but *their* brains weigh on average 1300 g. This is why humans are so brainy; it is because the brain mass is so much larger than would be expected for mammals of our size.

For consideration

1. Explain why brain size in mammals varies approximately as the cube root of body mass during development, while in comparing one species with another the adult brain size varies approximately as the cube root squared.

2. The brain of a sparrow is much larger than that of a small mammal of the same mass. Why?

3. Supposing humans grew to an adult body mass of 700 kg, what sort of head to body ratio would they have?

Further reading

R. McNeill Alexander, *Size and Shape* (Studies in Biology series no. 29, Arnold, 1971)
A useful introduction.

S.J. Gould, *Ever since Darwin* (Penguin, 1980)
Essays 9 and 21–24 are on scaling topics.

S.J. Gould, *The Panda's Thumb* (Penguin, 1983)
See essays 29 on lifespans and 9 on Mickey Mouse.

T.A. McMahon, 'The Mechanical Design of Trees' (*Scientific American* vol. 233, no. 1, 1975)
Trees seen from an engineering point of view.

The formation of plant cell walls

The cell walls of plant cells are of immense economic value. They make up cotton, fibres, paper and timber. Almost indigestible to humans (see page 42), they also contribute considerably to dietary fibre. Many cell walls which have escaped digestion have been compressed to form coal, oil, gas or peat. Yet despite the fact that cell walls make up a third of the dry mass of plants, little is known about how they arise. In this topic we shall discuss what is known about the structure, formation and functions of cell walls in higher plants.

THE STRUCTURE OF CELL WALLS

Analysis of the primary cell walls of higher plants reveals that there are three main components: **cellulose**, **hemicellulose** and **pectins**. The strength and rigidity of the wall is due to cellulose, a polymer of β-glucose. Thousands of parallel cellulose molecules are linked into **microfibrils**, which are fibres 10–30 nm in diameter. These fibres criss-cross in the wall and are linked together by hydrophilic hemicelluloses, which are polymers of five carbon sugars. In the zone where the cell walls of adjacent cells meet, the walls are glued together by calcium pectate, the calcium salt of a polymer of glucuronic acids, which forms the **middle lamella**. Running through this whole structure, from one side to the other, are **plasmodesmata**, which are cytoplasmic connections between cells through which water moves and compounds can be exchanged. How is this complicated structure formed?

CELL WALLS ARISE AT THE MERISTEMS

New cells originate, and new cell walls are created, at the meristems during the aftermath of mitosis. Think of a plant cell in the final stages of cell division. In shoot or root meristems, at telophase of mitosis, identical diploid sets of chromosomes have reached the opposite ends of the dividing cell. Although nuclear membranes form around each chromosome set, the cytoplasm in which these daughter nuclei are embedded is still continuous. At this stage, the microtubules which orientated the chromosomes during their journey to the poles of the spindle at anaphase are still intact, running from one daughter nucleus to the other. It is in the central zone between the nuclei that the new cell wall begins to form, as the **cell plate**.

The following description of what happens is based on the interpretation of a large number of electron and light micrographs of sections through the meristems. In the central zone between the nuclei, Golgi vesicles accumulate. When they join together, their membranes combine to form the new cell membranes of the two daughter cells. The Golgi vesicles contain pectins, which are released into the space between the newly formed membranes to generate the middle lamella. Throughout this process, the microtubules which run from one cell to the other remain intact. They probably come to lie in the plasmodesmata which connect the adjacent cells.

CELLULOSE SYNTHESIS

From the very start, cellulose is synthesised by each of the daughter cells and deposited in the cell wall between the pectin layer and the cell membrane. Like most products secreted from the cell, it is generated by the Golgi body. This was first noted in the mid 1960s, when radioactively labelled glucose was fed to roots and the roots were sectioned and autoradiographed after different lengths of time. The label was first found in the Golgi body and then inside the Golgi vesicles. Thirty minutes after application, the label appeared in the cell walls (figure 26.5).

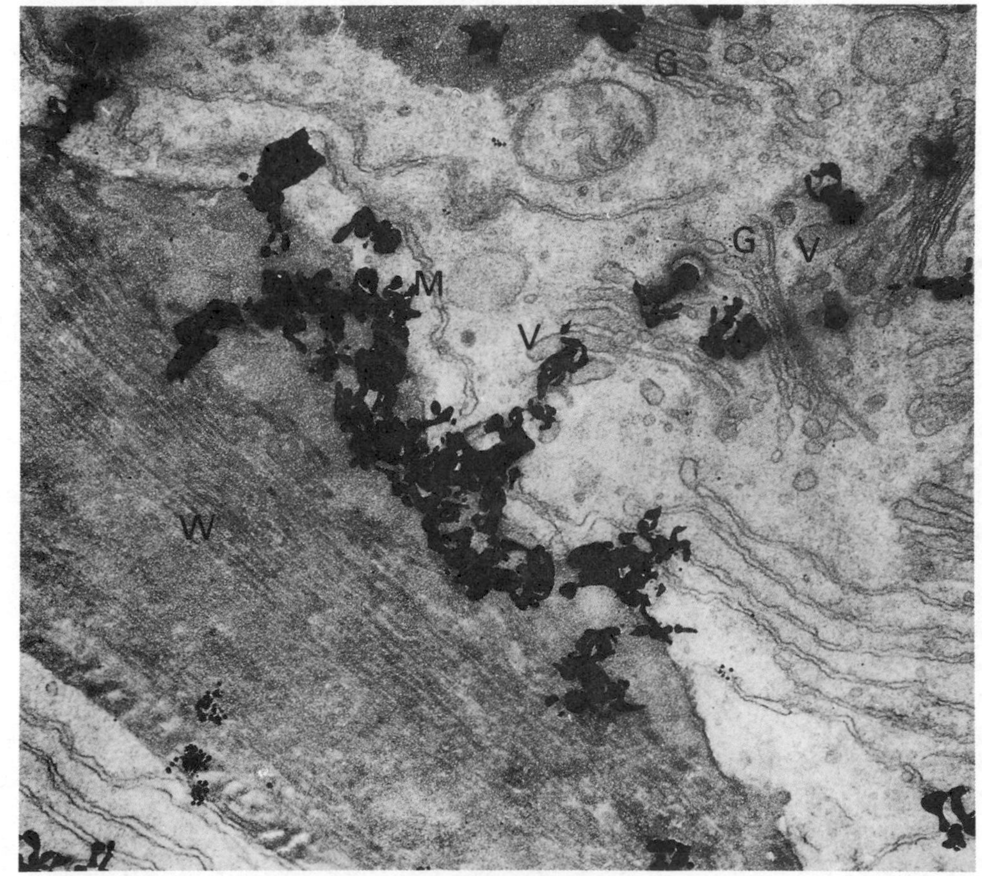

Figure 26.5 Autoradiograph of a root cap cell of wheat (*Triticum*) which was exposed to ^{14}C-glucose for 30 minutes. The black silver grains have accumulated in the Golgi bodies (G), the Golgi vesicles (V) and the space between the cell membrane (M) and the cell wall (W). After longer exposures to radioactive glucose the radioactivity appears in the wall itself.

In root cells Golgi vesicles, containing the compound which provides glucose for cellulose synthesis (uridine diphosphate glucose, UDPG), are pinched off the Golgi bodies. These vesicles move across towards the cell membrane and combine with specific parts of it. Guided by microtubules, the Golgi vesicles move towards particular membrane proteins known as **rosettes**. These are the complexes which make cellulose.

Each rosette consists of six membrane proteins with a hole in the middle. We can imagine a continuous microfibril of cellulose emerging from the hole, lurching in a random direction (figure 26.6). If you were standing on the cell membrane at this stage the microfibrils being extruded would resemble a forest of rapidly growing bramble stems.

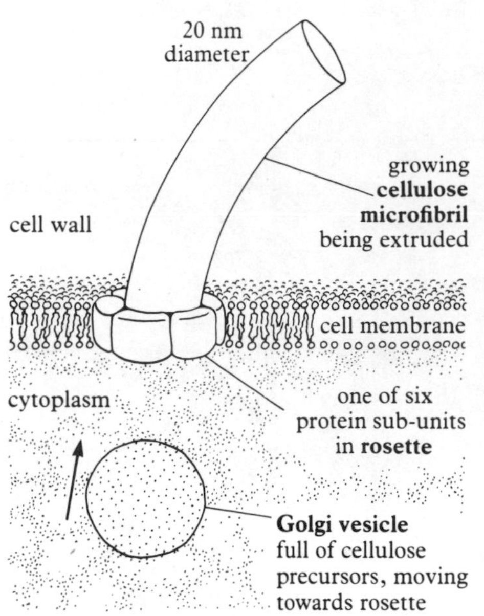

Figure 26.6 Hypothetical view of a *rosette* in a cell membrane, extruding a cellulose microfibril.

CELL EXPANSION

It is not yet known how and where the hemicelluloses are made. At first they link the cellulose microfibrils into a complex and rigid framework. Soon, however, the cell begins to elongate. In shoots this may occur under the influence of the hormone auxin (indole-3-acetic acid). Auxin is believed to stimulate the release of hydrogen ions across the cell membrane into the cell wall. These reduce the pH to about five. At this pH the enzyme **hemicellulase** becomes active, splitting the hemicellulose links between the microfibrils which cross one another. Under pressure from the cell contents, which are turgid and press on the cell wall, the cell elongates in one direction. This creates a parallel pattern amongst the microfibrils, which are now able to slide past one another.

CELL SPECIALISATION

The cell wall is usually wet, since pectins and hemicelluloses are strongly hydrophilic and bind water molecules. The wall is fully permeable to water; water flows across it, down water potential gradients, in channels about 40 nm in diameter (about 200 times the diameter of a water molecule). This water carries ions, which constantly bombard the cell membrane and may be taken into the cell if they encounter the active site of an ion-pumping protein (see page 57).

At this stage all newly expanded cells look similar to one another. By virtue of their position, newly expanded cells may specialise into any one of nine or ten cell types. In many cases, cellulose deposition will continue even after the cell has differentiated.

The new cellulose wall which is deposited is known as the **secondary wall**, to distinguish it from the primary wall, which was deposited before and during elongation. In an epidermal cell above ground, the Golgi body will begin to accumulate the precursors of waxes (cutin and other esters of monohydric alcohols). These will be extruded through the pores in the cell wall onto the plant surface, where they become the **cuticle**. These waxy scales reduce water loss and prevent the easy entry of fungi, bacteria and viruses. In some cells, such as those created by the cork cambium or the endodermis of the root, the Golgi body will make **suberin**, a water repellent wax which blocks the pores in the cell walls. The most dramatic transformation of the cell wall occurs, however, in lignification.

All xylem vessels, tracheids and fibres have in their cell walls large deposits of **lignin**, a polymer of substituted phenylpropane units. The phenylpropane units are made in the Golgi apparatus from glucose, when the set of enzymes associated with lignification begins to work. The Golgi bodies, held in position by microtubules, release lignin precursors into the cell wall at specific places. When polymerised into the cell wall, the lignin fibres replace those of the hemicelluloses. They encrust the cellulose microfibrils throughout the primary and secondary cell walls, not just on the inner surface. In areas known as **pits**, the cell wall does not become lignified. These pits, which often develop in places formerly occupied by plasmodesmata, allow water to freely enter and leave the xylem cell.

For consideration

1. What is the value of a cell wall to a plant cell and to the plant as a whole?

2. What are the selection pressures in plants which promote the evolution of (a) thicker cell walls (b) thinner cell walls? How do these selection pressures differ in different parts of a tree?

3. How permeable are cell walls to bacteria and viruses? What is the importance of this in the lives of plants?

Further reading

P. Albersheim, 'The Wall of Growing Plant Cells' (*Scientific American* vol. 232 no. 4, 1975)

J. Brett, 'A Wall Should not a Prison Make' (*New Scientist*, vol. 99, no.1374, 1983)
A brief account of cell elongation.

D.H. Northcote, *Differentiation in Higher Plants* (Carolina Biology Reader no. 44, 2nd edition, 1980)

27 . The Control of Growth

Sleepy seeds: seed dormancy

A plant can be regarded as the vehicle which enables one seed to produce more seeds. When the seeds are released, they may encounter a hazardous environment. They may be shaded, deeply buried, or exposed to frost and herbivores. Individual seeds which germinate in unfavourable conditions or at the wrong time of year are eliminated by natural selection. We are left only with seeds which tend to germinate when the resulting plant is likely to leave most offspring. Some seeds do not germinate even when exposed to water, oxygen and a reasonable temperature. This is because they are **dormant**. They can usually monitor the environment, and germinate when a suitable environmental stimulus is applied.

Most herbaceous species have seeds which survive for a long time in the soil. The seed of greatest known age to have germinated was of the tropical water lily *Nelumbo nucifera*, which was 237 years old. Long-term experiments suggest that many species have seeds viable for well over 50 years. The soil under arable fields, grasslands and secondary woodlands contains a 'bank' of 3000 to 50 000 dormant but living seeds in each square metre. These seeds provide a buffer against extinction. For example, in the hot and dry British summer of 1976, about ten times as many plants of Thorn-apple (*Datura stramonium*) as usual appeared. Possibly the hot, dry conditions stimulated the germination of buried seeds.

Some seeds are born dormant and others have dormancy thrust upon them. Seeds have dormancy thrust upon them when they lack one or more of the germination essentials, such as water, oxygen or a reasonable temperature. Seeds which are born dormant require some external or internal signal apart from water, oxygen and a reasonable temperature before they can germinate. They are usually only sensitive to this signal when their tissues have taken up water. We shall concentrate on seeds which are born dormant.

INNATE DORMANCY

Sometimes the embryo in the seed is immature. Such seeds will not germinate for days or months after shedding, whatever treatment is applied. In some woodland species, such as Ash (*Fraxinus excelsior*) and Lesser Celandine (*Ranunculus ficaria*), this mechanism may prevent germination in the deep woodland shade of summer.

The seeds of legumes (Leguminosae) are usually hard, with testas impermeable to water and oxygen. Their testas must be damaged to allow water and oxygen to enter the seed if germination is to take place. In nature the movements of soil particles during frost-heaving, and the activities of micro-organisms, abrade the seed coats. Left to natural events like these, germination may be staggered over several years. Agricultural seedsmen overcome this problem by treating seeds in rotating drums lined with carborundum.

RAINFALL

When it rains the desert blooms. Many of the plants which emerge belong to small herbaceous species with short life cycles (in the case of *Boerhaavia repens*, only eight days from seed to seed!). Such species are known as **ephemerals**. In some desert ephemerals, including the Californian poppy (*Eschscholzia californica*), a little rainfall is not enough to stimulate seed germination. Nevertheless, if the seeds are refreshed by a heavy shower of more than fifteen millimetres of rain, or if they are continuously soaked in running water, they can germinate. If the seeds had germinated in response to a weak shower, the plants might have been desiccated before they were able to flower.

The basis of this mechanism is a water soluble **germination inhibitor**, probably abscisic acid (ABA) in some species, which is slowly but continuously made in the testa. When it rains heavily, enough of this inhibitor is washed out for germination to begin. In some other species the inhibitor is nonanoic acid, $CH_3(CH_2)_7COOH$. Some species of desert ants keep their stored seeds dormant by spraying them with myrmacetin, which is 1-hydroxynonanoic acid.

TEMPERATURE

The seeds of most species only germinate within a relatively narrow temperature range and have a precise optimum temperature. This may enable them to detect suitable places or times of year to germinate. For instance, seeds of the common dock (*Rumex crispus*) are stimulated to germinate by fluctuating temperatures, such as those which might occur in spaces between adult plants. In the laboratory, germination of ling seeds (*Calluna vulgaris*) occurs best after the seeds have been exposed to short temperature shocks of 45 °C. This may be why a dense flush of heather seedlings appears as soon as a grouse moor is burnt.

In many species of temperate climates and more northerly latitudes, germination occurs in spring and a period of chilling is required before the seeds are able to germinate. This occurs for instance in the birches (*Betula*) and roses (*Rosa*). The cold requirement is usually about six weeks at 0–5 °C. It is known as **stratification** because gardeners used to perform the chilling process by leaving seeds outside between layers (strata) of sand in the winter. It can be substituted for by soaking the seeds for a day or two in gibberellic acid, which suggests that the cold period may somehow induce gibberellic acid synthesis.

LIGHT

Many seeds have mechanisms which sense whether they are above or below the surface, and in that case, whether conditions are suitable for germination. The light-sensing pigment is **phytochrome**. Three different types of light-sensing mechanism exist.

Firstly, many seeds have a light requirement, for instance tobacco (*Nicotiana*) and several varieties of lettuce (*Lactuca*). Buried seeds presumably reach the surface by frost-heaving or through the activities of humans, worms, moles or larger burrowing animals. Only then will they germinate. Other species, such as love-in-a-mist (*Nigella*), even produce seeds which only germinate in the dark.

Secondly, hundreds of species have seeds which, although they can germinate in light or dark, will not germinate on the soil surface beneath vegetation. This phenomenon is common in woodland herbs and amongst small-seeded grassland plants. Presumably these seeds lie dormant on the surface until the shading leaves disappear. Then their growth is less likely to be restricted by competitors.

Thirdly, seeds buried at depth often play a waiting game. They defer germination until they are moved to the surface. In many arable weeds, burial of the seeds at a depth greater than fifteen centimetres induces in the seeds a light requirement for germination. The necessary light flash may be as short as a tenth of a second. For many seeds this light requirement is only met during ploughing, when the soil is turned over.

As a result of these mechanisms, the plants which you see in a community represent only the tip of the iceberg. Beneath them lurk vast populations of sleepy seeds, ready to burst forth if and when a suitable opportunity arises.

For consideration

1. The ratio of far-red (735 nm) to red (660 nm) wavelengths of light is higher under vegetation than in full sunlight. Why is this so? What are the implications of this difference?

2. The germination of some seeds appears to be inhibited by high levels of carbon dioxide. Why might such carbon dioxide sensitivity be a selective advantage to seeds of some species?

Further reading

M. Black, *Control Processes in Germination and Dormancy* (Carolina Biology Reader no. 20, 1973)
A readable summary of germination mechanisms, with plenty of examples.

P.F. Wareing and I.D.J. Phillips, *Growth and Differentiation in Plants* (Pergamon Press, 3rd edition, 1981)
An advanced book, useful for reference.

The physiology of seed germination

A seed develops from an ovule after fertilisation. Normally it is released into the environment to fend for itself. It contains the embryo, the new plant, surrounded by maternal tissues which protect and sometimes nourish the embryo. Seeds often also contain a food storage tissue, the **endosperm**, but whether an endosperm is present or not, the seed contains energy foods such as starch, fats and proteins. These ultimately provide both the energy and building blocks from which new cells in the seedling are created.

The co-ordinated processes involved in the breakdown of the stored food reserves, their transfer to the embryo in soluble form, and their processing in the new seedling have all been examined closely in the barley grain (*Hordeum* spp.) because of its economic importance in brewing beer.

THE BARLEY SEED ABSORBS WATER

Barley seeds, like those of all other grasses, are endospermic (figure 27.1). The endosperm contains thin-walled cells packed with starch grains. When the seed is released it is dry, with a water content of about fifteen per cent of the total mass and a water potential of perhaps $-100\,000$ kPa. The cell walls are corrugated, and the mitochondria are shrivelled. The respiration rate is so low that it is difficult to measure. It is water uptake which triggers the germination of the seed.

When the micropyle makes good contact with a moist soil surface, water is absorbed first into the dry cell walls of the endosperm and embryo, and then into the cells themselves. It moves into the cell walls of the grain down the water potential gradient, a process known as **imbibition**. Then it moves into the cytoplasm of the cells across the selectively permeable cell membranes, by osmosis. The cell walls, mitochondria and enzyme systems become hydrated and, within a day, the respiration rate increases a thousand-fold.

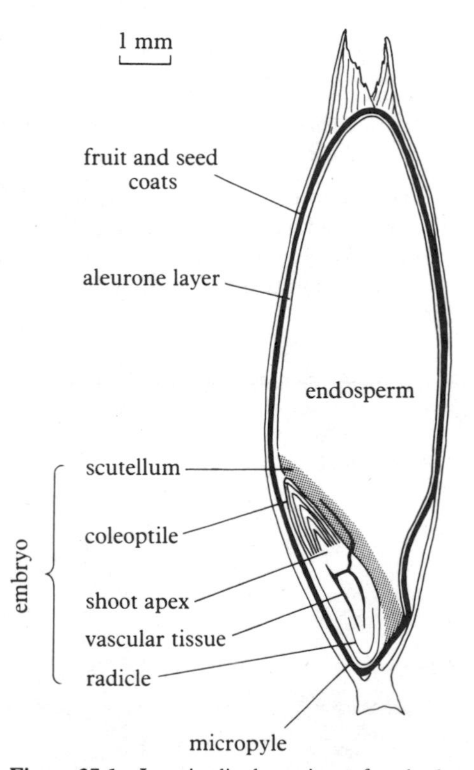

Figure 27.1 Longitudinal section of a barley 'seed'. Notice the position of the aleurone layer. The 'seed' illustrated here is really a fruit, since it is surrounded by a testa fused to the ovary wall. The scutellum is a shield-shaped segment of the embryo, the closest part of the embryo to the endosperm. (*After Galston, Satter, Davies*)

GIBBERELLIC ACID AND ENZYMES

In the first two days auxin and cytokinins produced by the shoot apex are pumped down into the embryo and cause the xylem and phloem to begin differentiating. The shoot apex then releases gibberellic acid (GA) which is transported down the phloem to the scutellum and released into the endosperm. When GA encounters the aleurone cells around the endosperm it makes them release digestive enzymes. In fact the digestion of the endosperm is triggered by as little as 5 ng per dm^3 of gibberellic acid, equivalent to one gram dissolved in two hundred million litres of water.

There is convincing evidence that this gibberellic acid somehow triggers the synthesis in the aleurone cells of all the enzymes which pour into the endosperm. This evidence is of several types. When radioactively labelled amino acids are added to isolated barley aleurone tissue in the presence of added gibberellic acid, all the enzyme molecules released carry the label. In the equivalent control experiment without added gibberellic acid, no enzymes are released and none are labelled. Inhibitors of DNA transcription in the nucleus, and of translation at the ribosome, prevent the enzymes from being produced. Under the influence of gibberellin a messenger RNA is formed which can support the synthesis of the starch-splitting enzyme, amylase, in a cell-free preparation derived from aleurone cells. In control preparations without gibberellin no such messenger RNA can be isolated.

The enzymes liquefy the endosperm by converting large, insoluble storage molecules to smaller, soluble ones. Between them, the enzymes hydrolyse fats into fatty acids and glycerol, hemicellulose into five carbon (pentose) sugars, starch into maltose, proteins into amino acids, and DNA into nucleotides.

THE EXPANSION OF THE EMBRYO

The small molecules which result from this digestion are absorbed by the scutellum of the embryo and transferred in particular to the shoot and root apices. There the manufacture of polymers from monomers requires energy from ATP hydrolysis (see page 33). The ATP is usually provided by the aerobic respiration of glucose. In the first few days after wetting, this respiration reduces the dry mass of the seed, since glucose is lost as water and carbon dioxide. The dry mass of the seed continues to decline until the seed dies unless, in the meantime, the seedling has placed its cotyledon or stem into the light and has started to photosynthesise.

The continual uptake of water softens the testa and the absorption of water by the cells in the embryo causes them to expand rapidly. This exerts pressure on the testa, which ruptures, allowing the radicle to protrude. Once the radicle protrudes it can absorb both water and nutrient ions, increasing the rate of hydration of the seedling. At this point, germination has taken place.

For consideration

1. Assess the relative value to seeds of lipids and carbohyrates as storage reserves.

2. The respiratory quotient of a tissue is the ratio of the carbon dioxide it produces to the oxygen it absorbs. The respiratory quotient of a germinating barley seed may vary from 0.7 to 1.0, and then rise to infinity before settling back to 1.0 again. Account for this.

Further reading

M. Black, *Control Processes in Germination and Dormancy* (Carolina Biology Reader no. 20, 1973)
The author summarises the whole range of dormancy mechanisms and germination triggers.

J.A. Bryant, *Seed Physiology* (studies in Biology no. 165, Arnold 1985)
This summarises the biochemical events within a germinating seed.

Geotropism in roots

In several areas of plant physiology, such as phloem transport and the mechanism of stomatal opening and closure, you may have noticed the multitude of conflicting hypotheses to explain the phenomena, and the lack of sound evidence on which to decide which hypothesis is correct. This is especially true of theories to explain the positive geotropism of roots. Indeed, the current hypotheses are almost the opposites of the previous ones. Thus it is worthwhile to discuss the history of ideas on root geotropism, as an illustration of the scientific method in action.

SOME PRELIMINARY OBSERVATIONS

When roots are placed horizonally they bend downwards. It was shown by Sachs in 1887 that this is not merely the roots sagging under their own mass; his roots grew straight down even when supported by mercury. In Maize (*Zea mays*) the bending occurs 1.5–3.0 mm behind the tip, remarkably close to the cap. Some reports suggest that a horizontal root continues to grow at the same rate on the lower surface but grows faster on the upper surface. Other measurements suggest that growth is slowed down on the underside as well as being speeded up on the upper side.

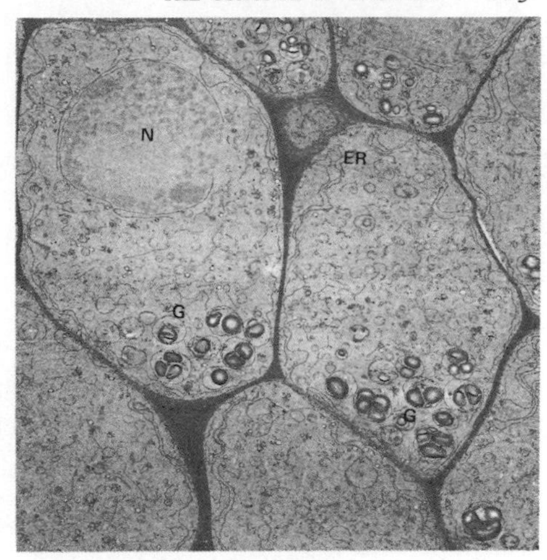

Figure 27.2 Electron micrograph through root cap cells of maize which have been exposed to gravity. Notice how the amyloplasts (G) have fallen to the base of each cell and the endoplasmic reticulum (ER) has been displaced towards the top of the cells. N is the nucleus.

THE ROOT CAP SENSES GRAVITY

If the root cap is removed, a root is unable to respond to gravity for a few days. The starch grains (**amyloplasts**) in the cap seem to be essential for the perception of gravity. Starch grains have been removed from the roots of cress, white clover and red clover by incubating the roots with gibberellic acid and kinetin at 30 °C for thirty-five hours. These roots continued to grow straight when placed horizontally. The starch grains began to reappear a day later and the roots regained their capacity to respond to gravity. Furthermore, electron micrographs show clearly that the amyloplasts move down to the bases of the cells when exposed to gravity, displacing the endoplasmic reticulum to the tops of the cells (figure 27.2). Over a range of species, the slower the starch grains fall, the longer a root has to be exposed to gravity before the root responds to it. Nevertheless, the mechanism by which the root cap transmits a signal to the cells two or three millimetres behind it remains obscure.

THE CHOLODNY-WENT HYPOTHESIS

In 1926, on the basis of their famous experiments on phototropism in coleoptiles, Cholodny and Went put forward the hypothesis that positive geotropism in roots might be caused by a hormone. When **auxin** was isolated and characterised and its role in phototropism was realised, it was reasonable to suspect that it might play a part in root geotropism.

Since auxin is produced by the apex of the shoot, perhaps, by analogy, it was produced by the root cap in the root. One hypothesis was that it accumulated under gravity on the lower side of a horizontal root. If this high concentration on the lower surface inhibited cell elongation, a horizontal root would bend downwards.

One piece of evidence was consistent with this idea. When roots were bathed in auxin solutions, growth was stimulated by those solutions containing about 10^{-8} moles of auxin per litre and inhibited at higher concentrations. Roots were 10 000 times more sensitive to auxin than shoots, and concentrations high enough to inhibit growth seemed likely to occur in the lower surface of an intact root. Unfortunately, however, the standard coleoptile test cannot be carried out successfully with agar blocks exposed to root tissue because the concentrations of auxin in the root are too low.

This hypothesis to explain positive geotropism in roots appeared in most school and university textbooks, often stated dogmatically as if fully proven, from the 1950s right up to the 1980s. Yet it was not demonstrated until 1973 that auxin occurs in roots at all! By that time, the attention of researchers was switching to **abscisic acid** (ABA), because it had been clearly shown that auxin in roots is transported not from the root cap towards the stem, but in the other direction, from stem to cap.

Experimental results supporting this downward movement of auxin had been published by Scott and Wilkins in 1968 (figure 27.3). They cut root segments of maize six millimetres long and placed them vertically. On top was placed a donor block containing radioactively labelled C^{14}-auxin, and at the base was placed a receiver block in which the accumulation of radioactivity could be monitored. It was found that the auxin moved towards the root tip but not towards the stem. It is now established that auxin transport in the intact root takes place only in the vascular tissue and toward the root cap, at the rate of about 1 cm h^{-1}. All the auxin in the root is probably derived from the tips of the stem and leaves and this may be why its concentration in roots is so low.

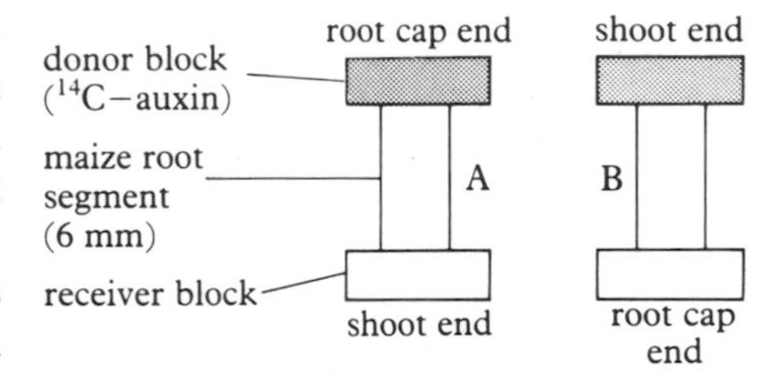

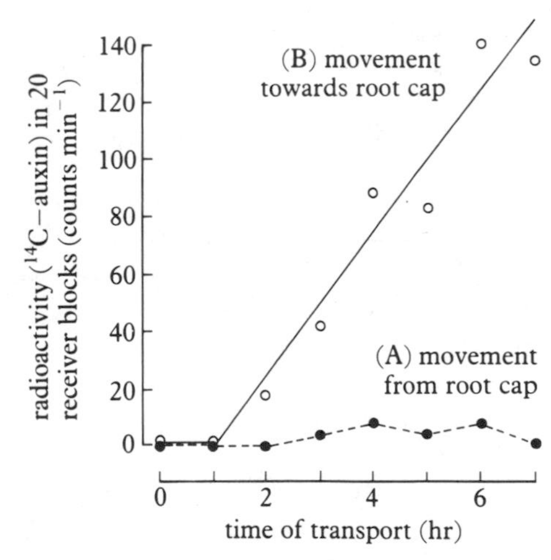

Figure 27.3 Experiment which indicated that auxin in root tissue moves towards the apex but not away from it. It makes no difference whether the donor block is at the top or the bottom; the auxin moves from the shoot to the root end, but not in the opposite direction. (*After Scott and Wilkins, 1968*)

THE POSSIBLE ROLE OF ABSCISIC ACID

The plant hormone abscisic acid was unequivocally identified in the root caps of maize in 1973/4. It inhibits elongation in maize roots at concentrations as low as 10^{-8} to 10^{-4} moles per litre. If the cap is removed and abscisic acid is applied to the underside of a horizontal root at concentrations as low as 10^{-9} moles per litre, the root grows downwards. Furthermore, a series of experiments carried out by Wilkins and his co-workers in which they partially removed root caps and/or inserted mica barriers in the apices have yielded results which support the hypothesis that root caps produce a growth inhibitor (figure 27.4).

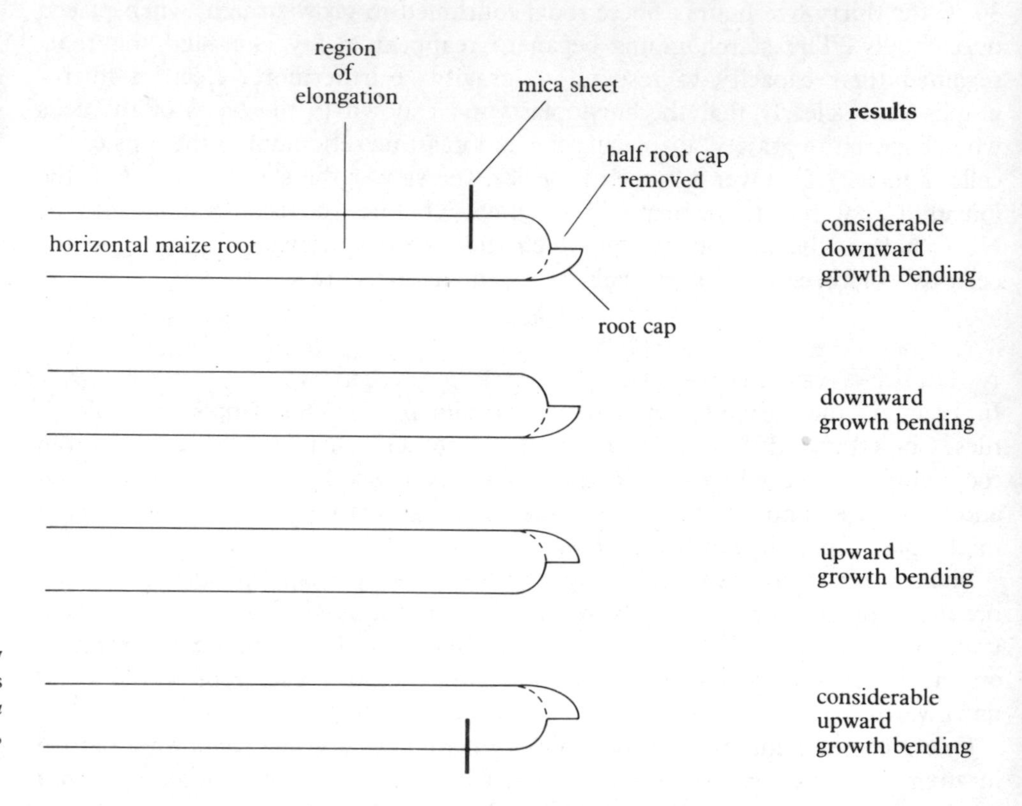

Figure 27.4 Results of some experiments by Wilkins which are consistent with the hypothesis that the root cap produces an inhibitor. (*Data from M. B. Wilkins*, Advanced Plant Physiology, *Pitman, 1984*)

However, several hurdles remain before we can regard the role of abscisic acid in positive geotropism as proven. We must explain why analyses suggest that there is more ABA in the upper than lower halves of the root caps of horizontally grown root tips. We shall need to prove that ABA is actually made by the root cap, that it moves from the top to the base of the root cap in a horizontally grown root, and that it is transported from the cap to the region of elongation. Even then, we shall have to show how the downward movement of the amyloplasts affects the synthesis or secretion of the hormone. Maybe in the end we shall find that abscisic acid is not involved at all!

For consideration

1. Interpret the results of each of the experiments shown in figure 27.4.

2. Make a list of all the hypotheses featured in this topic. Briefly describe how each hypothesis was tested, and state whether the test supported or refuted the hypothesis.

Further Reading

H.E. Street and H. Opik, *The Physiology of Flowering Plants* (3rd edition, Arnold, 1984)
Chapter 10 deals with growth movements, including geotropism.

M.B. Wilkins (ed.), *Advanced Plant Physiology* (Pitman, 1984)
A detailed account of recent research into geotropism is given in Chapter 8.

---

28 . Mendel and the laws of heredity

Mendel under the magnifying glass

No one doubts the immense contribution made to genetics by that well known schoolmaster and monk, Johann Gregor Mendel (figure 28.1). But there is something odd about his work.

In the 1930s the statistician R.A. Fisher had a close look at Mendel's experiments. His conclusions were disturbing. It is not that Mendel's results are inconsistent with his theory; on the contrary, they are not only consistent but too good to be true.

Figure 28.1 An Austrian postage stamp celebrating the centenary of Mendel's death. 'Entdecker der Vererbungsgesetze' means 'Discoverer of the law of heredity'.

DID MENDEL FIDDLE HIS RESULTS?

Take for example the experiment in which he crossed heterozygotes of the gene which leads to either smooth or wrinkled seeds. The offspring are expected to produce three smooth seeds for every wrinkled one. Mendel reported getting 5474 smooth and 1850 wrinkled seeds, compared with the ideal result of 5493 and 1831.

Of course in this kind of experiment the exact prediction is never expected in practice. That is because the alleles in each seed are the result of two random events. The chance of a dominant (or recessive) allele entering the zygote from the male gamete is 50 per cent, just like getting a head (or tail) in tossing a coin. The same is true for the female gamete. Because Mendel's experiments measure random events, the results are expected to deviate from the ideal in just the same way as a coin-tossing experiment would.

It is easy for a statistician to calculate how large these deviations ought to be. In particular the median deviation is a useful figure: if the experiment were performed many times, half of them should have deviations less than the median, and half greater. Mendel's actual deviation was 19 (5493 minus 5474), that is less than the median value.

If this happens with a single experiment, it is nothing to worry about. When it happens time and time again, our suspicions are aroused. In nearly all Mendel's experiments the deviation is less than the median value, whereas it ought to be less in about half of them. Taking all the experiments together Fisher calculated that, assuming Mendel's theory is true, the chance of obtaining results so close to the ideal is only 0.007 per cent, or 1 in 14 000.

HOW DID IT HAPPEN?

Of course it is just possible that Mendel had an assistant working for him in the monastery garden at Brno, helping with the tedious counting of pea plants and seeds (figure 28.2). If the assistant had known what result Mendel was expecting, he might have doctored the results to please his superior.

Although Fisher considered this possibility, another explanation seemed more likely. Any clear thinker in the mid-nineteenth century, he suggested, could have worked out Mendel's theory if a few simple assumptions were made. Two of these were that inheritance is caused by particles passing from parents to offspring, and that the male and female parents make equal contributions to the zygote. In Fisher's view Mendel probably worked out the whole theory in his head at an early stage. Then the impressive experiments were merely a way to check that his ideas were correct, and to provide a demonstration for the benefit of other people.

Figure 28.2 The garden of the monastery where Mendel carried out his genetic experiments. Did he do the counting himself or did he have an assistant to do it for him?

But there may be another explanation. The trouble with the characters Mendel studied is that a few peas or plants are intermediate. This is because each phenotype varies, and sometimes the variations of the two phenotypes overlap. For instance some of the peas turn out to be half way between round and wrinkled – they may be genetically round (or genetically wrinkled) but modified by other factors such as the environment. These peas cannot be classified objectively. However, Mendel may have tried to classify them, and in doing so was swayed by knowing the phenotype ratio he expected to see.

But even if Mendel did doctor his results it hardly detracts from his great achievement in describing so clearly the principles of genetics. It is one of the tragedies of science that this talented amateur received no recognition before his death in 1884. He deserved a Nobel Prize, but didn't even get a pat on the back.

For consideration

1. Can you think of any other respect in which Mendel's results seem almost too good to be true?

2. If you invented a theory which turned out to be correct, but you had originally supported it with biased evidence, would you deserve full credit for the theory?

Further reading

R. Dunbar, 'Mendel's Peas and Fuzzy Logic' (*New Scientist*, vol. 103, no. 1419, 1984)
A short and interesting article.

R.A. Fisher, 'Has Mendel's Work Been Rediscovered?' (*Annals of Science*, vol. 1, no. 2, 1936)
An advanced appraisal of Mendel's experiments.

Genetic counselling

Muscular dystrophy is not one disease but several. In all of them muscles waste away and weaken. This is because the muscle fibres degenerate and are replaced by fat and connective tissue. There is no effective remedy.

In general, any disease can be said to have both environmental and genetic causes. For instance you can only get malaria if the parasite enters your body *and* you lack genetically determined resistance. In some cases everyone is exposed to the environmental causes, and whether you contract the disease depends solely on the genes you happen to carry. This is what we mean by calling a disease 'genetic'. Most muscular dystrophies are genetic.

One rare form of muscular dystrophy is caused by a dominant allele on one of the autosomal (ie. non-sex) chromosomes. A heterozygous person married to someone without the disease may wish to have children; but there is a 50 per cent chance that a child will also be affected. Another rare form of the disease is caused by a recessive allele. In this case a heterozygous person is called a **carrier**, because although not diseased he or she is carrying the allele. So it may be passed on to a child. In this case, if both parents are carriers, there is a 25 per cent chance that a child will get the disease.

In fact the commonest form of muscular dystrophy is caused by a sex-linked recessive allele, transmitted like the allele for red-green colourblindness.

Any responsible parents who know there is a genetic disease in one of their families are anxious about the risk to their children. But, as you can see from the muscular dystrophy example, working out the risks may be no easy task. One disease may have several different genetic causes. This is where genetic counselling comes in.

COUNSELLING

Genetic counsellors do not tell parents whether they can have children. That decision is for the parents to make. What the counsellor does is to help them decide by explaining the facts and indicating the options available.

Since it is important not to make mistakes, careful work goes into preparing for a counselling session. Usually the family doctor knows of a genetic disease in the family, and refers the case to the counsellor. Then more information is collected. The counsellor needs to find out which relatives have the disease, so that a pedigree or genetic chart of the family can be constructed. This may make it possible to identify whether either parent is a carrier.

Medical tests may reveal further information. In some cases it is possible to tell whether a person is a carrier even though the disease is not present. For instance, in muscular dystrophy, the muscle breakdown leads to high levels of the enzyme creatine kinase in the blood. Surprisingly the carriers also show this feature, so that a blood test can usually identify them.

WHAT PARENTS CAN DO

Sometimes counsellors can reassure parents. Partially genetic diseases and those caused by several genes at once often involve low risks to the child – say, less than one in twenty. This is not much worse than the risk of some serious malformation arising in any pregnancy, which is about one in forty. In this case parents often accept the risk and have children.

Diseases caused by a single gene may involve higher risks, which can nevertheless be estimated accurately. In these circumstances some parents opt for artificial insemination, while others may adopt a child. Since counsellors realise the disappointment parents may experience in this situation, they try to minimize any psychological problems that might arise.

There is an old Chinese saying: 'a poor doctor cures, a good doctor prevents'. This is a bit hard on all those excellent doctors who can cure, but it does emphasize the wisdom of preventing diseases where possible. This point is often made, for example, in the case of cancer; it applies just as much to genetic diseases, many of which are serious and incurable. So it is fortunate that genetic counselling is becoming an increasingly important branch of medicine.

For consideration

1. What are the arguments for making preventive medicine a more important part of health care? Are there any arguments against?

2. Why is it surprising that carriers of muscular dystrophy allele have a high level of creatine kinase in their blood?

Further reading

C.A. Clarke, *Human Genetics and Medicine* (Studies in Biology no. 20, Arnold, 1977)
A general introduction.

P.S. Harper, *Practical Genetic Counselling* (John Wright, 1981)
A detailed account for the medical profession.

29 . Chromosomes and genes

Mapping human genes

Humans are not as good as fruit flies for mapping genes. To measure linkage between fruit fly genes, two flies must be mated and a large number of offspring counted. In humans of course we cannot set up mating experiments, and insufficient offspring would be produced even if we did.

For humans quite a different method is now available. Its main use is to identify which chromosome any particular gene is situated on, but in a modified form it can also determine the order of genes on a single chromosome.

SOMATIC CELL HYBRIDS

There is nothing unusual about a cell hybrid. This is a cell, derived from the fusion of two unlike cells, in which all the chromosomes have been collected into a single nucleus. Every time a sperm fuses with an egg, the two sets of chromosomes come together at the zygote's first mitosis; so all the zygote's descendants are hybrid cells.

The breakthrough in human gene mapping occurred when **somatic cell hybrids** were first produced. These are single hybrid cells made from ordinary body cells (not gametes) by the process of *cell fusion*. One surprise was that the original cells need not be from the same individual, or even the same species. In human gene mapping it is usual to fuse human with mouse cells.

Let's look briefly at the principle behind the technique. Suppose a hybrid cell could be made in which there was a complete set of mouse chromosomes but only one from the human set. Now imagine that you could identify which of the 23 human chromosomes you had in the cell, and could also discover whether any particular human gene was present in the cell. Clearly this method would allow you to work out which genes are on which chromosomes.

For many years human chromosomes could not be reliably identified, because they look so alike. Now, though, they can be stained so that each acquires a distinctive banding pattern, making it look different from the 22 others (figure 29.1).

There are also methods for detecting individual genes. If a gene codes for a protein, all you need to do is look for that particular protein in the cell. However a problem arises when the protein from the human gene is similar to a protein from a mouse gene. For instance you may be looking for the human gene for cytochrome c, but you know that every hybrid cell also has the mouse's cytochrome c. However, the corresponding proteins in the human and mouse cells are often slightly different, and can be separately detected using electrophoresis.

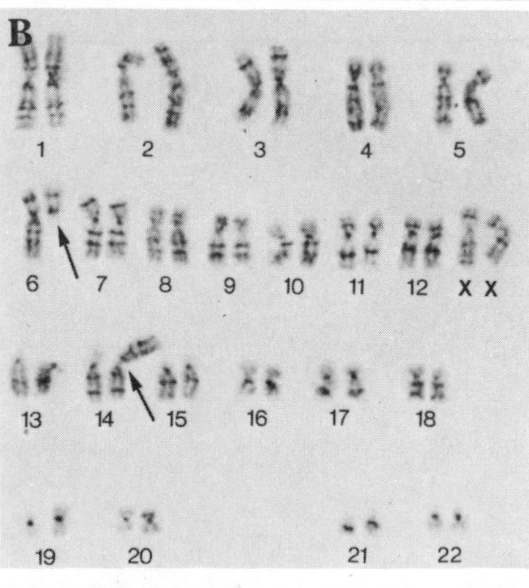

Figure 29.1 The chromosomes of a human female. **A** The appearance of a single squashed nucleus; **B** The chromosomes have been identified and arranged in homologous pairs. Note the distinctive banding patterns revealed by the ASG (acetic/saline/Giemsa) method of staining, developed at the Western General Hospital in Edinburgh. Note also the translocation of part of a chromosome 6 to a chromosome 14; this had no ill effects on the woman in question.

PROCEDURE

A culture of mouse cells is mixed with a culture of human cells – often lymphocytes or fibroblasts (figure 29.2). The cells are then encouraged to fuse, under the influence of polyethylene glycol or an inactivated virus. Only one or two in a thousand cells fuse, and some of these fuse with their own kind. However there are ways of killing all the cells that are not needed.

The cells left each have two separate nuclei, one from the mouse and the other human. It is only when the cell next undergoes mitosis that the nuclei join so that all the chromosomes come within a single nucleus. In the course of mitosis the mouse chromosome set remains intact, but most of the human ones are lost. The result is a collection of hybrid cells each with only a few human chromosomes – but different cells have different human chromosomes.

So the next stage of the procedure involves separating the cells and culturing each on its own to produce a clone of genetically identical cells. Only rarely does a cloned cell contain just one kind of human chromosome; usually there are at least two or three. However it is still possible to identify which chromosome carries a particular gene. We might find, for example, that the protein from the gene is only found in a clone containing human chromosomes 3 and 7, and in another clone containing 2, 7 and 11. Then clearly the gene is carried on chromosome 7.

To work out the order of genes along a chromosome, the chromosome in question must be damaged. Ionising radiation is used to knock pieces off it, and the loose pieces are soon lost from the cell lineage during cell division. If a pair of genes are close together, the chromosome will only rarely be broken between them. So the two proteins, from the two genes, will nearly always be both absent or both present in the hybrid cell. However, between two distant genes, the chromosome will be broken more often; in this case, quite a few clones will have only one of the proteins. Just as in crossing-over, the frequency of breaks between the two genes indicates the distance between them.

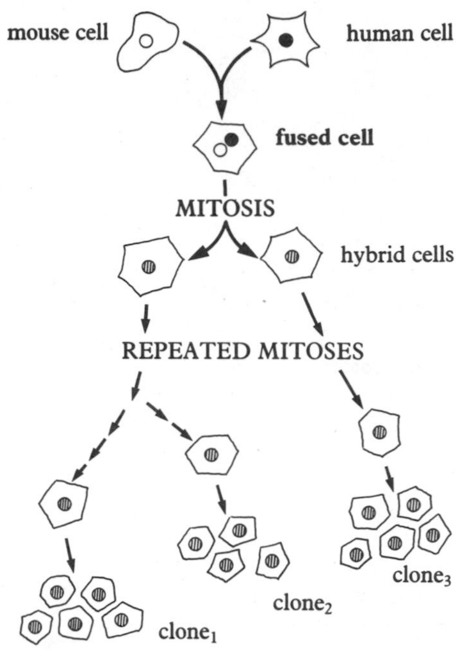

Figure 29.2 Somatic cell hybrids: a mouse and a human cell fuse. At the next mitosis the chromosomes in each cell are brought together in a single nucleus. In this and the next few divisions, most of the human chromosomes are lost. Finally individual cells are isolated to form genetically different clones.

For consideration

1. Why is it important to be able to map human genes?
2. What other methods can you think of for discovering which chromosome carries a particular gene?

Further reading

E. Sidebottom and N. Ringertz, *Experimental Cell Fusion* (Carolina Biology Reader no. 102, 1984)
Cell hybrids have more uses than just gene mapping.

J.R. Warr, 'The Use of Cell Cultures in Human Genetics' (*Journal of Biological Education* vol. 16, no. 1, 1982)
A general account of the whole field.

Arranging one's sex

Are males any use? In many organisms they merely add sperm to the female's eggs, and then do nothing further to help raise the offspring. Indeed some females get along very well without males. For example water fleas (*Daphnia*) and aphids can, for several generations, lay parthenogenetic diploid eggs which turn into viable female offspring.

For some reason though, most species do have males. So at some stage it must be arranged what sex each individual is to have. There are several ways in which this is done.

SEX CHROMOSOMES

The best known method involves sex chromosomes. In the X–Y system, the female is XX and **homogametic** – that is, all her gametes will carry an X chromosome. The male is XY and **heterogametic** – half the gametes carrying an X and half carrying a Y. The X–Y system is widespread throughout the animal kingdom, and is also found in some flowering plants which have separate sexes, such as the Hop. Certain insects including grasshoppers lack the Y chromosome, so that the female and male are XX and XO respectively.

autosomes	**XX**		♀
autosomes	**X**		♂
autosomes	**XX**	Y	♀
autosomes	**X**	Y	♂

autosomes	**XX**		♀
autosomes	**X**		♀
autosomes	**XX**	**Y**	♂
autosomes	**X**	**Y**	♂

Figure 29.3 Sex determination in fruit flies and humans. The chromosomes determining sex are printed in bold type. In the fruit fly each set of six autosomes, i.e. non-sex chromosomes, can be counted as having 2 units of maleness genes, and each X chromosome has $1\frac{1}{2}$ units of femaleness genes. The set of genes represented more strongly determines the sex of the individual. In humans, each X chromosome can be counted as having 1 unit of femaleness genes and each Y chromosome as having 3 units of maleness genes. The set of genes which is represented more strongly determines the sex of the individual.

It is not necessarily the male that has the two different sex chromosomes. In organisms where the female has them, the chromosomes are usually called Z and W; the female is ZW and the male ZZ. This system is found in butterflies, moths, some birds and fish, and a few flowering plants. In chickens the W is missing, so that males are ZZ and females ZO.

From the variety of sex chromosome arrangements, it might be suspected that they do not all function in the same way. Even within the X–Y system, fruit flies and humans arrange the sex of their offspring differently. Abnormal arrangements of sex chromosomes give a clue as to how they work. For instance XO fruit flies are male and XXY are female (figure 29.3). Even though normal males are XY, the Y chromosome has no effect on sex. It is believed that the X chromosomes carry genes for femaleness, while the autosomes have maleness genes. The sex of the individual is determined by the imbalance between the two sets of genes.

In humans however, XO individuals have a condition called Turner's syndrome and are sterile females. On the other hand, an XXY person is male. Unlike the fruit fly, the human Y chromosome does appear to carry genes for maleness. The main factor determining sex is the presence or absence of a Y chromosome.

Species using sex chromosomes usually produce equal numbers of male and female offspring. But is this biologically sensible? After all, in most species one male can fertilise many females, so it might be more efficient to produce more females than males. It is also a purely random affair whether an individual is male or female. Sometimes, perhaps when there is a greater need for one sex than the other, it would be more efficient for the mother to control the sex of her offspring.

OTHER WAYS OF DETERMINING SEX

Honey bees face the latter problem acutely. Thousands of female workers are required to run the hive, while the hundred or so male drones are merely a parasitic nuisance except at mating time. Accordingly the queen bee can determine the sex of each offspring. She carries a store of sperm from the drones she mated with, and can control whether sperm is released onto each egg before it is laid. Fertilised eggs become females, and unfertilised eggs drones. The system is known as **haplo-diploidy**, since sex is determined by whether the individual has a haploid (male) or diploid (female) set of chromosomes. Incidentally, there is only a single gene determining the sex; all heterozygotes become female, and all non-heterozygotes become male. Haplo-diploidy works because a haploid organism cannot be heterozygous, and so must be male.

Sex chromosomes and haplo-diploidy are both genetic methods of fixing sex. A third method is not genetic: sex is determined by the environment. For instance the marine worm *Bonellia* has a planktonic larval stage, and an adult male which attaches itself to the female in the same way that a parasite would. Any larva settling on the sea bottom out of contact with other individuals becomes a female. It grows a long proboscis with which it feeds. If however a larva by chance settles on the proboscis of a female, it becomes a male and develops into a dwarf 'parasite.' By this means every larva acquires the appropriate sex for its situation, and there is little wastage.

By contrast, other examples of environmental sex determination are more difficult to understand. Consider the various reptiles whose sex is fixed by the temperature at which the eggs are incubated. The eggs of a North American turtle, when incubated at 23–28 °C, produce only males, while incubation at over 30 °C gives only females. In the intermediate temperature range both sexes appear. It means that a female laying eggs in sunlit sand, or in the shade of vegetation, is effectively determining the sex of all her offspring. Warm or cool weather also probably influences the sex ratio in the new generation of turtles.

The reason for this extraordinary system is not understood. Some lizards and alligators are influenced by temperature in the opposite way: warm eggs produce males, and cool eggs females.

WHY HAVE SEX AT ALL?

Not only are scientists puzzled about some of the ways organisms fix their sex; the existence of sexual reproduction itself has caused many heads to be scratched. An important aspect of the sexual process is meiosis and fertilisation, during which genes are shuffled into new combinations. One likely advantage of this is that novel gene combinations arise in the new generation, making the individuals possessing them better able to cope with a changing environment. For instance a tree may inherit genes for resistance to a fungal disease, and for larger fruit, whereas none of its ancestors had both genes in the same plant.

So sexual reproduction can produce new improved varieties. But it can also destroy them. This tree will shuffle its genes when it produces its gametes. The two genes will often be separated again, and quite possibly none of the tree's offspring that survive to maturity will have the 'good' combination. If on the other hand the tree could reproduce asexually like the parthenogenetic aphid, all its offspring would be genetically identical to itself. They would all carry the favourable gene combination, as in the case of Cox's apple trees (see page 107).

In short, asexual reproduction is much more efficient at preserving new improved varieties of an organism. This is one reason why it is so difficult to answer the question posed at the beginning of this topic.

For consideration

1. Suggest one or more functions for the turtle's sex determination mechanism.

2. Considering the advantages and disadvantages of asexual and sexual reproduction, what do you think the ideal reproductive strategy would be? What species do you know of that use it?

3. What evidence supports the suggestion that in humans each X chromosome has 1 unit of femaleness genes and each Y chromosome has 3 units of maleness genes?

Further reading

J.J. Bull, *Evolution of Sex Determining Mechanisms* (Benjamin/Cummings, 1983)
An advanced text but contains some fascinating examples.

J. Maynard Smith, *The Evolution of Sex* (Cambridge University Press, 1978)
Chapter 1 summarizes the problems posed by the evolution of sex. The rest of the book is advanced.

---

30 . The nature of the gene

The evidence for a triplet genetic code

Each cell has, on its DNA, a library of instructions for proteins, written in code. Some of these instructions are transcribed by enzymes on to short lengths of RNA, which carry the same code. This code is translated during protein synthesis into the order of amino acids along a protein molecule.

You have probably learnt that each amino acid which appears in the final protein molecule is coded for on the DNA by a sequence of three successive nitrogenous bases, a **triplet** or **codon**. However, you may not realise that this is an hypothesis, deduced from the results of experiments by the 'code breakers'. In this topic we shall describe some of most important evidence which they collected.

On the following purely theoretical basis, we should expect a triplet genetic code. If two successive bases specified an amino acid the sixteen possible combinations would be insufficient to code for the twenty common amino acids. If three successive bases specify each amino acid the 64 possibilities are enough to code for twenty amino acids. A quadruplet code would be inefficient compared to a triplet code because of the 'energy costs' of making longer DNA and RNA, much of which would, presumably, be irrelevant. In fact the concept of a triplet code is backed up by the evidence obtained between 1960 and 1966 by Nirenberg and Ochoa in the United States and Crick and Brenner in Cambridge.

CELL-FREE SYSTEMS

Nirenberg found that a cell-free preparation from the bacterium *E. coli* could synthesize polypeptide chains in a test tube. The bacterial cells were gently broken by grinding them up in powdered alumina, and the cell wall and cell membrane fragments were thrown down by centrifugation. A suspension remained which contained *E. coli* ribosomes together with numerous enzymes and small molecules. Nirenberg added to this suspension a messenger RNA which had been made in his laboratory, a radioactively labelled amino acid and ATP. After incubating the preparation at 37 °C for an hour, he was able to detect whether the labelled amino acid had been synthesised into protein by precipitating the protein and testing it for radioactivity with a Geiger counter.

Some of Nirenberg's results with different synthetic messenger RNAs and the labelled amino acid phenylalanine in four different experiments are shown in Table 30.1. A synthetic messenger RNA with the base sequence UUUUU . . ., known as **poly**U, caused many radioactive phenylalanine molecules in the suspension to be incorporated into protein. However, messenger RNAs with base sequences such as AAAAA. . . (polyA) or CCCCC. . . (polyC) caused no more phenylalanine to be incorporated in proteins than in preparations to which no messenger RNA had been added. The code for phenylalanine was therefore U, UU, UUU, UUUU, UUUUU or some other U-multiple.

Table 30.1 This shows the result of adding different synthetic messenger RNAs to cell-free preparations from *E. coli* to which a radioactively-labelled amino acid, ^{14}C-phenylalanine, has been added.

Messenger RNA added to cell-free preparation	Counts per minute of phenylalanine in precipitated protein
None	44
PolyA	50
PolyC	38
PolyU	39 800

The real breakthrough did not come for another three years, when techniques were developed to join together nucleotides in a specific order into a chain of specific length. This allowed a much wider variety of synthetic RNAs to be used in cell-free systems. Nirenberg found that dinucleotides (two bases only) would not support the binding of specific amino acids to ribosomes. Trinucleotides (three bases), however, would do so. It was then easy to add different trinucleotides to cell free preparations and find out which labelled amino acid associated itself with the ribosomes in each case. In this way about 50 of the 64 codons in the genetic code were deciphered. It was already likely that the code consisted of trios of bases.

The rest of the codes were deciphered by Khorana in experiments which simultaneously established the triplet nature of the code. Khorana added to cell-free preparations synthetic RNAs with sequences of alternating bases (UGUGUGUG for instance) and analysed the polypeptide chains which were formed. PolyUG produced a polypeptide with alternating cystine and valine, polyAG made alternating arginine and glycine, and polyAC produced alternating threonine and histidine. When polyGUAA was added to the preparation, a mixture of peptides two or three amino acids long was formed. Therefore a codon not previously assigned, UAA, was found to be a stop codon.

CRICK'S MUTANTS OF THE T4 'PHAGE

Five years earlier, however, in 1961, Francis Crick had used a totally different technique to establish the triplet nature of the code. The T4 bacteriophage is a virus. When it is plated out on colonies of a strain of *E. coli* bacteria, it produces characteristic changes in the colonies. Mutants of the virus can be recognised by the altered responses of the bacterial colonies to infection.

Crick used a well-known mutagen, proflavin, to induce mutations in the virus. Proflavin is known to add or delete single bases to or from the DNA. Amongst the treated viruses, Crick found several strains which, on genetic analysis, seemed to have point mutations at different positions within the same gene. By infecting *E. coli* simultaneously with two different mutant strains of the virus, Crick was able to produce viruses which contained two or more of the point mutations within the same gene. He found that the effect of some mutations could be eliminated when certain other mutations were present in the same gene.

The original mutants were called 'minus' and the mutations which suppressed them were called 'plus'. Crick was able to identify several mutations of this type, to characterise them as + or −, and to combine them in different permutations. When he did so he found that the normal wild type phenotype was produced by +−, ++−−, +++−−−, −−−, and +++, but that combinations such as +, ++, ++++, −, −−, −−−−, ++−, +−− produced the mutant phenotype.

The explanation was as follows. Imagine that a − mutation is caused by a deletion of a base and a + mutation by the addition of a base. If one base is

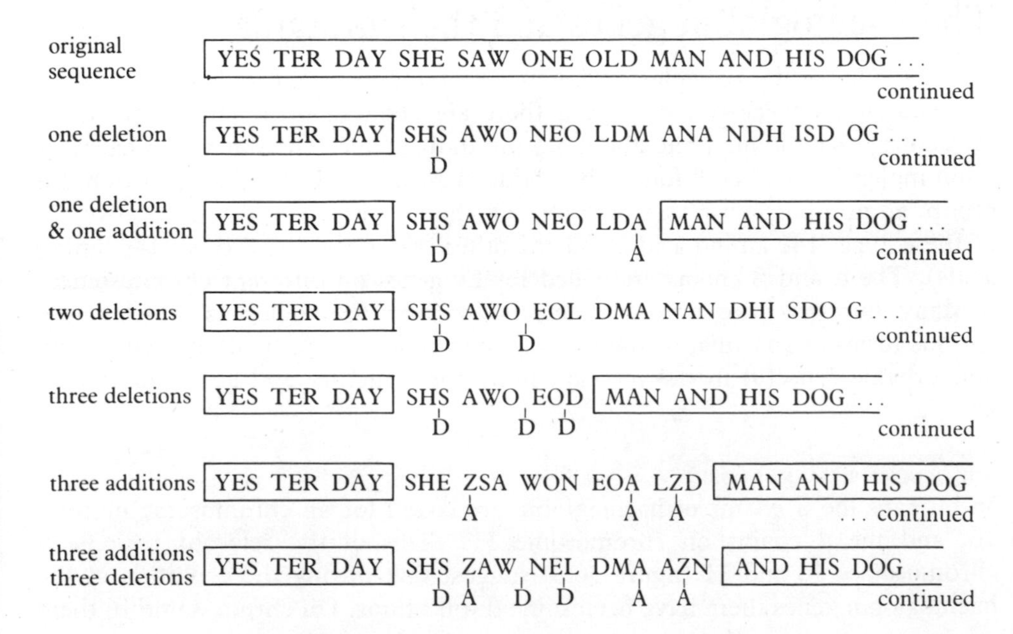

Figure 30.1 A sentence of three-letter words, showing that three additions or deletions can restore the reading frame.

removed but another is added nearby, the protein for which the gene codes will hardly be affected since the only stretch in which the 'reading frame' has been altered is between the positions of the addition and deletion (figure 30.1). The same happens each time the number of + mutations and − mutations is the same. On the other hand, one or two extra + or − mutations will alter the reading frame completely, so that the organism will produce a polypeptide chain with a greatly altered sequence of amino acids. Since +++ and −−− are exceptions to this rule, perhaps they merely alter one or two amino acids in the chain and leave the reading frame intact. If this is the case the code must be read in triplets.

The type of mutation which occurs in proflavin-induced mutants of bacteria is illustrated in figure 30.2. Here a base addition at one point in the chain was corrected by a base deletion at another point. Only the amino acid composition of the protein segment between the points was altered. If you examine these results closely, you will see that you can explain them on the basis of a triplet code.

Figure 30.2 Part of the base sequence of a gene in the food-poisoning bacterium *Salmonella*, showing the effect of one base addition and one base deletion on the sequence of amino acids in the protein. The mutations were induced by the mutagen proflavin.

normal gene (messenger RNA chain)

GUX	ACA	GCG	$^{C}_{U}$UA	CGC	GUC	ACC	CCU	GA$^{A}_{G}$

valine — threonine — alanine — leucine — arginine — valine — threonine — proline — glutamic acid

mutant gene (messenger RNA chain)

C deleted (in AAG); C added (in CCU)

GUX	AAG	CG$^{C}_{U}$	UAC	GCG	UCA	CCC	CCU	GA$^{A}_{G}$

valine — lysine — arginine — tyrosine — alanine — serine — proline — proline — glutamic acid

For consideration

1. Write out the sequence for a poly GUAA messenger RNA molecule − GUAAGUAAGUAA etc.. Using a table for the genetic code, and the fact that UAA is a stop codon, work out which peptides should result if you added this messenger RNA to a cell-free *E. coli* preparation. Predict the relative proportions in which the peptides should be formed.

2. How do the results shown in figure 30.2 support the idea of a triplet code?

Further reading

B.F.C. Clark, *The Genetic Code and Protein Synthesis* (Studies in Biology no. 83, 2nd edition, Arnold, 1984)

The haemoglobin genes and their mutants

Haemoglobin is the oxygen-carrying pigment found in the red blood cells of vertebrates. In every red blood cell there are about 250 million haemoglobin molecules, each globular and about 5.5 nm in diameter. The functional haemoglobin molecule consists of four polypeptide chains, each bonded to its own haem group, an iron porphyrin (see page 49). Of the four polypeptide chains, two are of the α type (141 amino acids) and the other two are of the β type (146 amino acids), The α and β chains are coded for by genes on different chromosomes.

Many humans have heritable defects in their haemoglobins which cause various forms of anaemia, so the haemoglobin genes are particularly well known and provide a useful model for explaining the structure and action of human genes.

THE GENES FOR HAEMOGLOBIN

In humans the α chains of haemoglobin are coded for on chromosome number 16, and the β chains on chromosome 11. Plans of the relevant regions of chromosomes 16 and 11 (figure 30.3) suggest that during the evolution of the haemoglobin genes there have been some duplications. On chromosome 16 there are two copies of the α chain gene. On chromosome 11 there are not only five different functional genes in the β chain cluster, but the DNA in between them contains the vestiges of two genes similar to the β gene which have been inactivated by mutations at some time in the evolutionary history of the species.

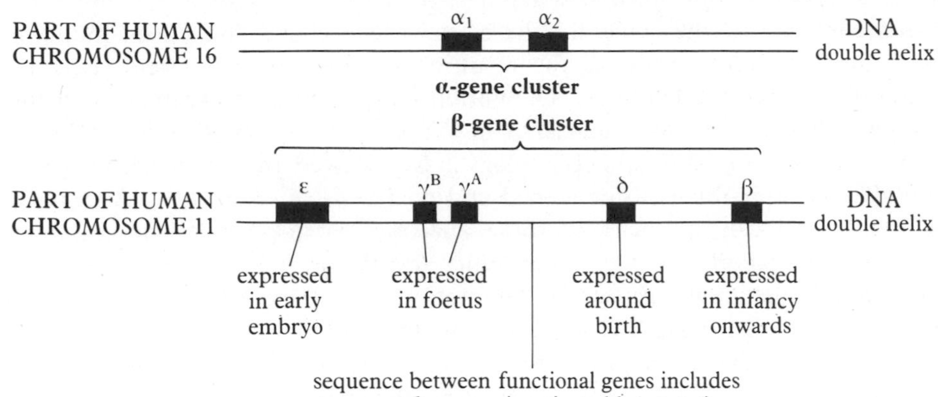

Figure 30.3 Plans of the α and β gene clusters on chromosomes 16 and 11 of human DNA. The regions between the genes, known as intervening sequences, do not code for useful information as far as is known. Over 95% of the β gene cluster (60 000) bases long) does not code for protein chains.

The genes along this β segment of chromosome 11 are switched on in turn during development. In the human foetus most haemoglobin molecules have two α chains and two γ chains (2α2γ). This provides the embryo's blood with a greater affinity for oxygen than the mother's blood, so that the foetus can take up oxygen from the maternal circulation. At about the time of birth the β chains begin to be made instead of the γ chains and the adult (2α2β) haemoglobin appears. The ∂ and ϵ genes in figure 30.3 are also involved in haemoglobin formation.

THE β CHAIN GENE IN DETAIL

Let us look further inside the gene which codes for the β polypeptide chain. A gene consists of a series of codons. A codon is a sequence of three successive bases along one strand of the messenger RNA. Each codon codes for a particular amino acid. The order of codons along the sense strand of the gene corresponds to the order of amino acids along the polypeptide chain for which the gene codes. So we should expect the β chain gene to begin with a start codon (TAC in the DNA), to continue for 146 codons for amino acids, and then to finish with a stop codon (ATT, ATC or ACT in DNA).

In general, however, eukaryotic genes are not continuous sequences like this. In the gene for the β chain of human haemoglobin, for example, the parts of the gene which code for parts of the polypeptide chain occur in three groups each known as an **expresson** or **exon** (figure 30.4). They are separated by two long regions of genetic rubbish called **interruptons** or **introns**. For the gene to be expressed a molecule of messenger RNA (mRNA) must be synthesized on its surface and the mRNA must have its genetic code translated into a protein by a ribosome. In this case it appears that mRNA is synthesized along the whole length of the gene and then 'spliced' in the nucleus so that the parts correspond-

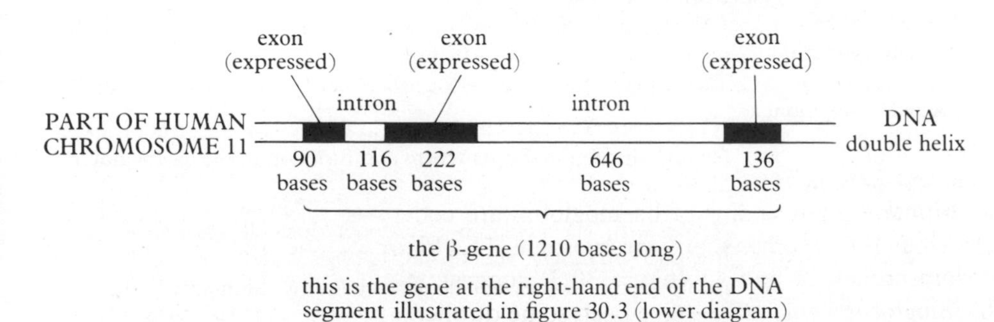

Figure 30.4 Plan of the structure of the gene for the β-chain of human haemoglobin on chromosome 11. This is a detailed view of the β gene at the right-hand end of the DNA segment of chromosome 11, illustrated in the lower part of Figure 30.3.

ing to the introns are cut out by enzymes. The shortened messenger RNA molecule leaves the nucleus and can be used as a basis for the synthesis of hundreds of β chain molecules in the cytoplasm.

THE MOLECULAR BASIS OF GENETIC ANAEMIAS

It seems likely that the genes for the α and the β chains have been mutating throughout evolution. Some of the resulting changes in amino acid sequence have probably been deleterious. They often cause inherited anaemias, diseases which are characterised by a deficiency in oxygen transport by the blood. Recent advances in recombinant DNA technology have allowed the haemoglobin genes from affected individuals to be cloned and analysed (see page 139). As a result the molecular basis of many of these anaemias is now understood. Let us consider two examples, sickle cell anaemia and thalassaemia.

Sickle cell anaemia is caused by a mutation from GUG to GAG in the sixth codon of the gene for the β chain of haemoglobin. This results in the substitution of a valine residue in place of a glutamic acid residue at the appropriate point on the β chain. This small change alters the structure and properties of the β chain so much that the oxygen-carrying capacity of the haemoglobin molecules is reduced. The allele is common in the those parts of the tropics where malaria occurs.

About 20 per cent of Africans and 9 per cent of American Negroes are carriers because they possess one copy of the sickle cell allele. This confers on them some resistance to malaria. Individuals with two copies of this allele, however, tend to be weak and anaemic and die in their early teens.

Humans who suffer from **thalassaemia** manufacture too little functional haemoglobin because the production of either the α chains or the β chains is greatly reduced. This disease is frequent in the tropics, probably because, as with sickle cell anaemia, carriers of the gene have some resistance to malaria. Sufferers from thalassaemia major, with two mutant alleles, have no haemoglobin at all in some of their red blood cells and die young from multiple symptoms, including severe anaemia, iron poisoning and heart and glandular defects. In thalassaemia Constant Spring (named after the area where the first case was described), the affected individual hardly produces any α chains at all. This is because of a mutation in the stop codon at the end of the gene coding for the α chain, from ATT (stop) to GTT (glycine). In the synthesis of messenger RNA on the surface of this gene, the mRNA continues past the altered stop codon at position 141, stopping only at codon 172. This abnormally elongated mRNA is unstable and unable to bring about α chain synthesis.

The understanding of the molecular basis of some of these genetic disorders raises the faint hope that in future the diseases may be corrected to some extent by genetic manipulation. For example, stem cells taken from the red bone marrow of an individual might be injected with genes to correct the defect and then reintroduced into the bone marrow to churn out normal cells. A safe procedure such as this is, however, a long way off.

For consideration

1. (a) With reference to figure 30.3, explain how the cells in the baby which produce red blood cells switch off the γ chain gene and switch on the β chain gene at birth? (b) Why might such knowledge be valuable in the treatment of individuals whose foetal haemoglobin persists into adult life?

2. Look up the structures of the amino acids valine and glutamic acid. Would it be possible to distinguish between normal haemoglobin and sickle-cell haemoglobin by electrophoresis?

Further reading

J.E. Darnell, 'The Processing of RNA' (*Scientific American*, vol. 249, no. 4, 1983)
This explains the evidence for RNA splicing and how it may occur.

N. Maclean, *Haemoglobin* (Studies in Biology no. 93, Arnold, 1978)
An introduction to the structure and function of the haemoglobin molecule.

N. Maclean, 'Haemoglobin – A Model Molecule' (*Journal of Biological Education*, vol. 15, no. 1, 1981)
An update of the previous book, containing a summary of the genetic anaemias.

31 . Genes and development

Plant breeding, cloning and cell fusion

The selection of improved strains of crop plants has been taking place intentionally and unintentionally for several thousand years. In the last twenty years, however, the pace of plant breeding has quickened. Most of the successes have been the result of the established techniques of hybridisation and selection. Recently, several other potentially valuable techniques have been devised to cross and to propagate plants. This topic describes some of the new methods.

A plant breeder has two main aims. The first is to create diversity by producing new strains or hybrids from which the best can be selected. The second is to minimise the variation in the desirable strains produced by obtaining plants with a high degree of uniformity.

HOW DIVERSITY CAN BE CREATED

One conventional way to create new strains of crop plants is to cross the crop species with a wild relative with a desirable feature, such as resistance to fungal and viral pathogens. This method was used to breed disease resistant strains of the potato grown in the Western world, *Solanum tuberosum*, by crossing them with a species from the Andes, *Solanum demissum*. *S. demissum* has hairs on its stems which secrete gum and trap aphids, the main vectors of virus diseases.

Amongst other notable successes were the production of rapidly maturing, short-stemmed varieties of rice which responded particularly well to fertilizers and gave an exceptionally high yield of grain. Rice yields in south-east Asia doubled or trebled as a result, creating the so-called 'green revolution'. In America, the basis of the crop of maize (*Zea mays*) for the past forty years has been F1 and F2 hybrids, which exhibit **hybrid vigour** and considerably outyield the parental strains.

Some crop plants, however, cannot be crossed to yield viable seeds. In such cases it may be possible to cross two species by **cell fusion**, that is fusing their somatic (non-reproductive) cells. Potatoes and tomatoes, for example, have been crossed in this way (figure 31.1). The first step is to create naked protoplasts, by removing the cell walls of plant leaf cells. The cell walls are dissolved by pectinase and cellulase enzymes. This leaves the spherical cell contents, nucleus and cytoplasm surrounded by the cell membrane, floating as a suspension in an isotonic sucrose solution. Then the mixture of protoplasts from the two species is treated with a substance called polyethylene glycol. Under its influence the cell membranes of adjacent cells combine, and later the nuclei inside the hybrid cell fuse. If the two species which have been fused are too dissimilar, the hybrid nucleus loses many chromosomes, as in hybrid cells of human and mouse cells (see page 128). Otherwise the nuclei remain intact and plantlets can be grown from individual cells in tissue culture. So far this technique has had no commercial impact.

It was in the 1950s that F.C. Steward first showed how complete plants could be grown in culture from individual cells derived from carrot tap roots. Since

A

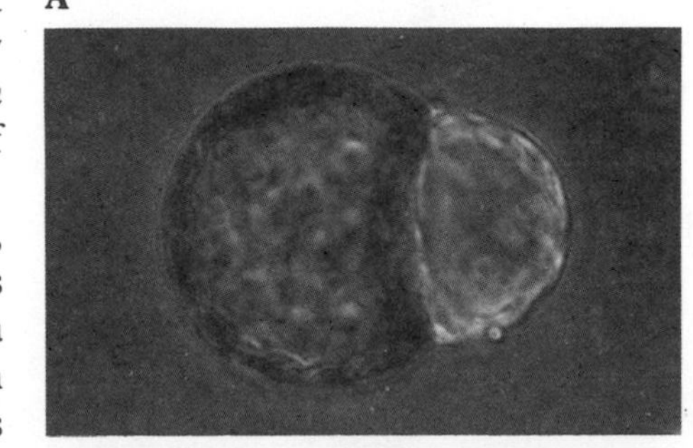

B

Figure 31.1 **A** The fusion of protoplasts from a potato (on the left) and a tomato (on the right) under the influence of polyethylene glycol. Several different hybrids have been produced in this way. If they possess chloroplasts derived from the potato they are known as 'pomatoes', whilst the ones with chloroplasts from the tomato are called 'topatoes'. As yet such hybrids have no commercial importance. **B** From left to right – potato leaves, hybrid leaves, tomato leaves. Notice how the hybrid shows characteristics of both the 'parent' plants.

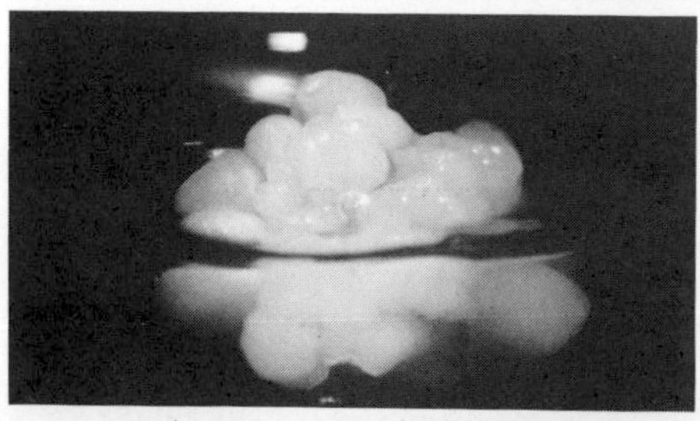

A

B

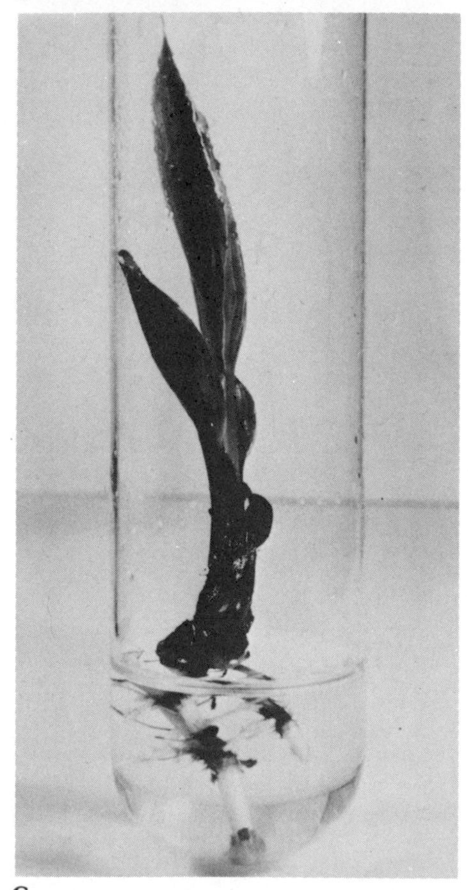

C

Figure 31.2 Three stages in the production of plantlets of coconut palm from small pieces of living callus tissue. **A** Development of embryoid structures from callus. **B** Germination of embryoids to form shoots. **C** Coconut plantlet ready for establishment.

then methods for culturing plant tissues have been considerably refined, though there is still an element of luck. Individual cells or groups of cells from the suspension are grown in sterile conditions on agar which contains sucrose, nutrient ions, vitamins and amino acids. Growth hormones such as auxins and cytokinins are required, and their timing and concentrations are vital to the success of the enterprise. Lumps of undifferentiated cells, known as **calluses**, form first, and plantlets emerge from them later.

Even when the calluses are derived from the cells of the same plant, there is considerable phenotypic and genotypic variation amongst the plantlets which are formed. They often differ in chromosome number, exhibit chromosomal aberrations or possess single somatic gene mutations. This itself is a potentially valuable source of diversity. For instance, to select plantlets of sugar cane resistant to the eyespot fungus *Helminthosporium sacchari*, a large number of genotypes in tissue culture were dosed with the toxin normally produced by the fungus. The ones which survived were the basis of the new resistant strains.

THE CREATION OF UNIFORMITY

There are three well-tried methods of developing uniformity in a crop strain. One is to **inbreed** the plants for several generations. In inbreeding, half the heterozygosity is lost in each generation, and the result is a homozygous strain of the crop plant. A second technique is to cross two inbred strains to produce F1 hybrids. There is little genotypic or phenotypic variation amongst the hybrids. Thirdly, an age-old ploy is merely to take **cuttings** from a valuable individual of a perennial species. Such cuttings should be genetically identical, and ultimately may be grown over a wide area. Plants propagated in this way include King Edward potatoes and Bramley's seedling apples.

Some crop plants, however, exhibit considerable genotypic and phenotypic variation between individuals, and because of incompatability self-pollination may be impossible (see page 107). Taking cuttings may also be impossible. How can the best individuals be propagated?

The oil palm and the coconut palm, for instance, have only one growing point. In this case tissue culture is the answer. Cells taken from a particularly productive individual are grown in tissue culture and the calluses are split to yield as many individuals as possible (figure 31.2). This promises to revolutionise the coconut industry, which produces nuts, oil, and copra of tremendous economic importance.

Another way of producing uniformity quickly might be to generate new plants from pollen grains. Imagine that you have bred a perennial plant with desirable characteristics. It may be highly heterozygous, so that its offspring may be very variable. This would make the strain unsuitable as a long term crop plant. If the generation time was lengthy, a programme of inbreeding and selection might take too long. Much quicker would be to collect its haploid pollen grains, treat them with the drug colchicine, which induces chromosome doubling, and grow the cells individually in tissue culture. Whole plants can be grown from pollen grains in this way. We should expect the resulting diploid plants to exhibit a range of phenotypic variation because as products of meiosis, each pollen grain and the plant arising from it should be different in genotype. Those plants which exhibited the desirable characteristics of their parent could be selected. Each plant is no longer heterozygous; because of the chromosome doubling which has taken place it should be homozygous at all its loci.

Techniques like this are becoming more and more important in conservation and horticulture. Rare orchids, for example, have been propagated for decades from segments of apical meristem in tissue culture, yielding a healthy profit for the growers and ensuring at the same time the survival of the species.

For consideration

1. What are (a) the advantages and (b) the disadvantages of genetic uniformity within a crop?

2. You have two haploid plants of the same species with different but desirable characteristics. Suggest two ways of producing a diploid plant with a combination of these characteristics.

3. Suggest, in detail, one reason why a hybrid between two different inbred strains of a crop plant is usually more vigorous than either of its parents.

Further reading

J.B. Land and R.B. Land, *Food Chains to Biotechnology* (Nelson, 1983)
This contains a clear introduction to the principles and practice of plant breeding programmes in agriculture.

C. Tudge, 'The Future of Crops' (*New Scientist*, vol. 98, no. 1359, 1983)
This and other well illustrated articles in the same issue survey the potential for increasing crop production by modern techniques.

Genetic engineering

The genetic code is universal, that is, it is identical in all organisms. The same piece of nucleic acid should code for a protein of the same amino acid sequence in whatever species it is expressed. It should therefore be possible to add foreign DNA to a cell and make it manufacture a protein characteristic of a different species. This is known as **genetic engineering** or **recombinant DNA technology**.

Genetic engineering has been most successful so far with recipient organisms such as bacteria and yeast. Both these species are well known genetically, they both have plasmids, they reproduce quickly in simple culture, they are easy to cultivate and can rapidly produce large quantities of proteins for medicine or research. The value of this approach is illustrated by the fact that the isolation of 5 mg of the protein somatostatin required half a million sheep's brains, but the same mass of this protein can be produced in a week by nine litres of genetically engineered bacteria.

In this topic we shall discuss the problems encountered by scientists trying to express foreign genes in bacteria by explaining how human insulin went into large scale production.

There are three main problems in recombinant DNA technology. The first is preparing the donor DNA. The second is to introduce the DNA into bacteria in such a way that the foreign gene is expressed. The third is selecting those bacteria which successfully express the gene, and eliminating the others.

PREPARING THE DONOR DNA

It is often difficult or impossible to extract the appropriate gene from the chromosomes of the organism which produces the protein. Attacking the chromosomes with **restriction endonuclease enzymes** may produce tens of thousands of DNA fragments of a variety of lengths. The only reliable way of finding the gene you want is to expose its complementary RNA on a column and 'fish' for it by flushing the gene fragments down the column.

Genes derived directly from eukaryotes may be unsuitable for expression in bacteria. Most genes of eukaryotes contain **introns** (see page 135), lengths of genetic rubbish which are transcribed onto the messenger RNA but cut out of the RNA during 'splicing'. Bacteria have no splicing enzymes and so the rubbish is translated, producing an elongated polypeptide chain of unpredictable structure.

The best way to prepare donor DNA is therefore to manufacture the gene which is to be incorporated in the bacterium. The correct base sequence may be known from sequence analysis of the DNA. Alternatively, as was the case in the manufacture of human insulin, the amino acid sequence may be known. Molecular biologists could work out from the known genetic code what sequence of DNA bases would code for the correct sequence of amino acids in the insulin chain.

Insulin, which is secreted by cells in the human pancreas, is a protein of 51 amino acids whose primary structure was first worked out by Sanger. It consists of an A chain (21 amino acids) bonded to a B chain (30 amino acids) by two disulphide bonds. The research team who began to modify a bacterium to synthesize human insulin decided to make the genes for the A and B chains separately, and to join the polypeptide chains together after manufacture. It took two months to assemble the A-chain gene and three to make the B-chain gene.

INTRODUCING DONOR DNA INTO BACTERIA

Bacteria can absorb fragments of DNA from the medium in which they are growing. There is no guarantee that if a bacterium takes up some DNA it will become incorporated in the genotype and reproduce along with the bacterial genes. However, if the donor gene is placed within a **plasmid** there is a strong chance that it can be carried for generation after generation in the bacterial cell.

Several plasmids have been synthesised by molecular biologists for use in genetic engineering. The value of using a plasmid such as the pBR322 plasmid (figure 31.3) is twofold. Firstly, it possesses two genes which confer antibiotic resistance on any bacterium which contains the plasmid. Those bacteria which lack the plasmid can rapidly be selected out of a culture by adding the appropriate antibiotic. Secondly, the exact sites at which the plasmid can be cleaved by each of the common restriction endonuclease enzymes are known (figure 31.3). This enables the donor gene to be incorporated at a particular place in the plasmid.

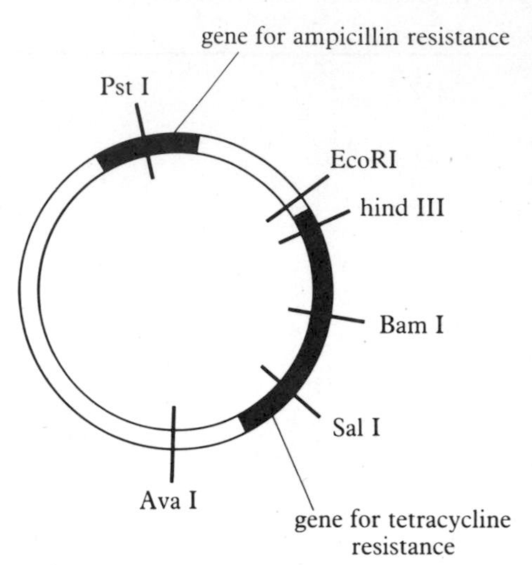

Figure 31.3 The pBR322 plasmid, showing its genes for drug-resistance and the positions at which it can be cut by different restriction endonuclease enzymes. (*After Gibson*)

To produce synthetic insulin, certain restriction endonuclease enzymes were used to split the plasmid and to produce complementary 'sticky ends' on both plasmid and donor gene. With the aid of the enzyme **DNA ligase**, which joins strands of DNA, the donor gene was incorporated into the gene for tetracycline resistance in the plasmid, thus inactivating the tetracycline-resistance gene.

SELECTING THE RELEVANT BACTERIAL STRAINS AND OBTAINING THE POLYPEPTIDE

Millions of plasmids were then mixed with a population of *E. coli* strain K12. Some of the plasmids were picked up by the bacteria. Which bacteria contained the plasmids with the suite of insulin genes in the correct position? These bacteria were recognised because they continued to divide in a medium containing the antibiotic ampicillin, having the gene for ampicillin-resistance on their plasmids, but did not divide in tetracycline, having an inactivated resistance gene.

One potential problem in these new, genetically engineered strains of bacteria is that although the gene for the relevant protein might be present, it might be switched off. For this reason, before insertion in the plasmid, each donor gene was bonded to a segment of DNA which carried the gene for the lac operon in *E. coli*. This is the suite of genes, present on the bacterial chromosome, which synthesize messenger RNA when lactose is added to the surrounding medium. When lactose is added to the genetically engineered strain, it uncovers the operator gene (figure 31.4) and a messenger RNA molecule is synthesized for both β-galactosidase and the donor gene. The result is that the cell secretes not just the desired protein, but the desired protein bonded to β-galactosidase.

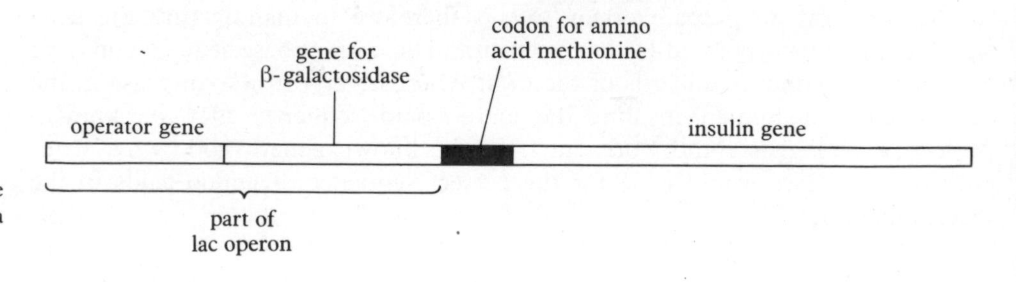

Figure 31.4 Sequence of codons attached to the pBR322 plasmid during the production of a human insulin chain in *E. coli*.

The research team producing synthetic insulin had, however, thought of that problem. Between the β-galactosidase gene and the donor gene they had incorporated in the plasmid a codon for the amino acid methionine (figure 31.4). Methionine happens to be attacked and dissolved by cyanogen bromide. When this chemical was added to the genetically engineered protein it split in half, releasing a free insulin peptide.

The strains containing the genes for the A and B chains of insulin were grown in separate cultures. Having purified the A and B chains, the researchers worked out a method to bond them together. Successful union was achieved in 1978. Since then the American company Eli Lilly has scaled up the process, and carried out clinical trials of the new product in hospitals. Factories specifically designed for insulin production were built in Indianapolis and Liverpool (figure 31.5), and the product is now being used routinely to depress the blood sugar levels of four million diabetics in the developed countries.

Figure 31.5 A technician checks the specific gravity of a fermentation solution at Dista Products Ltd., Liverpool. The solution contains genetically engineered bacteria which produce human insulin. The 'key workers' in the factory are these microscopic bacteria – yielding insulin on a macroscopic scale!

SOME OTHER USES FOR GENETIC ENGINEERING

Other strains of genetically engineered *E. coli* have been useful in producing large quantities of particular genes for base sequencing. The genes from humans are cloned in *E. coli* and sequenced by a rapid modern technique. In this way the sequences of over a hundred mammalian genes have already been determined, including those for insulin, haemoglobin and its mutants (see page 134), interferon and cytochrome c.

One of the most promising potential uses of genetic engineering is in the production of vaccines. At present a heat-killed or attenuated strain of a pathogen is injected and the host organism manufactures antibodies against the foreign proteins on the injected material. The success of recombinant DNA technology will make it possible to manufacture vaccines containing antigens, produced in micro-organisms, of any desired amino acid sequence.

For consideration

1. What are the advantages and disadvantages of using yeast, rather than *E. coli*, to express foreign genes in genetic engineering?

2. DNA from woolly mammoths and Egyptian mummies has been cloned in *E. coli* to produce enough identical DNA for its base sequence to be determined. Of what value is this kind of research?

3. In what circumstances might genetic engineering be unsafe?

Further reading

S.N. Cohen, 'The Manipulation of Genes' (*Scientific American*, vol. 233, no. 1, 1975)

W. Gilbert and L. Villa-Komaroff, 'Useful Proteins from Recombinant Bacteria' (*Scientific American*, vol. 242, no. 4, 1980)

R. Gibson, 'Genetic Manipulation, Principles and Practice' (*Biologist* vol. 29, no. 4, 1982)
An intelligible introduction, highly recommended.

R.P. Novick, 'Plasmids' (*Scientific American* vol. 243, no. 6, 1980)

S. Petska, 'The Purification and Manufacture of Human Interferons' (*Scientific American*, vol. 249, no. 2, 1983)

Transferring genes to higher plants

Many dicotyledonous plants, in fact over six hundred different species, can be affected by a cancerous growth known as **crown gall tumour** (figure 31.6). The proliferation of cells on the stems or at the shoot tip is stimulated by a soil bacterium, *Agrobacterium tumiefaciens*. This apparently mundane form of parasitism is potentially valuable to humans, for the way in which the bacterium infects plants suggests a way by which we might be able to genetically manipulate higher plants.

A. tumefaciens cells invade a stem through a wound and stick to the pectins in the middle lamella of the cell wall. Each bacterium contains a plasmid (see page 140) consisting of about 20 000 base pairs. On this plasmid there are three main sets of genes. One set is essential for the plasmid to inject itself through the cell

Figure 31.6 A crown gall tumour on a sugar beet plant. The tumour causes shoots and roots to grow in a disorganised fashion.

membrane of the host cell. Once inside, the plasmid travels to the nucleus, where a second set of genes incorporates itself into a random position in the host cell chromosomes.

There are thirteen genes in this particular set. They code, amongst other things, for auxin and cytokinins, which promote the cell divisions which cause the cancerous tumour to develop. There is also a gene which codes for an **opine**, an unusual derivative of amino acids which occurs in crown gall tumour but not in normal cells. The metabolism of the host cell is diverted to making opines, which pass across the cell membrane to be taken up by the descendants of the original bacterium in the cell wall. These bacteria digest the opines as a source of organic nitrogen compounds. The third set of genes on their plasmids code for opine-digesting enzymes (figure 31.7).

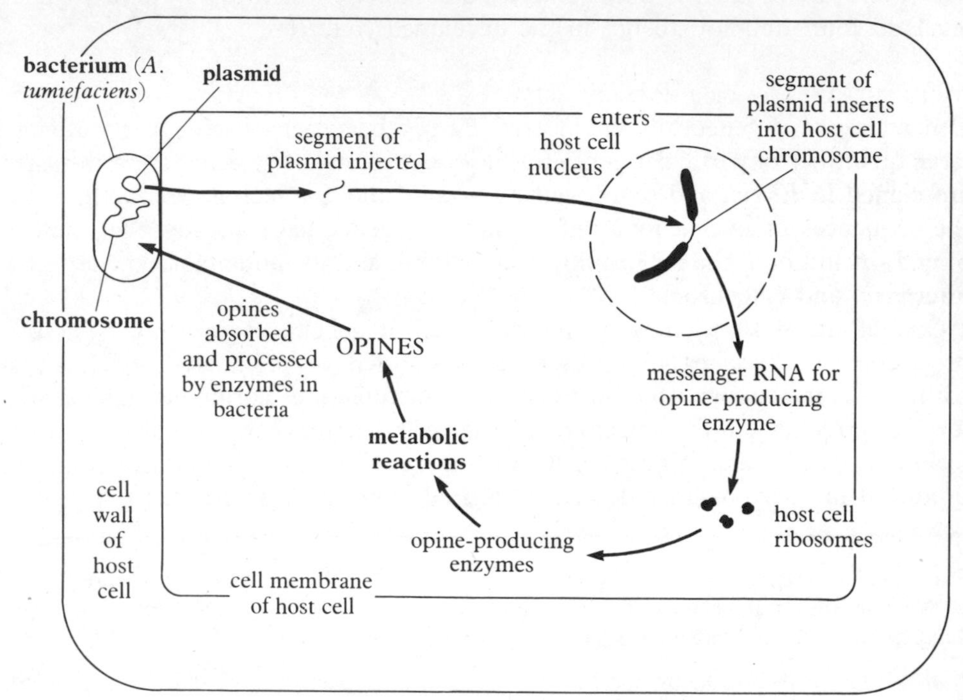

Figure 31.7 By inserting some of its own genes into the chromosome of the host cell, *A. tumiefaciens* can alter the metabolism of the host cell and absorb and process the new compounds (opines) which are made.

The striking thing about the bacterial DNA, when it has inserted itself into the host cell DNA, is that it can be transmitted through the egg and the pollen like a normal Mendelian gene. If, by genetic engineering, a useful gene could be joined to the part of the plasmid which invades the host cell, the useful gene might be incorporated in the host cell chromosomes and inherited by the plants' offspring.

This has already been done. Genes cloned in *E. coli* have been transferred to the plasmid and passed to the host cells. The tumours have been removed and new plants, all identical in genotype, propagated from the fragments. The most spectacular success so far, using crown gall, is the successful transfer of a gene from a fruit fly into a petunia! If this is possible, most gene transfers must surely be possible once our knowledge of plant genes improves.

For consideration

1. Suggest some commerical applications of our ability to insert new genes into plants by using *A. tumefaciens* as a vector.

2. Plan a programme to transfer the genes for nitrogen fixation (see page 47), carried on a plasmid, from the bacterium *Klebsiella* into a potato, using *A. tumefaciens* as a vector. What problems might you encounter in trying to manufacture nitrogen-fixing potato plants in this way, and how might you overcome them?

Further reading

M.D. Chilton, 'A Vector for Introducing Genes into Plants' (*Scientific American*, vol. 248, no. 6, 1983)

A. Hepburn, 'Mother Nature Got There First' (*New Scientist*, vol. 99, no. 1377, 1983)

J.D. Watson, J. Tooze and D.T. Kurtz, *Recombinant DNA – A Short Course* (Freeman, 1983)

32 . The organism and its environment

The energetics of food production

In his essay on population in 1778, Malthus predicted that famines were inevitable because although population increases geometrically, food supply only increases arithmetically. His alarmist predictions have not yet come true, at least on a global scale, largely because improved crop plants, fertilizers, pesticides and mechanization of farms have allowed food production to keep up with the increase in the population. Nevertheless, the land area on which crops can be grown and livestock raised is finite. An understanding of the main limiting factors in the conversion of solar energy to food is essential if we are to make the best use of the land available.

THE EFFICIENCY OF PRIMARY PRODUCTION

Primary production is the accumulation of dry mass by plants, some of which can be harvested as food. Imagine a crop plant growing in ideal environmental conditions at such a density that none of the leaves overlap, so there is no shade, and most of the light is absorbed. How efficient is it?

Of the total solar radiation (wavelengths 200–10 000 nm) only about half is in the visible wavelengths (400–700 nm) which can be absorbed and used by the photosynthetic pigments in the leaves. Some of this energy, particularly in the green wavelengths, is reflected from leaves or passes straight through them. Thus only about forty per cent of the total solar energy can enter the light reaction of photosynthesis.

Furthermore, the light reaction is inefficient. As the electrons energised by light move around the circuit of electron carriers in non-cyclic photophosphorylation, and the proton pumps operate in chemiosmosis across the thylakoid membranes (see page 43), much energy is lost as heat. At the most, only a quarter of the energy trapped by the pigments ends up in the potential chemical energy of glucose and other molecules. This represents only ten per cent of the total solar energy falling on the plant.

The glucose formed, however, is not all devoted to growth and starch production. About a quarter to a third is expended on metabolic reactions such as respiration, which provides the ATP required for cell division, growth and maintenance. Moreover, plants are ectotherms and their rates of respiration and photorespiration are strongly temperature dependent. C4 plants, with their additional Hatch–Slack pathway of photosynthesis and lack of photorespiration, have net production efficiencies higher than those of C3 plants.

Bearing in mind all these considerations, the optimum rate of conversion of solar energy to biomass is about seven or eight per cent of total solar radiation. Efficiencies of this magnitude have been achieved in cultures of unicellular protists, agitated in large plastic tubes at high temperatures, and supplied with abundant light, carbon dioxide and nutrients. In these cultures, as distinct from higher plants, every cell photosynthesises. This highlights the fact that in crops, with an equivalent annual efficiency of 1.5 per cent for wheat and 2.5 per cent for maize, production is limited by environmental factors which, if modified, might allow crop production to be increased.

Some of these limitations are as follows. The leaves of a wheat or maize crop do not cover the ground surface for the whole year, and growth is often limited by lack of light, low temperatures, lack of water, nutrient deficiency, or pests. Irrigation, proper soil management, the addition of fertilizers, the breeding of crop strains responsive to high levels of fertilizer, the use of pesticides and the breeding of pest-resistant crop strains all help to overcome these limitations. Unfortunately the manufacture and transport of nitrogenous fertilizers and pesticides depends mainly on energy provided by the combustion of fossil fuels. This makes them too expensive for many non-industrialized countries, and it is the less-developed countries in which the imbalance between population and food supply is most marked.

THE EFFICIENCY OF SECONDARY PRODUCTION

Another potential source of food is secondary production, the harvesting of animal material from natural or artificial ecosystems, such as fish from the sea or bullocks from a pasture. Since the animals waste so much of the energy provided by the plants, this represents a loss of energy to humans, although the nutritional quality of the proteins from animals is somewhat higher than from plant food. The animals do not eat all the plant biomass, and much of it dies before it can be eaten. In fact it is unusual for the animals to eat more than forty per cent of the plant biomass in an ecosystem (Table 32.1). A high proportion of the ingested energy is egested, even in ruminants and termites despite their mutualistic gut symbionts which increase the efficiency of cellulose digestion.

Table 32.1 Percentage of net primary production consumed by primary consumers in various ecosystem studies.

Plant community	Consumers	Production consumed (%)
Grass and herbs	Invertebrates	0.5
Grass	Invertebrates	9.6
Marsh grass	Invertebrates	7.0
Sedges and grasses	Invertebrates	8.0
Aquatic plants	Bivalve molluscs	11.0
Algae	Zooplankton	25.0
Phytoplankton	Zooplankton	40.0

The rest of the energy passes across the gut wall and is assimilated by the cells. Some is ultimately lost as heat in respiration and other metabolic reactions, and some is expelled from the body as nitrogenous waste such as urea. The rest is added to the biomass of the animal; this latter represents the material which we can eat. You will appreciate that it is even less efficient to harvest secondary consumers, such as fish from the sea, than it is to harvest primary consumers, such as bullocks.

Several factors influence the rates at which animals in ecosystems use energy. Herbivores egest a higher proportion of the energy in their ingested food than carnivores. Of the energy passing across the gut wall into the organism, a higher proportion is used in respiration in endotherms than ectotherms of the same mass, since endotherms need to respire faster to maintain their high body temperatures and metabolic rates. Small endotherms, like pygmy shrews or young birds, have higher respiration rates per unit mass than large endotherms. One reason for this relationship could be that small endotherms have a large ratio of surface area to volume over which heat can be lost. Of course the age of an organism also influences its energy budget. The graphs of increase in mass against time for most organisms are sigmoid. This means that during the growing stage of an animal's life it devotes a higher proportion of its assimilated energy to growth than when it has reached full size.

Our emerging understanding of the factors limiting primary and secondary production should enable us to increase food production still further. We must hope fervently that now that birth rates are declining rapidly in many less-developed countries, the supply and distribution of food will increase fast enough to avert a Malthusian catastrophe.

For consideration

1. Comment on the likely value of earthworms, grown on household rubbish, as a potential food source.

2. Why have plants not evolved to use the energetic ultra-violet wavelengths of light in photosynthesis?

3. Place these organisms in order from the most efficient to the least efficient in converting ingested food energy to secondary production during their growing stages: pygmy shrew, elephant, cabbage white butterfly, snake. Explain the reasoning behind your sequence.

Further reading

J.M. Anderson, *Ecology for Environmental Sciences: Biosphere, Ecosystems and Man* (Arnold, 1981)
This contains concise and interesting discussions of primary and secondary production.

D.O. Hall, *Biological Solar Energy Conversion* (*Biologist*, vol. 26, no. 17, 1979)
A brief quantitative summary of the energetics of crop plants.

Energy flow and nutrient cycling in ecosystems

The concept of an ecosystem is valuable because it allows energy flow and nutrient cycling in different ecosystems to be compared in the same units. In this topic we shall compare a temperate marine bay with a temperate forest at the same latitude and try to account for the differences.

ENERGY FLOW

Figure 32.1 illustrates several important differences between the two communities. The major difference is that a much higher proportion of the net primary production is eaten by herbivores in the marine bay than in the forest. The zooplankton in the bay appear to be efficient grazers. The herbivores in a forest, however, are so inefficient that about eighty-eight per cent of the net primary production becomes detritus. Let us try to explain this phenomenon.

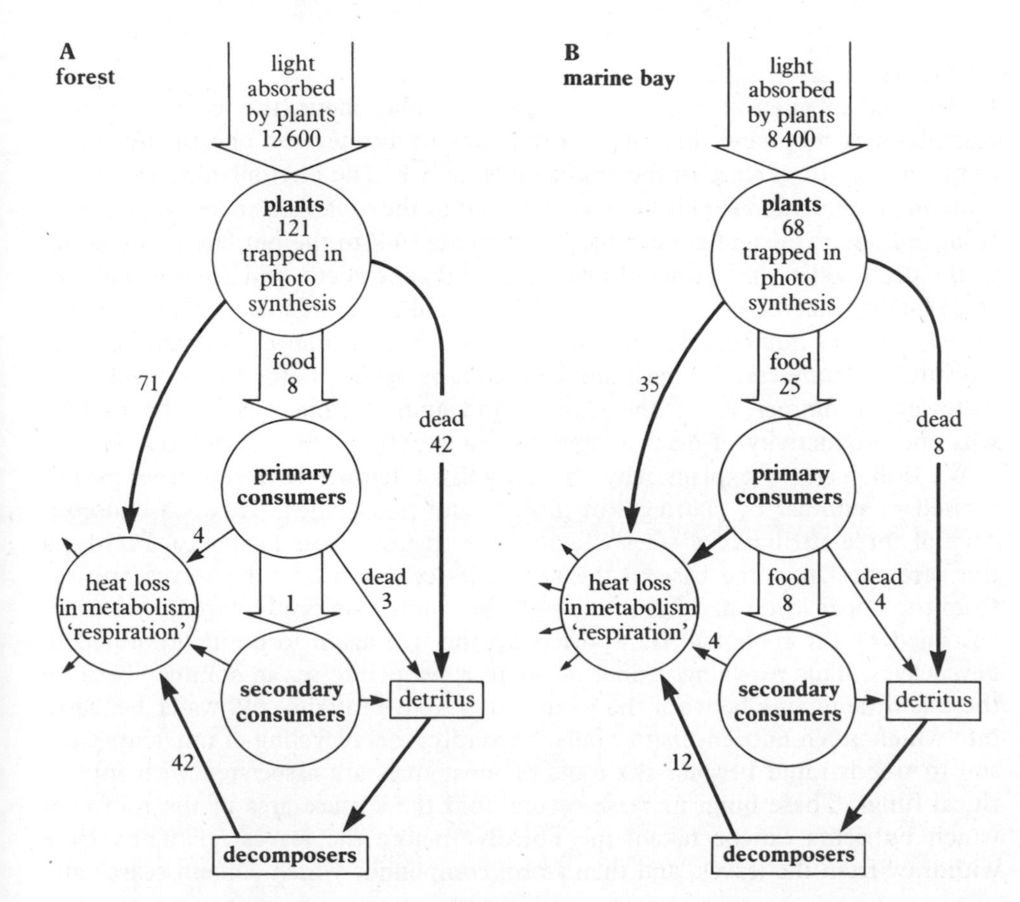

Figure 32.1 Energy flow diagrams for (**A**) a temperature deciduous forest and (**B**) a marine bay. All the estimates are in kJ m^{-2} day^{-1}. The plants in the marine bay trap a much higher proportion of the solar radiation in photosynthesis than in most open seas. (*Data from E. P. Odum, Japanese Journal of Ecology, 1962*)

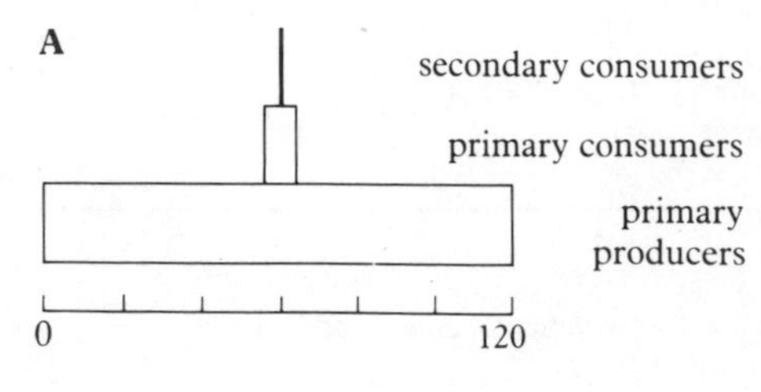

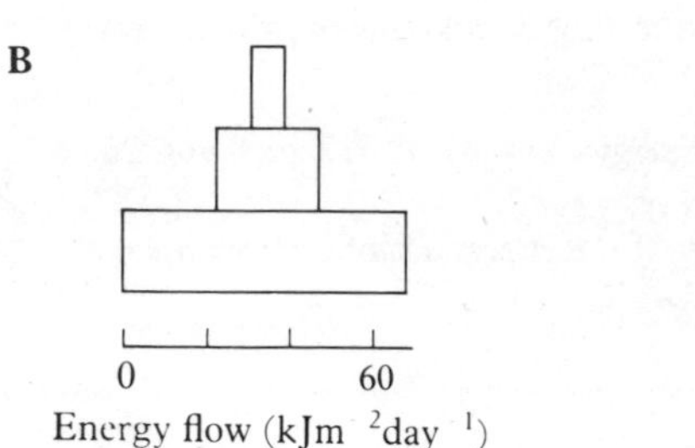

Figure 32.2 Pyramids of energy flow for (**A**) the temperate deciduous forest and (**B**) the marine bay, calculated from the data in Figure 32.1.

In a deciduous temperate forest, tree leaves are exposed to the sun for only half the year. Nevertheless, about one per cent of total annual solar radiation is trapped in plant mass. In a climax forest there may by as much as eight square metres of leaf for every square metre of ground surface. Why don't the herbivores benefit from this vast quantity of herbage? Why are the trees not stripped of their leaves each year by insects which invade forests like a swarm of locusts?

Insect herbivores in forests are between the devil and the deep blue sea. Their plant food is largely inedible because of the plants' chemical defences, and they are continually preyed upon by voracious predators, which regulate their population sizes. A high proportion of tree biomass is dead xylem, low in water and nitrogen but rich in indigestible lignin. The leaves are often protected by chemical compounds such as tannins and alkaloids, which cost the tree much energy and nitrogen to manufacture. The result is that the pyramid of energy flow for a forest tapers rapidly between the primary producers and the primary consumers (figure 32.2).

Why hasn't much unpalatability evolved amongst phytoplankton? Most zooplanktonic individuals are filter feeders which eat hundreds of protists of several species simultaneously, without discrimination. Their method of feeding is unlikely to result in natural selection for unpalatability. Most forest herbivores, however, are far smaller than the plant on which they feed, and specialize on a single host tree species. Palatable trees will be defoliated and this will cause selection for unpalatability.

Why is the sea not carpeted by large overlapping seaweeds, trapping solar energy with almost the same efficiency as leaves in a forest? We really do not know. Of course less light reaches the phytoplankton in the sea than the leaves in a forest, since the water surface reflects some solar radiation and absorbs some of the rest. But given that the phytoplankton consists mainly of organisms with one or very few cells, why isn't the sea bright green like pea soup? Part of the reason may lie in nutrient deficiency.

NUTRIENTS

In polluted inland waters rich in nitrates and phosphates the water may well resemble pea soup, and this suggests that lack of nitrates and phosphates might limit the growth of algae in the sea in midsummer. The phytoplankton begin to undergo a population explosion in spring, but as the organisms take up nutrients from the water, die and are eaten, the nutrients sink to the bottom of the ocean in the dead bodies and egesta of producers and consumers, and are not replaced from below. Starved of nutrients, the phytoplanktonic organisms cannot reproduce, and their numbers decline, mostly through predation by zooplankton. As a result, the numbers, dry mass and productivity of the photosynthetic cells is at its lowest in summer, just at the warmest and sunniest time of year. This may be why the productivity of deep temperate seas resembles that of deserts.

We still have to explain why the growth of temperate forest trees is not limited in summer by shortages of nitrates and phosphates. This is a consequence of three attributes of trees which all result from their longevity. Firstly, a tree stretches from the base to the top of its ecosystem. If the leaves are lost from the upper layers and fall to the soil, their nutrients can be rapidly released, absorbed by the roots and transported up the tree again to be incorporated in new leaves. This recycling cannot occur in a temperate sea in summer because there is little mixing between the warm water above the the cold water beneath, into which much nutrient matter falls. Secondly, this recycling of nutrients from soil to tree is rapid because the roots of most trees are associated with mycorrhizal fungi. These fungi increase several-fold the surface area of the root over which nutrients can be taken up. Thirdly, before the leaves fall many trees withdraw from the leaves, and then store, compounds which contain scarce and valuable elements such as nitrogen. These are used when the next crop of leaves is produced.

Thus in the final analysis most of the differences in energy flow and nutrient cycling between temperate forests and temperate seas result from their different types of plant inhabitants. Long-lived trees reproduce slowly and devote most of their net primary production to vegetative growth, whereas short-lived algae duplicate themselves by binary fission every few minutes.

For consideration

1. The net primary production (photosynthesis minus respiration) of the plant plankton is often measured by the light and dark bottle method. In this, samples of sea water are suspended just below the surface in two bottles, one with clear glass and the other painted black. The concentration of oxygen is measured in both bottles at the beginning and end of a twenty-four hour period. Suggest how you could use these figures to calculate the net primary production of the phytoplankton, and discuss possible sources of error in the technique.

2. In most temperate seas the biomass of the phytoplankton, having fallen in the summer, increases again in autumn. Suggest some reasons why this autumn peak occurs.

Further reading

T.J. King, *Ecology* (Nelson, 1980)
A more detailed discussion of energy flow and nutrient cycling in ecosystems.

P. Collinvaux, *Why Big Fierce Animals Are Rare?* (Penguin, 1975)
The essay entitled 'Why the sea is blue' discusses why the sea is not clogged up with giant seaweeds.

G. Monger (ed.), Revised Nuffield Advanced Science: *Biology Study Guide II* (Longman, 1986)
Chapter 29 contains some particularly useful data and study exercises on energy flow and nutrient cycling.

Succession

In the 1890s, a young ecologist, Frederick Clements, organised a mule train to take him around Nebraska to look at vegetation. What he saw revolutionised plant ecology. On land abandoned by recent settlers, and in places where buffalo used to graze, he observed that one set of species appeared to follow another in a predictable sequence. Eventually the species composition of the vegetation appeared to stabilise, reaching a climax stage characteristic of the area.

Clements was the first person to point out forcefully that plant communities change with time. He thought of a plant community as a unit which, like an organism, goes through predictable stages of development until, eventually, it reaches adulthood. In this topic we shall discuss whether Clements was right or wrong.

Figure 32.3 Stages of secondary succession in an oak-hornbeam forest in southern Poland. The sequence **A** to **E** illustrates the vegetation 0, 7, 15, 30 and 150 years after the forest was felled. C–E are shown overleaf.

DIFFERENT TYPES OF SUCCESSION

When plants invade an area which has not previously supported plants, a **primary succession** takes place. This occurs on rocky screes, sand dunes and in lakes, and probably starts the formation of soil. As plants become established they increase the amount of soil humus and create microhabitats suitable for the invasion of animals and micro-organisms.

On a global scale, however, **secondary successions** cover a much greater proportion of the land area than primary successions. Secondary succession occurs on soils in which plants used to grow, where the vegetation has been removed by fires, felling or overgrazing. The various stages of secondary succession in an oak–hornbeam forest in Poland over 150 years are shown in figure 32.3.

A successional sequence like this can be **deflected** at any stage by grazing, coppicing or fires. For example, in the absence of intense grazing, a herb-rich community on chalk or limestone rock is invaded by birches, willows, hawthorn, privet, Blackthorn and other shrubs and slowly reverts into a climax forest of Beech, Ash or Yew. If sheep, cattle or rabbits begin to graze the vegetation early in this succession, the usual course of succession is deflected and a grassy, species-rich vegetation develops typical of the grassland of southern

C

D

E

Figure 32.3 (continued) Can you describe, and explain, the sequence of changes shown in photographs A–E?

Britain. In many areas this was maintained by rabbit grazing, until the epidemic of the virus disease myxomatosis in 1953–5, which almost eliminated rabbits. Secondary succession then resumed, and areas of scrub thirty or more years old are now frequent on the chalk.

ENERGY AND NUTRIENTS IN SECONDARY SUCCESSION

In a developing forest some of the net primary production of the plants is stored in branches, trunks and roots, and some accumulates in the humus in the soil (figure 32.4). For this reason a developing forest is described as a **storage ecosystem**. Similarly, the whole ecosystem accumulates carbon and nitrogen from year to year. The gain in nitrogen from rainfall, bedrock decomposition and nitrogen fixation by prokaryotes outweighs the loss in nitrogen by denitrification and leaching.

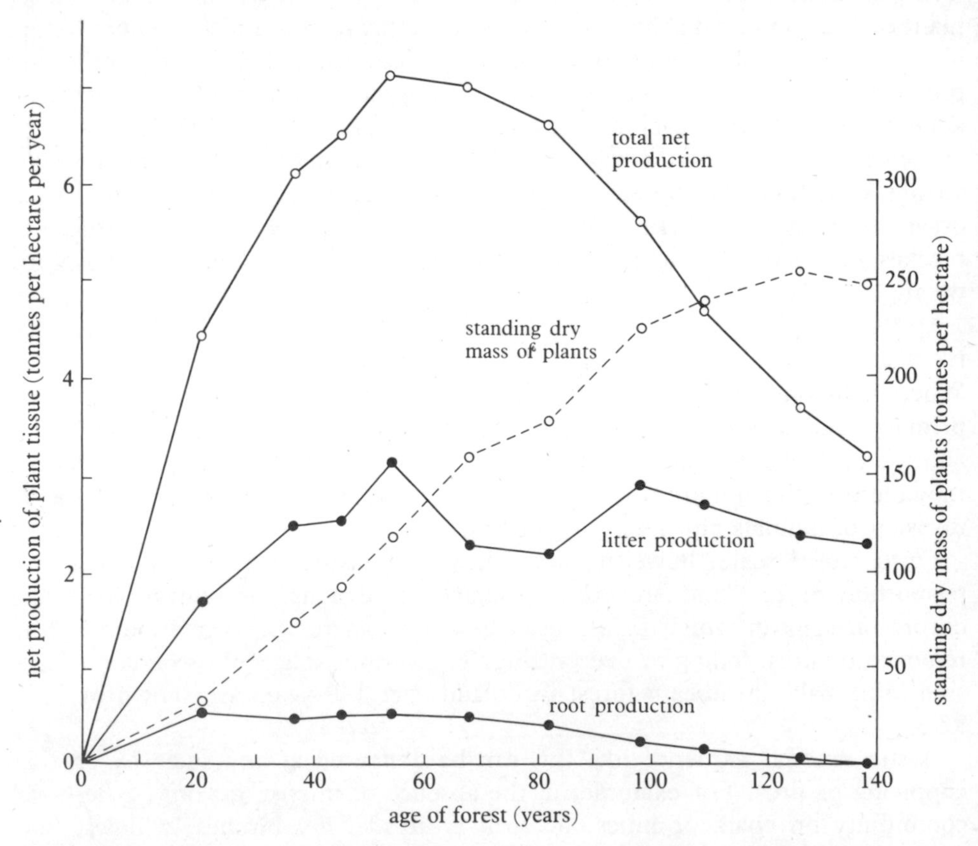

Figure 32.4 Changes in production and biomass with age in stands of Norway Spruce (*Picea abies*) of different ages in Russia. All the stands were in the same climatic region and grew on similar acidic soils. (*Data from Kazmirov and Morozova*, Dynamic Properties of Forest Ecosystems *Cambridge University Press, 1970*)

In a climax forest, however, the inputs of energy, carbon and nitrogen should be about the same as the outputs. Although the decomposition of dead trees causes the ecosystem to lose energy and carbon dioxide, the young trees which grow in their place accumulate biomass and nutrients.

A CRITICAL LOOK AT SUCCESSION

Clements' concept that species follow one another in a predictable sequence, resulting in a climax of constant composition, has been challenged by later research. One main objection is that each species has its own particular life history and physiological tolerance range. The other is that in many parts of the the world the climax state is never reached. Let us discuss both these objections in turn.

Different plant species have different micro-environmental preferences. As the conditions of climate and soil differ from place to place, so should the abundances of different plant species. Thus the species compositions of successions should differ in adjacent areas of the same habitat. Similarly, different plant species have different dispersal abilities. Whether or not a species is included in a succession depends on its ability to disperse its seeds into the area or to maintain dormant seeds in the soil. Different species also have different tolerances to others; for example plants show varying tolerance of the shade cast by other plants. The presence of one species may be promoted or inhibited by the presence of a mature tree of another species. That will influence the course of succession. The result of all these processes, operating simultaneously, is that in any one place the course of succession can never be precisely predicted.

The 'climax' may simply consist of areas in different stages of secondary succession. When a tree dies or falls the gap may become occupied by a sequence of successional species. For example, in many American forests it has been shown that species A grows beneath and replaces species B, whilst species B grows beneath and replaces species A. The result is a **stable mosaic** (Table 32.2).

Table 32.2 Probability of replacement of one species by the other in a forest dominated by Sugar Maple (*Acer saccharum*) and the beech *Fagus americana* at Warren Woods, south-west Michigan. Saplings of each species, which were more than one metre tall, were counted under adult trees.

	Percentage of saplings under mature trees of: Sugar maple	Beech
A Sugar maple saplings	36%	77%
B Beech saplings	64%	23%

In many areas the so-called climax trees are very long-lived and the climax state may take a long time to be reached. If environmental catastrophes are frequent the climax may not have time to develop. Evidence is accumulating that many tropical and temperature forests have been frequently devastated by hurricanes, cyclones, and fires. Human activity has also subtly changed the relative abundances of climax species in many parts of the globe.

In short, it seems that Clements was right in pointing out that communities change with time, but wrong in stereotyping the process.

For consideration

1. Explain the shapes of the curves in figure 32.4.

2. Imagine that a tropical climax forest, in which the annual input of carbon dioxide equalled its output, was felled, and secondary succession was allowed to occur on the site.
(a) What should be the effect on the carbon dioxide concentration in the atmosphere, and why?
(b) Reconcile your answer with the current view that the felling of tropical rain forests contributes largely to the increasing level of carbon dioxide in the Earth's atmosphere.

Further reading

P. Collinvaux, *Why Big Fierce Animals Are Rare* (Penguin, 1975)
Compulsive reading. It contains an entertaining and elegant essay on succession and several other relevant articles.

H. Horn, 'The Mechanism of Secondary Succession' (*Scientific American*, vol. 232, no. 90, 1975)
This article explains the dynamic nature of the 'climax'.

33 . Associations between Organisms

Intimate associations

The German mycologist Anton de Bary brought the term **symbiosis** into scientific use. In 1879 he defined it as any close or intimate association between two species. In this sense, parasitism, commensalism, and mutualism are all forms of symbiosis.

Mutualism is a symbiosis in which both partners benefit. Until recently many biologists were using 'symbiosis' to mean the same as 'mutualism', but now most textbooks have gone back to de Bary's original meaning and that is the sense in which the term will be used here.

INTIMATE ASSOCIATIONS

Consider parasitism. It is a nutritional association between two living organisms in which one benefits and the other is harmed. Usually we think of such associations as being pretty intimate, such as the tapeworm in a mammal's gut, but less intimate associations are also included in parasitism. A flea for instance spends only a tiny fraction of its time on the host, maybe only a few seconds each week; yet it is called a parasite.

There are two important ideas to bear in mind. One is the degree of intimacy in the relationship, and the other concerns how much harm or benefit each partner receives (figure 33.1). Judging by the way mutualism, commensalism and parasitism are defined, attention is normally focussed on the harm/benefit idea.

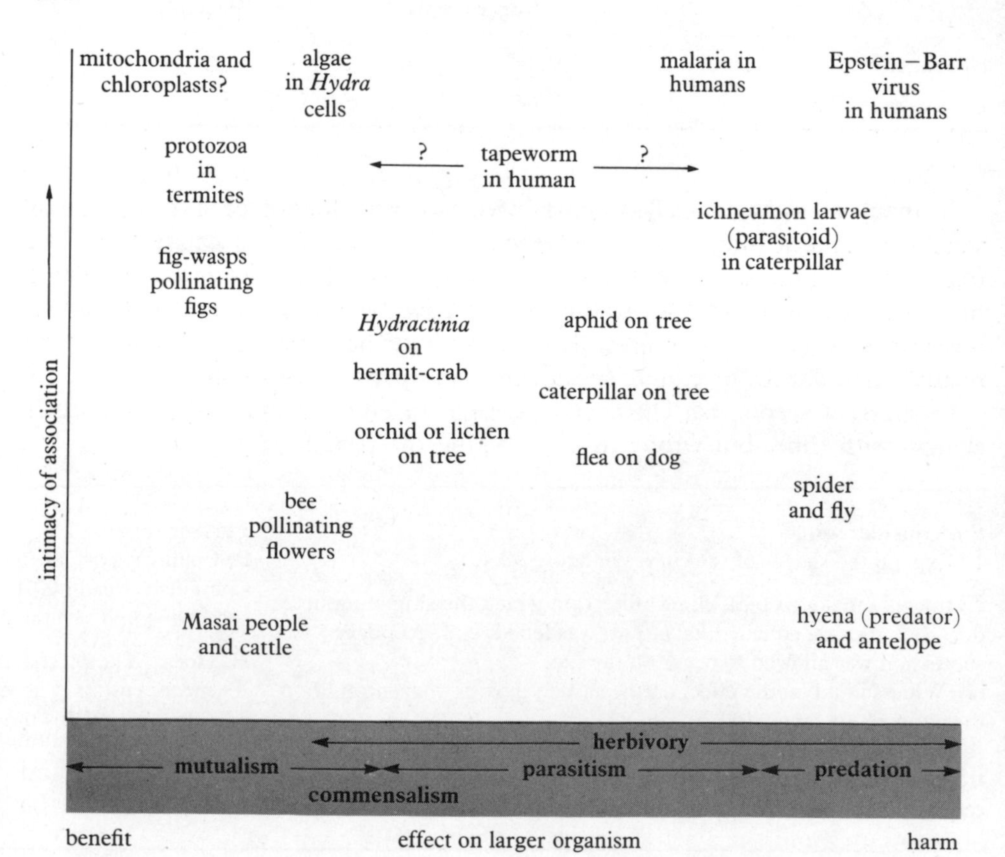

Figure 33.1 Associations between two organisms in which the smaller benefits. Only a selection of associations is shown to illustrate how all the degrees of intimacy and harm/benefit are found. The reason for including mitochondria and chloroplasts in this scheme is explained on page 166. (*Courtesy of Journal of Biological Education*)

HARM OR BENEFIT

At least one organism in an association benefits, usually the smaller. The larger organism may also benefit, and in extreme cases it is entirely dependent on its partner. For instance termites eat wood, but they cannot use it without the assistance of cellulose-digesting protists in their gut. At the other extreme the larger organism is killed, as in predation. Most predators kill their prey before they eat it, but this has the drawback that the food must be consumed before it decays.

Much more sophisticated are the **parasitoids** which eat their prey, or at least most of it, before it dies. The best known examples are the ichneumon wasps that lay their eggs in or on caterpillars and other insects (figure 33.2). The ichneumon larvae eat the prey from the inside, inessential organs first, thus prolonging the caterpillar's life. Only when the ichneumon larvae are near to pupation do they attack the caterpillar's nervous system and other essential organs, causing it to die. Not having invented refrigerators, this is how the parasitoids keep their food fresh.

Figure 33.2 The ichneumon wasp lays its eggs inside a caterpillar. Here the larvae have emerged from an almost completely consumed caterpillar. They are now ready to pupate.

In these examples it is clear whether the larger organism is harmed or benefitted. In other cases it is not. A tapeworm in a thin person can cause undernourishment in the host; but the same tapeworm in a fat person may prevent the host getting even fatter. There is a similar situation in some mycorrhizas – fungi that grow on living tree roots. In poor soils, the mycorrhiza takes organic compounds from the tree but also enables the tree to absorb more phosphate from the soil than it could on its own. This is mutualism. However, in rich soil, the same tree–mycorrhiza association may be parasitism: the tree can obtain all the nutrient ions it needs on its own, and the fungus gives nothing back for the nutrients it receives.

Quite apart from the variability of the nutritional relationship, it is sometimes difficult to discover what is actually going on. Drawing up a balance sheet of the harms and benefits each organism receives can in some cases be almost impossible. The next topic on lichens provides an example of this. In fact, it is not always terribly important whether an association is mutualism, commensalism, or parasitism. Something else is biologically more interesting.

DEGREES OF INTIMACY

Just as there are all degrees of harm or benefit to the larger organism or host, so also there is a continuum from close intimacy to no association at all. Several kinds of intimacy can be distinguished. One is a matter of timing: the bee, flea, and spider in figure 33.1 spend only a short time in association with the partner, whereas the protists in a hydra are there for the whole life cycle. On the other hand, intimacy can also be spatial. The bee does not enter the plant at all, and the flea only penetrates its host with its mouthparts. At the other extreme, the protists of the hydra are right inside the host's cells – and one cannot get more intimate than that.

But the most interesting thing about these associations is the adaptation of one organism to the other. In general, the degree of adaptation is related to the intimacy of the association. For instance the Masai people of Kenya subsist largely on milk and blood from their cattle, and the cattle gain protection. Neither species is obviously adapted for this mutualistic association. Slightly more intimate are the bee, flea, and spider examples; their most obvious adaptation is in the mouthparts. Any partner inhabiting a host's gut must be able to resist being digested and tolerate anaerobic conditions. Organisms in the host's bloodstream must be able to survive antibodies, macrophages, and other host defences. Intracellular partners similarly need to combat lysosomes and proteolytic enzymes. Any of these organisms in the host's gut or body must be able to enter and then get their progeny out safely.

In short, most of the adaptations shown by these organisms can best be explained by the closeness of the association, rather than by the benefit or harm done to the organisms.

For consideration

1. Add further examples to figure 33.1. Does any pattern emerge, or do the examples form an even scatter across the diagram?

2. Construct a diagram, like figure 33.1, in which it is the larger organism that benefits.

Further reading

P.J. Whitfield, *The Biology of Parasitism: An Introduction to the Study of Associating Organisms* (Arnold, 1979)
This undergraduate text follows the approach used in this topic, and is not just about parasites.

R.A. Wilson, *An Introduction to Parasitology* (Arnold, 1979)
The physiology of parasitism is discussed, as well as all the major groups of parasites.

Getting on well together: lichens

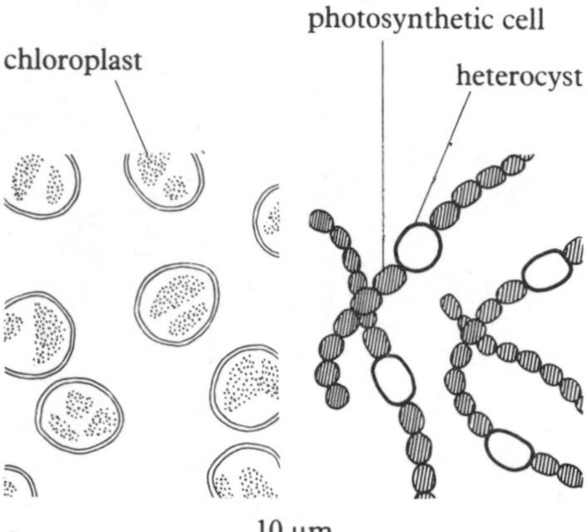

Although a lichen looks like a single organism, it is in fact an association between a fungus and a photosynthetic organism. Lichens have several remarkable features. One is their hardiness. It is well known that lichens are usually the first colonizers of a lifeless landscape, such as a new volcanic island. They also survive permanently on mountain tops and in arctic and antarctic environments which support no other terrestrial organisms.

This tolerance of extreme conditions depends on at least three properties. Lichens photosynthesise of course, so do not depend on a source of organic matter; they can survive temperatures as low as −268 °C and as high as 66 °C; and frequently their moisture content falls below ten per cent with no ill effects. The individual organisms making up the lichen association are nothing like as hardy when growing without the partner. Thus it seems to be the close association itself that gives lichens their unusual qualities.

chloroplast
photosynthetic cell
heterocyst
A
10 µm
B

Figure 33.3 Photosynthetic organisms in lichens: **A** the protist *Trebouxia;* **B** the cyanobacterium *Nostoc*, with a single large heterocyst in a chain of ordinary cells.

STRUCTURE AND GROWTH

About a quarter of all known fungi can form part of a lichen; they mainly belong to the *Ascomycetes*. Many contain the bright green cells of a unicellular protist, often *Trebouxia* (figure 33.3). Other lichens harbour instead the blue-green cells of a cyanobacterium such as *Nostoc*, and a few species have both kinds of partner. In most lichens the fungus accounts for more than nine tenths of the thallus mass, and it is for this reason that lichens are often referred to as lichenized fungi.

In many cases the fungus and photosynthetic partner can be separated and each grown in pure culture. Usually the protist and cyanobacterium change little on being isolated. However the fungus is different. When grown in pure culture it is a fairly shapeless mass of hyphae – a strong contrast to the intricate form of the same fungus in its lichenized state (figure 33.4). So the protist or cyanobacterium seems to have a powerful influence on the way the fungus constructs the lichen thallus.

Figure 33.4 A lichen that grows on tree trunks. It measures 5 cm across. This particular species is *Evernia prunastri*, a foliose ('leafy') lichen with a branching frond-like body form.

Lichenized fungi differ from ordinary fungi in other respects. In particular they have much slower growth rates, commonly only 2–3 mm yr^{-1}. Giving extra nutrients to a lichen in its natural situation may increase the growth rate by a small amount, but too much nutrient causes the association to disintegrate. The great age of many lichens is also legendary. Some in the Arctic are thought to be well over a thousand years old. Lichens have been used to estimate the age of the substratum on which they are growing – four hundred years in the case of some gigantic statues on Easter Island in the Pacific (figure 33.5).

Reproduction and dispersal are achieved in two main ways. It would seem most efficient for each propagule to contain a piece of every organism participating in the lichen association; then a complete new lichen thallus could grow from the single propagule. We do indeed find several kinds of composite propagule, the smallest of which is shown in figure 33.6A. All of these propagules are the result of asexual reproduction.

Sexual reproduction is also possible, but the strange thing is that only fungal spores are produced, just like unlichenized fungi (figure 33.6B). If a spore lands on a stone, tree bark, or soil, it can only develop into a lichen if the correct protist or cyanobacterial cells arrive independently. Where these are common free-living species, there is a fair chance that cells will eventually be blown or washed onto the spore or young mycelium. On the other hand *Trebouxia* has rarely, if ever, been found free-living, so how it gets incorporated into a new sexually reproduced thallus is a mystery.

Figure 33.5 The age of these extraordinary statues on Easter Island in the Pacific has been estimated using lichens. They are thought to be about four hundred years old. The lichens are the white patches on the surface of the statues.

WHAT SORT OF RELATIONSHIP?

There is now good evidence that the fungus benefits from the association. Both protist and cyanobacterium can photosynthesise, and much faster than one might expect from the lichen's slow growth rate. Tracer experiments using ^{14}C show that surplus photosynthetic products pass to the fungus in the form of glucose from *Nostoc*, and usually as alcohol from the protist. In addition, some lichens containing *Nostoc* can fix nitrogen. This probably occurs in the heterocysts, which are large cells lacking photosynthetic pigment (figure 33.3). The enzyme nitrogenase converts the nitrogen to ammonia which in turn is incorporated into small organic molecules.

But does the fungus help its partner in any way? Surprisingly there is no clear answer to this. It is easy to think up ways in which the fungus *might* be beneficial. It *might* protect its partner from physical damage or high light intensities, provide it with nutrient ions, water and carbon dioxide, or prevent its desiccation. There is however little or no evidence for any of these. For instance lichens appear to gain and lose water in just the same way that agar jelly does. In any case the protist and cyanobacteria can just as easily withstand desiccation in their free-living forms as in the lichen.

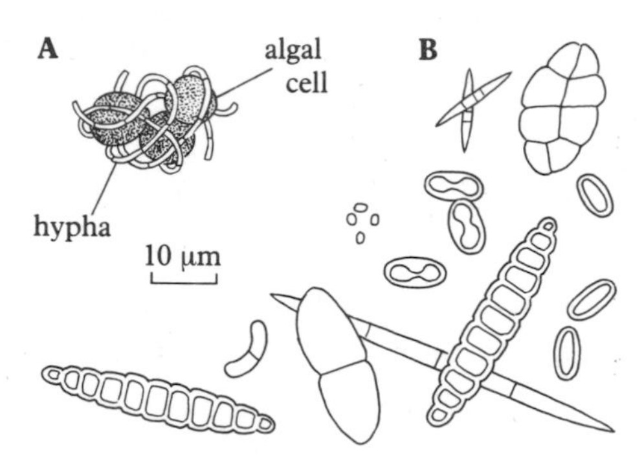

Figure 33.6 Lichen propagules: **A** Composite propagule (soredium) consisting of algal cells and hyphae; **B** Spores from various species of lichenised fungi.

The relationship between the organisms in the association probably varies from one lichen to another. In some the fungus may be a parasite on its partner. In others, particularly those consisting largely of *Nostoc* and its gelatinous matrix, it makes little difference whether the fungus is present or not; here the fungus may be a commensal.

Many lichens, though, are probably examples of mutualism. One reason is that lichens survive in many places where neither the unlichenized fungus nor its partner can exist alone; so presumably in those situations both partners benefit. Another reason relates to *Trebouxia*. Since it is very rare, even non-existent, in a free-living state, it is likely to be dependent on the fungus.

WHY STUDY LICHENS?

The main interest of lichens is the way novel characteristics are produced by the intimate association of two or more organisms. This opens up the possibility of creating new associations with valuable characteristics, for instance crop plants that can grow in conditions unsuitable for any other. Some lichens are already eaten by humans, and others support reindeer and caribou which in turn can be consumed by us.

Lichens are sensitive to atmospheric pollution. The number of species growing in an area can be used to measure how much pollution there is. By contrast, they are highly tolerant of radioactivity. So perhaps in the future, if there is a nuclear war, our descendants will be sowing and reaping the lichen fields on some distant planet, while back on Earth lichens will be the only survivors of the devastation.

For consideration

1. Design a crop 'plant' or domestic 'animal' with valuable new properties arising from an intimate association between two or more organisms. What particular agricultural problems might it solve?

2. Suggest an explanation for the slow growth rates of lichens: what mechanism might be involved, and what function might it serve?

Further reading

M.E. Hale, *The Biology of Lichens* (Arnold, 3rd edition, 1983)
A good general account.

D.L. Hawksworth and F. Rose, *Lichens as Pollution Monitors* (Studies in Biology no. 66, 1976)
Of particular interest to conservationists.

K.A. Kershaw and K.L. Alvin, *The Observer's Book of Lichens* (Warne, 1963)
A well illustrated guide to the British lichens.

Why bees dance

Karl von Frisch is best known for his theory of a dance language in honey bees. Once a forager bee has found some food, it can help other bees locate it, even if the forager never goes back there. In the 1920s von Frisch believed this was explained by the recruits learning the scents which the forager had picked up on its body. Later though, in the 1940s, he discovered correlations between the way the dance was performed and the food's distance and direction from the hive. So arose the famous idea that the dance was a 'language', telling the recruits where to go for the food.

In the 1960s, Adrian Wenner and his colleagues carefully examined von Frisch's experiments. So far as they could see, none of the experiments demonstrated that recruits used the language. There was no doubt that the dances, and in particular the sounds emitted during the dance, correlated with the food's position. But for communication to happen, it is not enough for the foragers to put information *into* the dance; the recruits must also be able to extract the information *from* the dance. What Wenner doubted was whether the recruits could 'read' the language.

The theory championed by Wenner's group was a modification of von Frisch's scent theory. There is plenty of scope for communicating information by means of scent. A forager may bring to the hive not only the scent of the food itself but also scents from the environment around the food source. So if recruits recognise the smells of clover, pine, and marsh plants on the forager, they will know the food source is a patch of clover near a pine wood and a marsh. This could pinpoint the site well enough. In addition the forager might have left its own scent at the food source, and this could help recruits locate it.

In some circumstances bees do seem to use only scents to locate a food source discovered by a forager. So the question is, do they ever use the dance language? There ensued a long dispute between Wenner and von Frisch, and it was only when an ingenious experiment was performed by another biologist that the issue was settled to most people's satisfaction.

GOULD'S EXPERIMENT

James L. Gould of Princeton University decided that a little trickery was needed. He chose to make the foragers tell the recruits, in dance language, that they had been to a place where in fact they had not been at all.

How could this be done? One obvious approach was to use a model bee, and make it transmit a message in dance language. But bees are not that easily fooled, and so far this method has not worked. Instead Gould painted over the ocelli of his foragers. The three ocelli are simple eyes located between the two large compound eyes, and painting over them reduces the bees' sensitivity to light. Getting these bees to deceive the recruits depended on another fact of bee behaviour: although bees dancing in the dark on a vertical comb use 'straight up' as a symbol of the sun's direction, if a light is provided they use that instead to indicate the sun.

Gould's bees were trained to take food from a table called the forager station, some distance from the hive (figure 33.7). They then returned to the hive. Having their ocelli painted, they could not see the lamp in the hive, and so danced using 'straight up' for the sun's direction. The recruits however were not ocelli-painted; they could use the lamp to interpret the dance. If the dance language theory was true, Gould should be able to adjust the position of the lamp and send the recruits off to any of the recruit stations he chose. And in fact this is just what happened. At least sometimes, bees do gain information from the dance but they use scents as well.

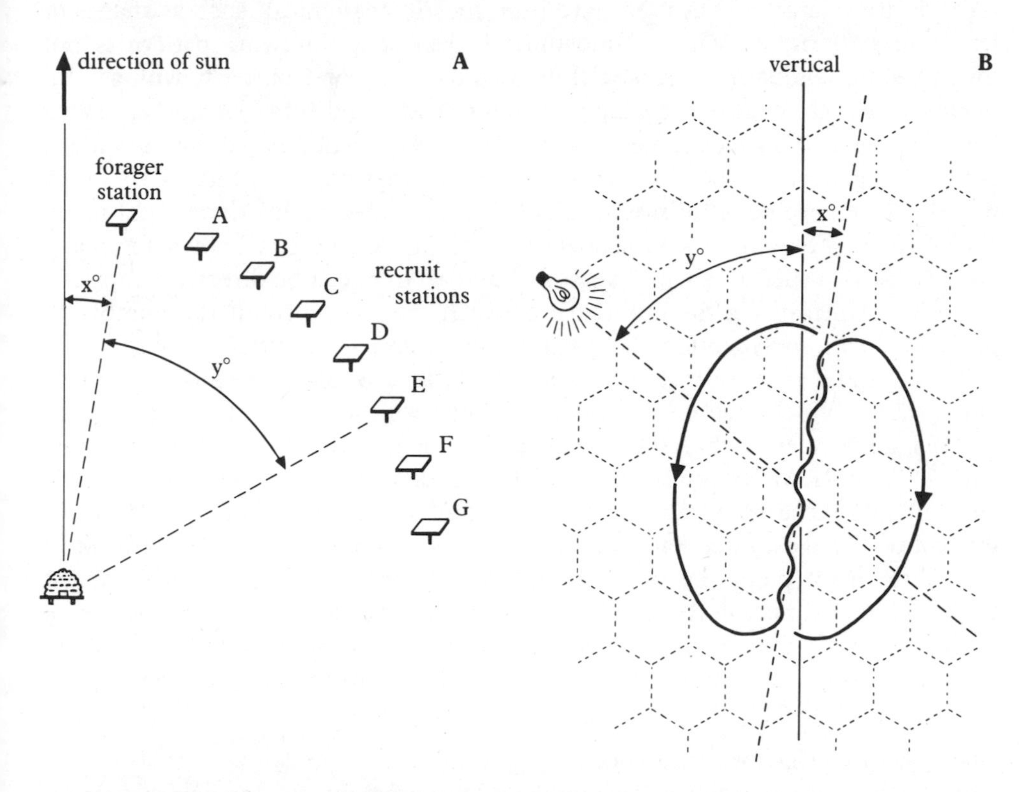

Figure 33.7 Gould's experiment on the dance language theory. **A** Plan of the experiment. Forager bees were trained to visit the forager station, x° clockwise from the sun, and then return to the hive where they danced. **B** The returning foragers danced on the vertical comb, x° clockwise from the vertical because they could not see the lamp and therefore behaved as if they were in the dark. However, because recruits following the dance could see the lamp, they interpreted the dance to mean the food was (x + y)° clockwise from the sun. So they flew to recruit stations D, E and F.

Research on bee communication has a long and chequered history. First there were von Frisch's scents, then his dances, then Wenner's sounds, and now scents again. All three types of stimulus seem to be involved in communication, but their precise roles and relative importance are still not understood. It is not an easy field in which to do research. The researchers are in the same position as a visitor to Earth from another planet who observes a group of human actors performing a play before an audience. All sorts of signals are passing from actors to audience (and from actor to actor), but which signals are significant, and to whom, – and when?

For consideration

Although Gould's experiment decided in favour of von Frisch's theory, what benefits do you think have probably resulted from Wenner's challenge?

Further reading

J.L. Gould, 'The Dance-Language Controversy' (*Quarterly Review of Biology* vol. 51, no. 2, 1976)
A detailed and fascinating account of the debate.

J.L. Gould, *Ethology: The Mechanisms and Evolution of Behaviour*. (Norton, 1982)
Chapters 13 and 24 deal with bees.

34 . Evolution in Evidence

A molecular clock

Biologists have been constructing evolutionary or phylogenetic trees for more than a century. They have been classifying organisms for much longer. Until recently the characteristics used have been mostly anatomical, such as shapes of leaves or patterns on wings. Unfortunately leaf shape or wing pattern is not much use for comparing a yeast cell with an oak tree, or a butterfly with a fish.

However these comparisons are now being made, and some intriguing results have emerged. The technique is to use a characteristic shared by as wide a range of organisms as possible, and what better than the structure of DNA or a protein? Cytochrome c, for instance, is found in nearly all eukaryotes and many prokaryotes. The amino acid sequence of its protein part is known for many species. So two species, such as yeast and an oak tree, can be directly compared. One way of measuring the difference between them is to count the number of positions in the polypeptide chain where the amino acids differ.

This enables us to estimate how many mutations would be needed to convert one species' cytochrome c into another's. Suppose the only difference between the two molecules is that the amino acid sequence lysine–glycine–alanine–leucine is replaced by lysine–proline–serine–leucine. From the genetic code you can see that substituting glycine for proline or *vice versa* requires at least two nucleotide changes; interchanging alanine and serine requires at least one. So at least three point mutations would be involved if one cytochrome were changed to the other. Estimating the number of mutations is of great interest in studying evolution. Now that nucleic acids themselves can be sequenced, we have in some cases direct measurements of the number of changed nucleotides.

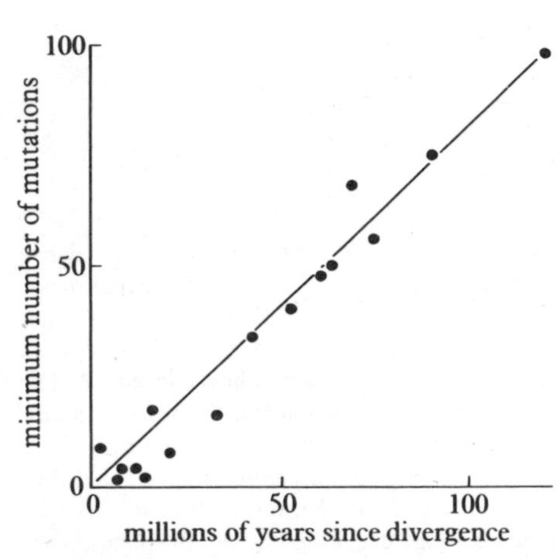

Figure 34.1 The molecular clock. Each point represents a pair of mammal species. The horizontal axis shows the date of the latest common ancestor of the species, estimated from the fossil record. On the vertical axis is the minimum number of mutations needed to convert one mammal's proteins into the other's. Based on seven proteins.

PHYLOGENY AND THE MOLECULAR CLOCK

This means that we can make two kinds of measurements on any pair of species: (a) the minimum number of mutations needed to convert one species' proteins into the other's; (b) the latest date when the two species had a common ancestor. The latter involves locating the probable point in the fossil record when the phylogenetic lines diverged.

When these measurements are plotted against each other, the points are remarkably close to a straight line (figure 34.1). The graph suggests that the minimum number of mutations is proportional to time – in other words they have been accumulating at a steady rate over hundreds of millions of years. So knowing the number of mutations enables us to estimate the time that has elapsed since the two lines diverged. Whatever the explanation for this extraordinary finding, it means that we can use some proteins as a **molecular clock**.

A molecule like cytochrome c can therefore be used, on its own, to construct a phylogenetic tree. Species with similar amino acid sequences are regarded as closely related, with a recent common ancestor; the more dissimilar the sequences, the earlier the two lines separated. Figure 34.2 shows part of a phylogenetic tree based on cytochrome c. You can see that it bears quite a close resemblance to the tree derived from the fossil record, but the correspondence is not exact. For instance, on the basis of cytochrome c the kangaroo – a marsupial mammal – is related more closely to most of the placental mammals than the human is. Either the orthodox theory is wrong, or the cytochrome c clock is not

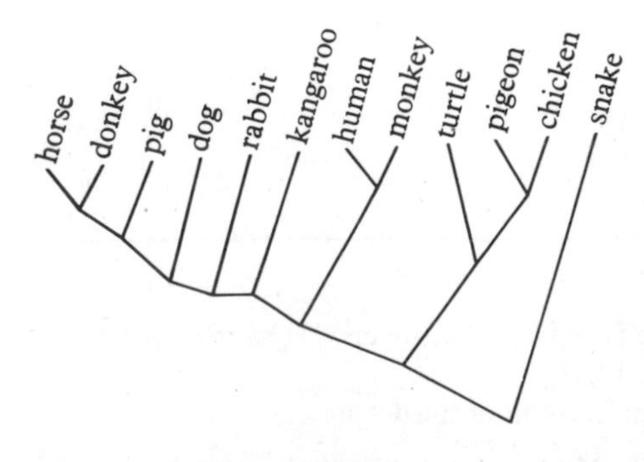

Figure 34.2 A phylogenetic tree based on cytochrome c. The greater the similarity of two species' amino acid sequences, the higher up in the diagram the species diverge.

entirely reliable – probably the latter. However, if the analysis were done on a variety of proteins and nucleic acids, we could probably construct quite an accurate phylogenetic tree.

NEUTRAL MUTATIONS

The molecular clock has not only provided a new way of constructing phylogenies. More controversially it has suggested a new theory of evolution, which does not involve natural selection.

The theory of **neutral mutations** has been developed by the Japanese biologist Motoo Kimura. He points out that, in the theory of natural selection, a mutation must confer a selective advantage on an organism if the mutation is to survive. Thus a fast evolving group should accumulate more changes in its amino acid sequences than a slow evolving group. However the molecular clock implies that amino acid substitutions accumulate at more or less the same rate in any group of organisms. To explain this, Kimura believes that most mutations which survive are 'neutral' – in other words the new form of the protein is no better and no worse than the original molecule. It is purely a matter of chance whether, after many generations, the new molecule has completely replaced the old one. On the other hand, harmful mutations are eliminated by natural selection.

There appears to be good evidence for this theory, even though it is surprising. It is surprising because it does not seem to matter which amino acids occupy most of the positions in certain proteins. This is in marked contrast with the well known example of sickle cell haemoglobin, where a single amino acid substitution has such drastic effects (see page 134). Those effects are to be expected if the amino acid is part of an enzyme's active centre, or in the part of the haemoglobin that binds the haem group. On the other hand, in some proteins part of the molecule may be merely a framework, which can be modified without seriously affecting the molecule's function. If that is so, random mutations could accumulate at a steady rate in those parts of the molecule, with natural selection playing no part in the evolutionary change.

For consideration

1. What kinds of macromolecule would you expect to function as molecular clocks, and which would give inaccurate results?

2. Although it has not been demonstrated that any visible phenotypic characters have evolved randomly, can you think of any which might change as a result of random evolution at the molecular level?

Further reading

R.E. Dickerson, 'The Structure and History of an Ancient Protein' (*Scientific American*, vol. 226, no. 4, 1972)
A detailed description of cytochrome c and its evolution.

M. Kimura, 'The Neutral Theory of Molecular Evolution' (*Scientific American*, vol. 241, no. 5, 1979)
A concise and readable account of the theory.

How good is the evidence for evolution?

Most people firmly believe that species have evolved from others in the past. It is frequently implied, if not actually stated, that evolution is a 'fact'. But can we be so sure? Has it been proved?

There are various kinds of evidence, and they all support the theory. At the same time most or all of them *could* have another explanation. And so long as that is true we cannot say the theory is proved; we have, at the most, good evidence for it.

EVIDENCE

Let us look briefly but critically at four specific lines of evidence:

1. **Fossils.** One problem with the fossil record is that it is lamentably incomplete. If it were complete, we would expect to find uninterrupted series of fossils linking every species with its distant ancestors. No series of this kind

has ever been found. The incompleteness of the record is not an argument against the theory of evolution; it merely reduces the value of fossils in deciding whether evolution has occurred.

Rather than demonstrating that evolution has occurred, fossils are used by biologists in a different way. On the *assumption* that evolution has occurred, fossils can indicate the probable stages any particular species went through as it evolved. This is something like route-finding – so long as you already know that the journey has been made.

2. **Classification.** Darwin himself thought classification offered strong evidence for evolution. If every group of related species has arisen from a common ancestor, then it should be possible to classify organisms as groups of groups of groups, and so on. This is what we find. But it does not demonstrate that evolution has occurred. When Linnaeus classified organisms he did the same to minerals, and geologists still use a similar classification. Soils too are arranged as groups of groups, but no one imagines they evolve – at least not in the sense that organisms may have done.
3. **Comparative anatomy**. It is often thought that only evolution can explain the widespread distribution of, say, the pentadactyl limb among vertebrates. But there may be a different reason. For any structure to exist in an adult, it must be possible for embryological processes to construct it. In a developing vertebrate embryo, the pentadactyl limb starts off as a fairly simple pattern of bone rudiments. It may be that, at this stage, there is only a very limited range of patterns which can be produced by chemical gradients, or whatever else is responsible for positioning the bones. So rather than pointing to a common ancestry, the pattern may simply mean there is only one efficient way to make a jointed limb in a vertebrate.
4. **Natural selection.** There are now impressive studies showing how natural selection can produce genetic changes in populations. The Peppered Moth is but one example. Does it necessarily follow, though, that natural selection turns one species into another? There is no reason why it can't. But in science we need evidence that something *does* happen, not just that it *could*.

   It is always possible that natural selection has its limits, just as artificial selection seems to. For instance, if we try to raise milk yields in cows by selecting high yielding individuals for further breeding, we find the milk yield can be pushed up so far but then no further. In the same way perhaps natural selection produces small scale evolutionary effects, such as subspeciation, but not larger scale changes.

There are in fact now several theories about how evolution might occur, and these are discussed in the next chapter.

A CRITICAL ATTITUDE

The next topic shows how questions have been raised about the Darwin–Wallace theory of natural selection. But virtually no biologist doubts that species *have* evolved. So why has this topic been written?

When everyone agrees that something is true, we are tempted to cease questioning it and become intellectually lazy. This happened to Newton's theory of gravitation. In the decades before 1900 hardly anyone doubted the theory, but it was soon to be completely overthrown by Einstein in the early years of this century. The same may happen to any well established theory in science, including the theory of evolution; if so the theory may at least need serious modification. Do *you* think that the evidence for evolution is so good that we no longer need to think about it critically? If so, then make sure you can answer the arguments you have just read.

For consideration

1. In how many ways does the phylogenetic tree in figure 34.2 differ from the phylogeny based on the fossil record? Suggest possible reasons for any discrepancies.

2. What other kinds of evidence support the theory of evolution? Can you think of alternative *scientific* explanations for them?

Further reading

S.J. Gould, *Hen's Teeth and Horse's Toes.* (Penguin, 1984)
Many essays in Gould's books are on evolution, but of particular interest is essay 19 in this one. He rails (probably quite rightly) against modern 'scientific creationists', and gives his own reasons for believing in evolution.

35 . The mechanism of evolution

Six ways evolution could occur

In 1858 Darwin and Wallace presented their theory of evolution by **natural selection**. Within a century, after much controversy, it was generally agreed that evolution was caused by two processes acting together: (a) the production of minor variations by random mutations and recombination, and (b) natural selection of the individuals which were best adapted to the environment.

Since 1960 significant developments have occurred in evolution studies. There are now several possible mechanisms for evolutionary change besides the one mentioned above. Here are some examples: three involve changes at the gene level and two arise due to the combination of species.

CHANGES AT THE GENE LEVEL

Neutral mutation: This has already been discussed in the last chapter. Neutral mutations change protein molecules in ways which do not make the mutant organism any more or less competitive. However these changes in protein molecules could have noticeable effects on the phenotype. Perhaps this theory may eventually explain features of organisms which do not appear to be adaptive – such as the complicated shapes of many plants' leaves, or some of the intricate patterns on the wings of butterflies and moths. Genetic drift is a similar mechanism that can operate when any population drops to a small size. Both mechanisms imply a randomness in evolution, rather than direction by natural selection.

Macromutation: Until the 1950s, the Darwin–Wallace theory assumed that evolutionary change was always slow and gradual. However a theory of macromutations is once more being taken seriously. Macromutations are sudden large changes occurring in a single generation. Even a simple mutation in one gene could have profound effects on the adult – if it is one of the genes that switch embryological processes on or off, or control their rate. For instance, a delay in switching off genes controlling brain growth could suddenly give a much larger adult brain size. Perhaps this kind of thing happened in our own evolution. Macromutations, in the form of gross chromosomal changes, may have played an important role in plant evolution.

Gene transfer: The above theories are variations on a theme. What they have in common is the idea that a new species has, as its immediate ancestor, just one other species. The remaining three theories depart from this. At the simplest level, it is now recognised that genes can be passed from one species to another. This would explain some curious facts. For instance, leghaemoglobin is a protein found in leguminous plants such as peas and clovers (chapter 10, 'Nitrogen fixation'). It is like myoglobin and combines with oxygen. One theory to explain leghaemoglobin is that it originally arrived in the legume from some other eukaryote, as part of a virus. Incidentally this kind of gene transfer may explain some of the inconsistencies in phylogenetic trees based on proteins (figure 34.2).

COMBINING TWO SPECIES

Interspecific hybridisation: Most hybrids are not new species, because they can interbreed with their parents. Occasionally, though, a reproductive barrier is set up, preventing them interbreeding. One example of this is the well known process of **allopolyploidy**, seen in the grass *Spartina townsendii*. In allopolyploidy a set of chromosomes from one species combines with a set from another. Then the chromosome number doubles, producing a polyploid.

Endosymbiosis: Another way in which species can combine is by means of the intimate associations described in chapter 33. One such association is the green protist *Chlorella* living inside the cells of a species of *Chlorohydra*. In this case the two species are clearly distinguishable; but if such an association became so close that neither could survive without the other we might eventually have difficulty saying whether it was two species or just one. In other words the intracellular partner – the protist in this case – would have become merely an organelle of the other. This **endosymbiosis** theory is thought to explain the origin of several organelles seen in eukaryotic cells, and is described more fully in the next chapter.

PAUSE FOR THOUGHT

The theories of evolution summarised above have not exhausted the possibilities. There are others, some too complicated to explain here. Where does it leave us?

None of the new ideas questions whether evolution has occurred. Nor do they deny that natural selection plays a role in evolutionary change. But we must now admit that we do not know how important Darwin and Wallace's concept of natural selection is. It could be responsible for nearly all evolution, or it might only cause minor changes. Further research may provide an answer.

For consideration

1. Which of the mechanisms outlined here can be regarded as slight modification of the Darwin–Wallace theory, and which are radically different?

2. Are any of the six theories incompatible, or could all of them be contributing to evolutionary change?

3. Can you think of yet further mechanisms?

Further reading

S.J. Gould, *The Panda's Thumb* (Penguin, 1980)
Chapter 18 'Return of the hopeful monster' is on macromutation.

R. Lewin, 'Can Genes Jump Between Eukaryotic Species?' (*Science*, vol. 217, no. 4554, 1982)
A short summary of what little is known about this.

C. Patterson, *Evolution* (Routledge, 1978)
This book is a good introduction to modern ideas on evolution.

What is a species?

Why do we think so much more about species than about varieties, races, genera, or families? One reason derives from the theory of evolution. A species can evolve independently of any other population; normally a part of a species cannot. For instance Wrens in Scotland cannot develop characteristics different from Wrens in England, because the two populations interbreed across the Scottish border. Wren genes are continually being swopped, preventing the populations diverging. On the other hand the St Kilda Wren is confined to its small group of islands 80 kilometres out in the Atlantic (figure 35.1). It has evolved independently to become noticeably larger with a different song. This has been possible because it cannot breed with the nearest Wrens in the Outer Hebrides.

It is the concept of interbreeding that gives us the standard modern definition of 'species'. *A species is a population of individuals that can breed with each other, but which is reproductively isolated from other populations.* 'Reproductive isolation' does not just mean that no interbreeding occurs; it means that no interbreeding *would* occur even if the two populations were mixed together in the wild. So, although geographical isolation has enabled the St Kilda Wren to evolve special characteristics, it is not a separate species. Mixture with the mainland population would allow interbreeding.

Figure 35.1 The wild, remote and treeless islands of St Kilda. They have their own subspecies of the Wren – *Troglodytes troglodytes hirtensis*.

A species defined in terms of reproductive isolation is called a **biological species**. However the term 'species' can be used in another way, as we shall see.

TAXONOMIC SPECIES

Taxonomists have been describing species for centuries, and were doing so long before anyone thought of biological species. Lichens and minerals are also classified into species. In this meaning of the word, a species is really just a taxonomic unit, or one kind of set into which specimens are grouped by similarity. These sets are called **taxonomic species**.

By and large, modern taxonomists try to make their taxonomic species the same as the biological species. But this is not always straightforward. For one thing, in many cases it is not known whether a taxonomic species interbreeds with other populations. Consider the Herring Gull *Larus argentatus* and the Lesser Black-backed Gull *L. fuscus*. In Europe these behave as two biological species, hardly ever interbreeding. It was only long after the species had been described that a chain of interbreeding populations was found to link them through Asia and America. Even so, for practical purposes they are still treated as two taxonomic species.

That is a case where two taxonomic species turned out to be the same biologically. Ignorance about whether different populations are capable of interbreeding may also allow the opposite to happen. For instance some biological species which look identical are liable to be classified together. There are three small brown birds, all widespread in England, called the Willow Warbler, Wood Warbler and Chiffchaff. To look at they are almost impossible to tell apart, and until the late eighteenth century all three were known as the Willow Wren. Then the great naturalist, Gilbert White of Selborne, noticed that Willow Wrens have three different songs. We now know that they are distinct biological species, and so they are described as three taxonomic species, even though they are still difficult to distinguish when not singing.

Ignorance about breeding systems is a particular problem in botany. One of the quickest ways of identifying populations that cannot interbreed is by counting their chromosomes. Unlike animals, many taxonomic species in plants have a wide range of chromosome numbers. For example Creeping Buttercup *Ranunculus repens* has diploid numbers ($2n$) of 16, 24, and 32; in White Clover *Trifolium repens*, $2n$ is 32 or 48. Presumably plants with different $2n$ numbers cannot interbreed. This is a common pattern; indeed it has been estimated that about half the taxonomic species of flowering plants are polyploid. There are many more biological than taxonomic species at present.

Yet another reason can be given why taxonomic species and biological species do not always coincide. Many plants do not reproduce sexually at all. For instance hawkweeds and most dandelions produce seeds, but not by meiosis and fertilisation. Each seed arises from the parent by mitosis, and is therefore genetically identical to it except when mutation occurs. So there are no biological species. Each taxonomic species is a clone, or a group of similar clones.

EVOLUTIONARY PROBLEMS

Evolutionary theory leads the taxonomist to expect two further problems. One indeed occurs, but the other does not.

The one that occurs revolves around subspecies. According to the theory of evolution, subspecies are sometimes populations that are on the way to becoming reproductively isolated, but have not yet achieved it. Thus the St Kilda Wren is a subspecies of our common Wren. If all intermediates between subspecies and full biological species are to be expected, the taxonomist may have to make arbitrary decisions about what are species and what are not.

You might expect the same kind of problem to arise in describing fossil species. Surely if a new species emerges from another by gradual modification, and the fossil record shows this happening, it would be entirely arbitrary where one chopped the fossil sequence: 'here the ancestor stops and the descendant starts'. However, to the taxonomists' relief, smooth fossil sequences are rarely found. To explain this, the theory of 'punctuated equilibrium' has been developed, the idea that species have changed, not gradually, but by sudden leaps.

For consideration

Suppose a taxonomist describes two closely similar taxonomic species with distributions which do not overlap. How would you set about discovering whether they are different biological species?

Further reading

A.R. Clapham, T.G. Tutin and E.F. Warburg, *Flora of the British Isles* (Cambridge University Press, 2nd edition, 1962)
'CTW' contains the standard key for identifying British seed plants and ferns. Although not exactly light bedtime reading, it is invaluable for reference and gives $2n$ numbers for almost every species.

Natural selection observed

One might think it would be easy to demonstrate natural selection in action. Surprisingly, though, there are comparatively few well documented cases. You are probably familiar with the studies of the land snail *Cepaea*, the Peppered Moth *Biston betularia*, the Meadow Brown butterflies on the Scilly islands, and perhaps others. Here we look at a less familiar example, involving a species of Darwin finch in the Galapagos archipelago.

GEOSPIZA FORTIS ON DAPHNE MAJOR

In 1975 a study of the finch *Geospiza fortis* (figure 35.2A) was started by Peter Boag and Peter Grant on the small, 40-hectare island known as Daphne Major. The birds were regularly caught so that they could be weighed and measurements made of their wing, leg, and beak – giving a 'body size index'. Before each bird was released, it was given a unique combination of coloured rings on its legs. These rings enabled the birds to be identified individually, even from a distance by using binoculars.

Besides measuring the birds each time they were caught, Boag and Grant also studied their feeding habits. During part of the year, the finches take a variety of food, including nectar and pollen from the cactus *Opuntia*. But in the dry season they are forced to subsist almost entirely on seeds. With their massive beaks they are well adapted for breaking these open.

So it was important to find out what kinds of seed were available to the birds at each time of the year. To do this, quadrats were randomly placed on the ground, and the number of each species of seed in each quadrat counted. From these counts, and from measurements of the size and hardness of each kind of seed, a 'size–hardness index' was calculated for each time of the year. The greater the number of large and hard seeds available, the higher the index.

A

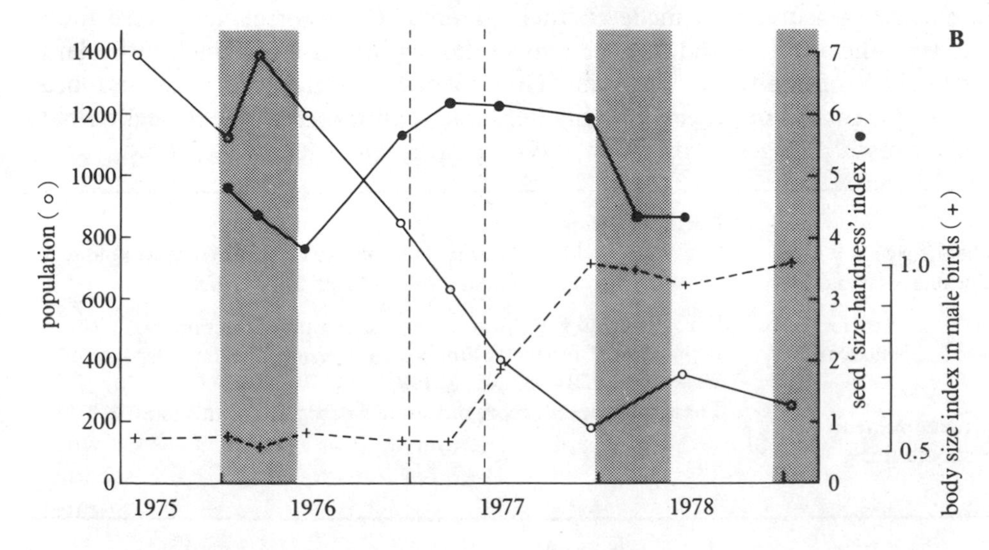

Figure 35.2 **A** *Geospiza fortis*. It is about the size of a sparrow; note the large beak. **B** Changes in the total population on Daphne Major of *G. fortis*, in its body size index (calculated from mass and various measurements of wing, leg and beak), and in the seed size-hardness index from 1975 to 1979. The breeding season in each year is indicated by shading, though in fact no breeding occurred in 1977. A change from 0.5 to 1.0 in the body size index represents about a 6% increase in mass and at most a 5% increase in linear measurements. (*After Boag and Grant, 1981*)

THE DROUGHT OF 1977

Figure 35.2B shows some of Boag and Grant's data. They extend from 1975 to 1979, and measurements were usually made just before, during, and just after the January–May breeding season, which coincides with the wet season.

In the early 1970s, the island received regular rains during the wet season. Consequently food was abundant and the finch population high. However in 1977 the rains failed, and only about one fifth of the usual amount fell. The finches did not breed at all, and of the 388 nestlings ringed in 1976 only one survived into 1978; 85 per cent of the total population died. Clearly in 1977 the birds were facing exceptionally hostile conditions.

Because regular records had been kept previously, the drought provided an excellent opportunity to observe what happened to the finches and their environment. For example, the seed size–hardness index rose to unusually high values. During a normal wet season, as in 1976, the index falls as the plants produce new seeds, many of them soft and small. As these small, soft seeds tend to be eaten first, the index rises during a normal dry season, but in 1977 it remained high throughout the year because new small soft seeds appeared only in relatively small numbers.

In the birds themselves average body size increased significantly. This happened within the males, and also within the females. However the effect was even more marked in the total population. The reason was that males are larger than females, and the sex ratio changed from approximately 1:1 in 1976 to about six males for every female in 1978. So in 1978 the birds surviving the drought were mostly those which were of above average size before the drought. Since smaller birds tend to eat smaller seeds, and only the large birds have beaks massive enough to cope with large and hard seeds, the smaller birds probably survived less well simply because their food was scarce.

HERITABILITY

Although this is a convincing case of natural selection, something more is needed to make it an example of evolutionary change. Evolution involves changes in gene frequencies – but how do we know that the gene frequencies changed in this case? They might not have done if the birds' sizes were determined merely by the environment, such as the amount of food they received as nestlings.

In figure 35.2B the body size index is based on mass and various measurements of the leg, wing and beak. The degree to which these are genetically inherited, in other words their **heritability**, was investigated on Daphne Major. Four hundred nestlings were ringed, of which 82 were recaptured when more or less full grown. Measurements made on these birds were then compared with the same measurements made on their parents. High correlations were found between the nestlings and their parents in the case of mass and the length, depth and width of the beak. Boag and Grant concluded that the mass and beak measurements were probably highly heritable, and therefore the drought would have caused changes in the population's gene frequencies.

For consideration

1. Why do you think there have been so few satisfactory demonstrations of natural selection causing evolutionary change?

2. If the measurements in figure 35.2B had been continued for another few years, what would you expect to have happened, and why?

3. In what circumstances would the heritability measurements described *not* indicate genetic inheritance? How should this possibility be ruled out?

Further reading

P.T. Boag and P.R. Grant, 'Heritability of External Morphology in Darwin's Finches' (*Nature*, vol. 274, no. 5673, 1978)

P.T. Boag and P.R. Grant, 'Intense Natural Selection in a Population of Darwin's Finches (*Geospizinae*) in the Galapagos' (*Science*, vol. 214, no. 4516, 1981)

These two papers report the studies described in this topic.

---

36 . Some major steps in Evolution

Origin of life

Nearly every modern textbook describing the origin of life gives one theory, and one theory only. A 'primordial soup' of organic molecules is thought to have formed from gases in the early atmosphere, such as methane and ammonia. Lightning or other energy sources could have combined these simple molecules into complex organic compounds, which then accumulated in ponds, lakes, or the sea. Indeed experiments have shown how, with no organism present, methane can easily be converted into amino acids and cyanide, formaldehyde (methanal) into sugars, and cyanide into adenine.

The original organism developed in this 'soup', and it is assumed to have possessed the features common to all extant forms of life. Most important of these are a genetic system of nucleic acids, and a collection of proteins. The proteins enable the genes to survive and replicate.

LIFE FROM SPACE?

Not everyone agrees with the 'soup' theory. Fred Hoyle and Chandra Wickramasinghe, both physicists, believe that life could not have arisen on this planet. They cite evidence to show that various organisms are adapted to conditions they could never have encountered on earth. For example, some bacteria survive when subjected to extremely low pressures, or to massive doses of x-rays. To explain this and other observations, they propose that life arrived (and still arrives) at the surface of the Earth from space, in the form of cells, viruses, or fragments of nucleic acids. According to Hoyle and Wickramasinghe, the main steps in evolution were caused by organisms absorbing ready-made extraterrestrial genes.

THE CRYSTAL THEORY

Another critic of the 'soup theory' is the chemist, Graham Cairns-Smith. It may be true that the latest common ancestor of extant forms was a cell with protein/nucleic acid biochemistry, rather like a bacterium, but Cairns-Smith thinks that the earliest organism was different. Nucleic acids are an excellent genetic material – in a cell that contains many complex enzymes. Proteins similarly function well – so long as there are nucleic acids to control their synthesis. Each needs the other. What the 'soup theory' fails to explain is how a solution of organic molecules could produce a living cell which depends on both proteins and nucleic acids.

Cairns-Smith's theory is perhaps even more startling than the extraterrestrial one. He suggests that life did start on Earth, but that the earliest living things contained no organic chemicals at all. They were crystals, in particular clay crystals made of silicates. Their atoms would have been mainly oxygen, silicon, and various metals. A major hurdle the theory must overcome is to show how clay crystals could be alive. What sort of genetic system did they have, how did they reproduce, and how could they evolve?

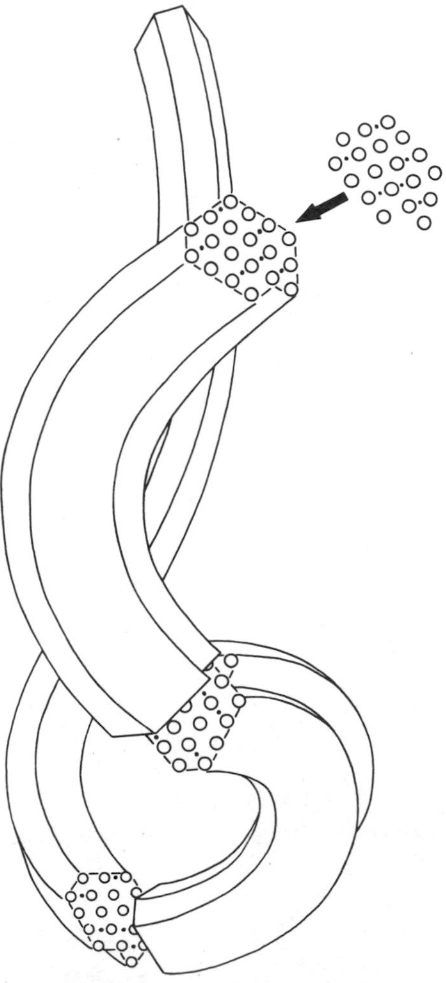

Figure 36.1 One kind of crystal organism. The genotype is the pattern of irregular atoms (dots) in the lattice of oxygen atoms (circles). Reproduction occurs when environmental forces break the crystal. Then each piece grows by adding new layers of atoms, with the same pattern, on the broken surface.

Figure 36.1 gives an idea of the principles involved. Some clay crystals are flexible rods, growing at their ends but not getting any thicker. In the cross-sections you can see that the rod is a regular crystal lattice, with most of the space occupied by oxygen atoms. Metal atoms in clay can be distributed between the oxygen atoms in an irregular pattern; and it is possible that each new layer of atoms making the rod a bit longer could copy the pattern in the layer next to it. This pattern of metal atoms could represent the 'genotype', and might determine the crystal's 'phenotype' – for instance its flexibility, cross-sectional shape, length, and how easily it breaks. The crystal could 'reproduce' whenever it got knocked into several pieces. One exciting aspect of the crystal theory of life's origin is that living clay crystals might be found today, or even manufactured.

For consideration

1. How would you set about investigating Hoyle's and Wickramasinghe's theory?

2. Suggest how even a completely inorganic crystal could be capable of evolving by natural selection. What stages might it go through to become an organic prokaryotic cell?

Further reading

J.D. Bernal and A. Synge, *The Origin of Life* Carolina Biology Reader no. 13, 1972)
A fifteen page pamphlet introducing the 'soup theory'.

A.G. Cairns-Smith, 'The First Organisms' (*Scientific American*, vol. 252, no. 6, 1985)
Like Graham Cairns-Smith's books, this article is entertaining and highly stimulating.

F. Hoyle and C. Wickramasinghe, *Evolution from Space* (Dent, 1981)
Despite some errors this book is worth thinking about.

S.L. Miller and L.E. Orgel, *The Origins of Life on the Earth* (Prentice Hall, 1974)
A fair knowledge of biochemistry is assumed.

Where did organelles come from?

It is sometimes stated that RNA is found in the nucleus and cytoplasm, while DNA only occurs in the nucleus. Although this general distinction can be made, it is not entirely true. Chloroplasts and mitochondria have their own DNA and RNA, and synthesise some of their own proteins. However, other proteins they need are synthesised on the cytoplasmic ribosomes from genes in the nucleus.

Simpler, surely, would be to have all the DNA in the chromosomes of the nucleus. So why does the eukaryotic cell have such an untidy arrangement?

ENDOSYMBIOSIS

One answer is provided by the **endosymbiosis theory** of organelles, which is summarised in figure 36.2. It is suggested that an anaerobic prokaryotic cell ingested some aerobic bacteria. By becoming endosymbionts – that is, symbionts within the host's cell – the bacteria made the composite organism aerobic and much more efficient at respiration. The bacteria evolved into mitochondria. Indeed mitochondria still bear remarkable similarities to bacteria. They synthesise DNA continuously, like prokaryotes, whereas eukaryotic nuclei do so only at one stage in the cell cycle (see page 101). They are sensitive to drugs that also affect prokaryotes but not eukaryotes. However the DNA loop of a mitochondrion is much smaller than the chromosome of a bacterium.

At a later stage, the composite organism is supposed to have acquired another symbiont. The new arrivals were motile bacteria, perhaps spirochaetes, that adhered to the surface as ectosymbionts. By making undulating movements they moved the organism along, thus allowing it to encounter more food.

You may think that is a bit far-fetched, but examples are known in modern protists. *Mixotricha* is one found in termite guts. It has four ordinary flagella, probably used only for steering. The protist speeds along in a dead straight line,

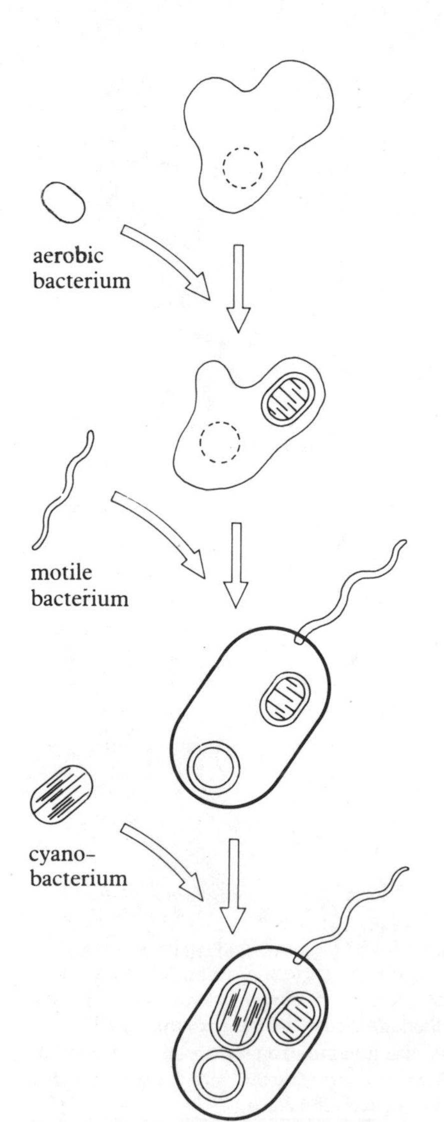

Figure 36.2 The endosymbiosis theory: **A** A prokaryotic cell with its chromosome; **B** It gains an aerobic bacterium and so develops mitochondria; **C** It gains a motile bacterium and then develops '9 + 2' organelles, mitosis and a nucleus; **D** Finally it gains a cyanobacterium and so develops chloroplasts.

which in itself is unusual for flagellates, by having thousands of undulating spirochaetes sticking to the surface. They act just like a coat of cilia. *Mixotricha*, incidentally, is even more remarkable than this, because it has altogether two kinds of ectosymbiont on its surface, two endosymbionts, and no mitochondria.

The eukaryotic cilium and flagellum possess the well known '9 + 2' pattern of microtubules (see page 7). According to the endosymbiosis theory, the cilium is the remains of the whole ectosymbiont cell together with several of its bacterial flagella. Other organelles are also supposed to be derived from the ectosymbiont. Among them are the basal bodies and centrioles, with a '9 + 0' structure, and the mitotic spindle and centromeres, since these depend on the centrioles (see page 103). Like the mitochondria and chloroplasts, centrioles and basal bodies cannot be manufactured by the cell using just the genes in the nucleus. Parent organelles must be present. This is further evidence that they were once independent organisms.

As for the chloroplasts, the evidence for their endosymbiotic origin is probably the strongest of all. Like mitochondria, they are sensitive to drugs that damage prokaryotes but not eukaryotes. The amount of DNA in a chloroplast is similar to that in a cyanobacterium. Indeed in some organisms, it is quite difficult to tell whether the cell contains endosymbionts or chloroplasts. Moreover chloroplasts can be transferred to animal cells, such as mouse fibroblasts, where they will continue photosynthesising for several days. This happens naturally when the mollusc *Elysia* feeds on algae: the chloroplasts land up in the hepatic tubule cells, and the mollusc benefits from the continuing photosynthesis. So if Martians really are green, there may be a good reason for it.

The endosymbiosis theory explains many features of the eukaryotic cell. But it is not entirely satisfactory. The mitochondria, chloroplasts, *and* nucleus are each surrounded by a double membrane, yet the nucleus is thought not to have developed from a symbiont. So where did it come from?

INVAGINATION

Figure 36.3 outlines an alternative theory. Imagine a prokaryotic cell with multiple copies of its DNA loop attached here and there to the cell membrane. The membrane contains all the enzymes for respiration and for photosynthesis. Invagination of the membrane could produce vesicles inside the cell, each bounded by a double membrane. Although all the vesicles or organelles would originally have possessed DNA (D), as well as the capacity to photosynthesise (P) and respire (R), specialisation could have occurred. So the nucleus lost P, R, and a little D; the mitochondria lost P and most of D; and the chloroplasts lost R and most of D.

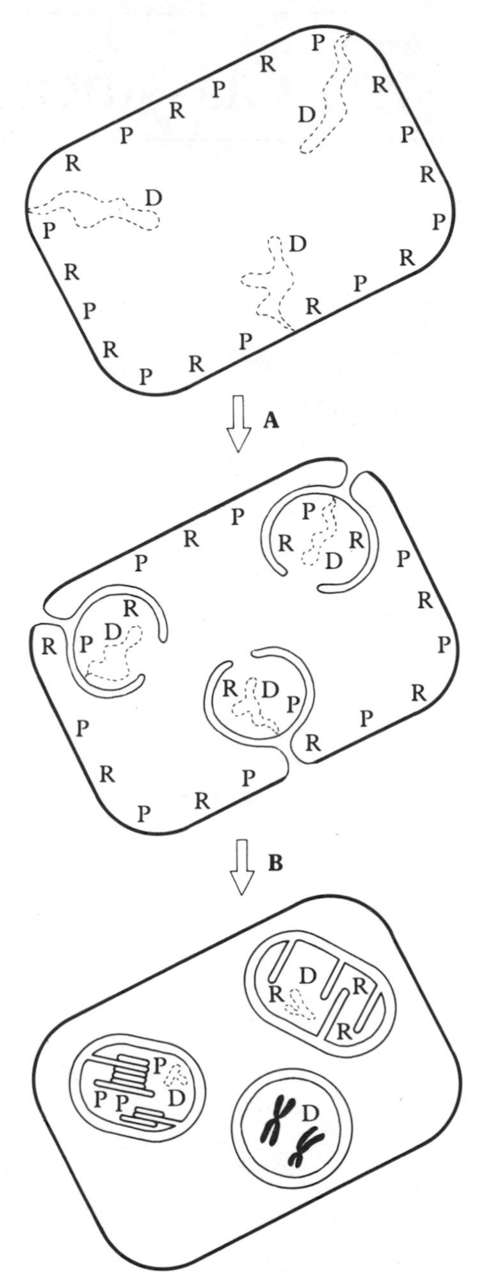

Figure 36.3 The invagination theory: **A** Each vesicle produced by invagination contains DNA (D), and enzymes for photosynthesis (P) and respiration (R); **B** Specialisation leads to nucleus, chloroplasts, and mitochondria.

For consideration

1. Although the endosymbiosis theory has gained widespread acceptance, can you think of any reason why an invagination theory must be ruled out?

2. What practical applications might there be for the principle of evolution by endosymbiosis?

Further reading

L. Margulis, 'Symbiosis and Evolution' (*Scientific American*, vol. 225, no. 2, 1971)
A well illustrated article.

L. Margulis, *Symbiosis in Cell Evolution: Life and its Environment on the Early Earth* (Freeman, 1981)
A detailed but readable account.

M. Tribe, A Morgan and P. Whittaker, *The Evolution of Eukaryotic Cells* (Studies in Biology no. 131, Arnold, 1981)
The authors describe the endosymbiotic theory and mention alternatives.

T. Uzzell and C. Spolsky, (*American Scientist*, vol. 62, no. 3, 1974)
This technical article discusses the invagination theory.

37 . Classification

Controversy over cladistics

Why do we group together the ectothermic animals whose embryos are provided with an amnion, and call them reptiles? Should we group them together at all? Such questions are central to the controversy surrounding the cladistic method of classification.

To gain an insight into the disagreement, it is worth considering a more general question: why do we classify organisms at all? Here are four possible answers.

1. To construct sets about which generalisations can be made. In the case of reptiles, we can say several things about them, other than that they are ectothermic amniotes. They have a scaly skin, rely on lungs for gas exchange, excrete uric acid, are absent from the coldest climates, and so on.
2. To make predictions about individual members of the set. Thus if several members of a plant family contain drugs or provide food, it makes sense to search for other useful drug and food plants in the same family.
3. To provide a filing system for handling biological information. Suppose, for example, you need information on the Hen Flea; you enter the name into a computer, and then it will print out for you a list of publications on that species.
4. To represent the way organisms have evolved. According to this view, the classification of organisms ought to be as nearly as possible identical with their phylogeny.

THE MAIN POINT OF CONTENTION

Orthodox taxonomists disagree with cladists on several points, but one stands out above the rest. It is best illustrated by those ectothermic amniotes. Orthodox taxonomists recognise the Reptilia as a class, while cladists refuse to recognise them as any sort of group at all.

Figure 37.1A depicts a hypothetical phylogenetic tree. It shows the possible evolution of the amniotes, much simplified. The amniotes are thought to be **monophyletic** – this means the group is derived from a common ancestor and includes all that ancestor's descendants. As the number of species increased, they formed the groups we now call birds, reptiles, and mammals. The reptiles are probably in some respects fairly similar to the ancestral amniote. By contrast birds have diverged widely, becoming endothermic and covered in feathers. Similarly mammals have become endothermic and hairy.

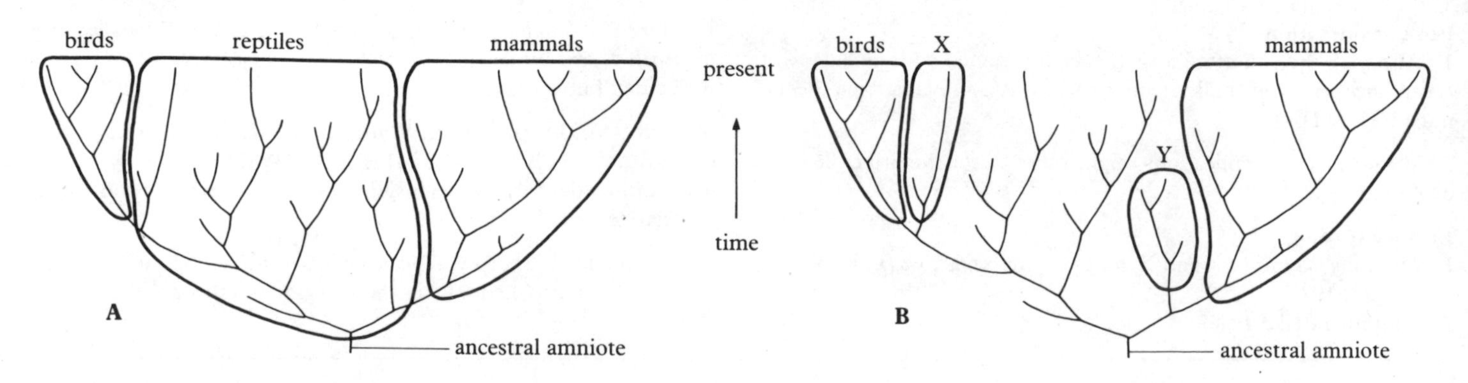

Figure 37.1 A simplified phylogenetic tree of the amniotes. **A** Orthodox classification groups organisms by similarity as well as by common origin. **B** Cladistic classification recognises only sister groups, such as birds and X (which might be the crocodile group), or mammals and Y (which might be extinct).

Cladists have special rules for constructing classifications (figure 37.1B). They attach names only to monophyletic groups, such as the birds and mammals. The reptiles are not named as a group; this is because they are **paraphyletic** – in other words, only one part of a larger monophyletic group. Moreover cladists give equal rank to the two groups coming from a single branching point on the tree. These are called **sister groups**. So if the birds are called the class Aves, the organisms in group X must also be a class, even if they may be almost identical with other reptiles.

If you were a participant in the cladistics controversy, you would no doubt have strong opinions about which approach is 'better'. Standing back from the debate, however, we can see that each has its merits. For the biologist principally interested in evolution, cladistics offers a powerful method. In addition to the fourth aim mentioned above, classification into sister groups may also be the best way to satisfy the first and second aims. This is because only monophyletic groups are named, and each will retain many of its common ancestor's characters.

On the other hand, some biologists are not so interested in evolution. They mainly want a classification satisfying the first three aims. It can be frustrating to have a useful group such as the reptiles abolished, leaving us with no easy way to make generalisations about them. Accordingly these biologists think it is more important to classify together organisms that resemble each other, than to recognise only the strictly monophyletic groups.

AND IN THE FUTURE?

It is not likely that either approach to classification will displace the other. Perhaps they will coexist, with cladistics as a means of unravelling evolutionary history, and the orthodox classification serving most other purposes. Although cladistics may become established in the lower ranks of the classification – from orders and families down to species – it will probably not be used much among the phyla and kingdoms. This is because working out evolutionary relationships between the larger groups has proved notoriously difficult.

Instead, quite a different development has taken place. The result is a classification of the phyla and kingdoms which is about as far removed from cladistics as one could imagine. This is the subject of the next topic.

For consideration

What other paraphyletic groups can you think of, besides reptiles? Would any biologists be seriously inconvenienced if they were abolished?

Further reading

C. Patterson, 'Cladistics' (*Biologist*, vol. 27, no. 5, 1980)
This article is by a leading exponent of the cladistic method.

Five kingdoms

More and more biologists are grouping organisms into five kingdoms. It is easy to see why the five-kingdom scheme is an improvement over the traditional arrangement based on just two kingdoms. Look again at the four aims listed at the beginning of the previous topic. The first aim is particularly important. How many generalisations could be made about 'plants' in the two-kingdoms scheme? Very few. This is because animals were defined as motile heterotrophs, while 'plants' formed a 'dustbin group' containing every living thing that was not an animal. No wonder 'plants' had little in common. How many similarities are there between green plants and bacteria, or green plants and fungi?

By contrast, a number of important generalisations can be made about at least some of the five kingdoms. Consider plants – remembering that the word now has a much more restricted meaning than in the two-kingdom arrangement.

Plants are multicellular and photosynthesise. In ecological terms, this means that they include all the larger producers. Fungi are multicellular and absorb their food in soluble form; so they include most of the larger decomposers. Animals are multicellular and ingest their food, and are therefore consumers. As Stephen Jay Gould has pointed out, it is a sobering thought that of all the five kingdoms only the animals, including ourselves, could be wiped out without much risk to the others.

Admittedly this scheme does leave the Protista as a 'dustbin group'. It contains everything not belonging to another kingdom. Despite all being unicellular they are rather a mixed bag, with all forms of nutrition represented. There is however another respect in which the five-kingdoms scheme is untidy.

PROTISTA OR PROTOCTISTA?

The green algae, or Chlorophyta, are an embarrassment. They contain both unicellular and multicellular forms which nevertheless do seem to be closely related. Similarly some organisms normally called fungi are unicellular. Yeast is an example. And the slime moulds cause confusion: they are unicellular for much of the life cycle, but then turn into a multicellular or multinucleate form (see page 15).

In other words, there are uncertain boundaries between the Protista and the plants and fungi respectively. For people who think this is a serious defect, one solution is to change the boundaries to make them more distinct. The green, brown, and red algae (Chlorophyta, Phaeophyta, and Rhodophyta) can be removed from the plants and grouped with the protists. Also the slime moulds and some fungi can join the protists. These six groups together make up what some biologists like to call the Protoctista.

The advantage of this arrangement is that now all the boundaries between kingdoms are clear cut. On the other hand the Protoctista is even more of a mixed bag than the Protista. There are few generalisations one can make about them. In particular they show an extreme variation in size, ranging from single-cell flagellates to gigantic multicellular seaweeds known as kelp, which can be a hundred metres long.

Of the two five-kingdom classifications, it does not matter much which is used. However, the Protista version has the practical advantage of putting all unicellular eukaryotes into one kingdom. If one is worried about the uncertain boundaries between the Protista and other kingdoms, one can console oneself with the thought that this is the only serious anomaly in a classification system which is otherwise based consistently on two fundamental characteristics of organisms, namely nutrition and cell structure.

For consideration

1. The five-kingdom scheme does not pretend to be phylogenetic – in other words each kingdom may have arisen from several different ancestral forms. Does this matter?

2. The system adopting the Protoctista should be called 'four kingdoms plus ragbag'. Discuss.

3. Do you agree that the five kingdom system is based more or less consistently on nutrition and cell structure? Explain your answer.

Further reading

S.J. Gould, *Ever since Darwin*. (Penguin, 1980)
Essay 13, 'The Pentagon of Life', is particularly relevant and, as usual, enjoyable reading.

C.S. Hutchinson, 'Biological Classification' (*School Science Review*, vol. 60, no. 215, 1979)
A compact presentation of the five-kingdom scheme; convenient for reference.

L. Margulis and K.V. Schwartz, *Five Kingdoms: An Illustrated Guide to the Phyla of Life on Earth* (Freeman, 1981)
A splendidly illustrated book that devotes a double-page spread to each of the 89 phyla. The Protoctista version is adopted.

Index